OXFORD STATISTICAL SCIENCE

OXFORD STATISTICAL SCIENCE SERIES

Celebrating Statistics

Papers in honour of Sir David Cox on the occasion of his 80th birthday

Editors:

A. C. DAVISON

Ecole Polytechnique Fédérale de Lausanne, Switzerland

YADOLAH DODGE

University of Neuchâtel, Switzerland

N. WERMUTH

Göteborg University, Sweden

OXFORD
UNIVERSITY PRESS

OXFORD
UNIVERSITY PRESS

Great Clarendon Street, Oxford OX2 6DP

Oxford University Press is a department of the University of Oxford.
It furthers the University's objective of excellence in research, scholarship,
and education by publishing worldwide in

Oxford   New York

Auckland Cape Town Dar es Salaam Hong Kong Karachi
Kuala Lumpur Madrid Melbourne Mexico City Nairobi
New Delhi Shanghai Taipei Toronto

With offices in

Argentina Austria Brazil Chile Czech Republic France Greece
Guatemala Hungary Italy Japan Poland Portugal Singapore
South Korea Switzerland Thailand Turkey Ukraine Vietnam

Oxford is a registered trade mark of Oxford University Press
in the UK and in certain other countries

Published in the United States
by Oxford University Press Inc., New York

© Oxford University Press 2005

British Library Cataloguing in Publication Data

Data available

Library of Congress Cataloguing in Publication Data

Data available

Typeset by SPI Publisher Services, Pondicherry, India
Printed in Great Britain
on acid-free paper by
Biddles Ltd., King's Lynn, Norfolk

ISBN 0-19-856654-9
ISBN 9780198566540

10 9 8 7 6 5 4 3 2 1

Preface

Sir David Cox is among the most important statisticians of the past half-century. He has made pioneering and highly influential contributions to a uniquely wide range of topics in statistics and applied probability. His teaching has inspired generations of students, and many well-known researchers have begun as his graduate students or have worked with him at early stages of their careers. Legions of others have been stimulated and enlightened by the clear, concise, and direct exposition exemplified by his many books, papers, and lectures.

This book records the invited papers presented at a conference held at the University of Neuchâtel from 14–17 July 2004 to celebrate David Cox's 80th birthday, which was attended by friends and colleagues from six continents. So many outstanding scientists would have wished to honour David on this occasion that the choice of invited speakers was extremely difficult and to some extent arbitrary. We are particularly grateful to the many excellent researchers who contributed papers to the meeting; regrettably it is impossible to include them in this volume.

The brief for the invited talks was to survey the present situation across a range of topics in statistical science. The corresponding chapters of this volume range from the foundations of parametric, semiparametric, and non-parametric inference through current work on stochastic modelling in financial econometrics, epidemics, and ecohydrology to the prognosis for the treatment of breast cancer, with excursions on biostatistics, social statistics, and statistical computing. We are most grateful to the authors of these chapters, which testify not only to the extraordinary breadth of David's interests but also to his characteristic desire to look forward rather than backward.

Our sincere thanks go to David Applebaum, Christl Donnelly, Alberto Holly, Jane Hutton, Claudia Klüppleberg, Michael Pitt, Chris Skinner, Antony Unwin, Howard Wheater, and Alastair Young, who acted as referees, in addition to those authors who did double duty. We also thank Kate Pullen and Alison Jones of Oxford University Press, and acknowledge the financial support of the Ecole Polytechnique Fédérale de Lausanne and the University of Neuchâtel.

Anthony Davison, Yadolah Dodge, and Nanny Wermuth
Lausanne, Neuchâtel, and Göteborg

February 2005

List of contributors

Ole E. Barndorff-Nielsen Department of Mathematical Sciences, University of Aarhus, Ny Munkegade, DK-8000 Århus C,Denmark; ...`oebn@imf.au.dk`

Sarah C. Darby Clinical Trial Service Unit and Epidemiological Studies Unit, University of Oxford, Harkness Building, Radcliffe Infirmary, Oxford OX2 6HE, UK; ....`sarah.darby@ctsu.ox.ac.uk`

Christina Davies Clinical Trial Service Unit and Epidemiological Studies Unit, University of Oxford, Harkness Building, Radcliffe Infirmary, Oxford OX2 6HE, UK; ....`christina.davies@ctsu.ox.ac.uk`

Anthony C. Davison Institute of Mathematics, Ecole Polytechnique Fédérale de Lausanne, CH-1015 Lausanne, Switzerland; ..`Anthony.Davison@epfl.ch`

Peter J. Diggle Department of Mathematics and Statistics, Lancaster University, Lancaster LA1 4YF, UK; ....`p.diggle@lancaster.ac.uk`

Yadolah Dodge Statistics Group, University of Neuchâtel, Espace de l'Europe 4, Case postale 805, CH-2002 Neuchâtel, Switzerland;
....`yadolah.dodge@unine.ch`

David Firth Department of Statistics, University of Warwick, Coventry CV4 7AL, UK; ....`d.firth@warwick.ac.uk`

Peter Hall Centre for Mathematics and its Applications, Mathematical Sciences Institute, Australian National University, Canberra, ACT 0200, Australia; ....`Peter.Hall@maths.anu.edu.au`

Valerie S. Isham Department of Statistical Science, University College London, Gower Street, London WC1E 6BT, UK;
....`valerie@stats.ucl.ac.uk`

Kung-Yee Liang Department of Biostatistics, Bloomberg School of Public Health, Johns Hopkins University, 615 N Wolfe Street, Baltimore, MD 21205, USA; ....`kyliang@jhsph.edu`

Peter McCullagh Department of Statistics, University of Chicago, 5734 University Avenue, Chicago, IL 60637, USA; ......`pmcc@galton.uchicago.edu`

Paul McGale Clinical Trial Service Unit and Epidemiological Studies Unit, University of Oxford, Harkness Building, Radcliffe Infirmary, Oxford OX2 6HE, UK; ....`paul.mcgale@ctsu.ox.ac.uk`

Amilcare Porporato Department of Civil and Environmental Engineering, Duke University, Durham, NC 27708, USA; ....`amilcare@duke.edu`

Nancy Reid Department of Statistics, University of Toronto, 100 St. George Street, Toronto, Canada M5S 3G3; .... `reid@utstat.utoronto.ca`

Brian D. Ripley Department of Statistics, University of Oxford, 1 South Parks Road, Oxford OX1 3TG, UK; .... `ripley@stats.ox.ac.uk`

Ignacio Rodríguez-Iturbe Department of Civil and Environmental Engineering and Princeton Environmental Institute, Princeton University, Princeton, NJ 85040, USA; .... `irodrigu@princeton.edu`

Andrea Rotnitzky Department of Biostatistics, Harvard School of Public Health, 655 Huntington Avenue, Boston, MA 02115, USA;
.... `andrea@hsph.harvard.edu`

Neil Shephard Nuffield College, Oxford OX1 1NF, UK;
.... `neil.shephard@nuf.ox.ac.uk`

Nanny Wermuth Mathematical Statistics, Chalmers Göteborgs Universitet, Eklandagatan 86, 41296 Göteborg, Sweden; ... `wermuth@math.chalmers.se`

Scott L. Zeger Department of Biostatistics, Bloomberg School of Public Health, Johns Hopkins University, 615 N Wolfe Street, Baltimore, MD 21205, USA; .... `szeger@jhsph.edu`

Contents

List of figures

List of plates

List of tables

David R. Cox: A brief biography

Yadolah Dodge

David Cox started his career as a statistician in 1944 at the age of 20. This chapter gives a brief history of his scientific life, with extracts from interviews he has given.

David Roxbee Cox was born in Birmingham on July 15, 1924. He studied mathematics at the University of Cambridge and obtained his Ph.D. from the University of Leeds in 1949. He was employed from 1944 to 1946 at the Royal Aircraft Establishment, and from 1946 to 1950 at the Wool Industries Research Association in Leeds. From 1950 to 1955 he was an Assistant Lecturer in Mathematics, University of Cambridge, at the Statistical Laboratory. From 1956 to

PLATE 1. David R. Cox.

PLATE 2. With F. J. Anscombe, Berkeley Symposium, 1960.

1966 he was Reader and then Professor of Statistics at Birkbeck College, London. From 1966 to 1988 he was Professor of Statistics at Imperial College, London. In 1988 he became Warden of Nuffield College and a member of the Department of Statistics at Oxford University. He formally retired from these positions on August 1, 1994.

In 1947 he married Joyce Drummond. They have a daughter and three sons.

Among his many honours, Sir David Cox has received numerous honorary doctorates. He is in particular Doctor Honoris Causa of the University of Neuchâtel and is an honorary fellow of St John's College, Cambridge, and of the Institute of Actuaries. He has been awarded the Guy medals in Silver (1961) and Gold (1973) of the Royal Statistical Society. He was elected Fellow of the Royal Society of London in 1973, was knighted by Queen Elizabeth II in 1985 and became an Honorary Member of the British Academy in 2000. He is a Foreign Associate of the US National Academy of Sciences. In 1990 he won the Kettering Prize and Gold Medal for Cancer Research.

At the time of writing this, Sir David Cox has written or co-authored 300 papers and books. From 1966 through 1991 he was the editor of *Biometrika*. He has supervised, collaborated with, and encouraged many students, postdoctoral fellows, and colleagues. He has served as President of the Bernoulli Society, of the Royal Statistical Society, and of the International Statistical Institute. He is now an Honorary Fellow of Nuffield College and a member of the Department of Statistics at the University of Oxford.

PLATE 3. D. V. Hinkley, M. Sobel, E. E. Bassett, S. Scriven, T. W. Anderson, P. A. Mills, E. M. L. Beale, A. M. Herzberg, E. J. Snell, B. G. Greenberg, D. R. Cox and G. B. Wetherill, London, 1968.

He prizes at a personal and scientific level the numerous friendships and collaborations built up over the years and especially those with postgraduate students. He taught on the MSc program at the University of Neuchâtel for over a decade. Each year the program brought together a mixture of students from many different branches of the applied sciences, and I asked him how one should teach statistics to a group of not-so-homogenous students. The following lines are David's random thoughts on teaching the course on survival analysis, which he was giving at the time:

'It is always a particular pleasure and challenge to teach away from one's own institution, the challenge arising in part because of uncertainty about the background knowledge that can reasonably be assumed, a key element in the planning of any course. The experience at Neuchâtel was notable especially for the impressive variety of subject fields of specialization, experience and interests of the students. This meant that a key issue was not, as so often, that of providing motivation, but rather that of finding a level of treatment that would enable topics close to the forefront of contemporary research to be addressed without making unreasonable demands on the students' theoretical and mathematical knowledge. Of course, it is hard to judge how successfully this was accomplished.

PLATE 4. With N. Reid, D. A. S. Fraser and J. A. Nelder, Waterloo, 1985.

'There are two extreme strategies that may be used in selecting material for such a course. One is to choose, perhaps rather arbitrarily, two or three topics and to develop these carefully in some depth so that hopefully some understanding of principles is achieved. The other, which I find personally more appealing both when I am a student at other people's lectures, and as a teacher, is to range more broadly and to omit detail.

'This raises the following general issue that arises in teaching any topic that is the focus of much research. Our subject has reached the state where in many fields the literature seems overwhelming. In survival analysis not only are there a considerable number of books, some technically very demanding, but also a very large number of research papers, some wholly theoretical, others concerned at least in part with specific applications, but often also containing new methodological points. Further new papers appear virtually daily.

'Yet somehow the implicit message, which I believe to be correct, must be conveyed that this volume of work is not to be thought overwhelming, that it should be regarded as evidence of the openness of the field and that by examining carefully new applications novel work of value can be done. Put differently there are a limited number of key principles and the rest is detail, important in its proper context, but not usually needed. Thus a central issue in each of the specialized areas of statistics is to separate essential principles from detail.

'Much current research in statistics is highly technical; look for example at any issue of the *Annals of Statistics*. It is sometimes argued: how much easier research must

PLATE 5. With Nancy Reid and David Hinkley, CIMAT, Guanajuato, Mexico, 1988.

have been all those years ago when simple and important results could be obtained by elementary methods.

'For myself, I very much doubt the correctness of this view, to some extent from personal experience, but more importantly from thinking about the history of key developments in our subject.

'Take, for example, the Cramér–Rao inequality. A simple elegant very brief proof is available requiring only the Cauchy–Schwarz inequality, itself provable by totally elementary methods. Does this mean that the pioneers had an easy task? It would be interesting to know Professor Rao's view about this. Certainly the earlier work of Aitken and Silverstone used the mathematically sophisticated ideas of the calculus of variations and was much more difficult.

'The tentative conclusion I draw is that statistics was and remains a field in which new contributions can be made from all sorts of backgrounds ranging from those with powerful mathematical expertise to those who have new ideas and formulations arising fairly directly out of specific applications. The intellectual vigour of the subject relies on there being a balanced mixture and thus on our subject attracting people with a wide range of interests. One of the strengths of the Neuchâtel course is precisely that.'

Another matter of interest for most statisticians is how our leaders think about our field. Here are David Cox's thoughts:

PLATE 6. O. E. Barndorff-Nielsen and D. R. Cox walking and contemplating at Rousham, 1990.

'I write some personal thoughts about working as a statistician. I do so largely from a UK perspective and from the viewpoint of a statistician, working mostly but not entirely in a university environment and involved with applications in the natural and social sciences and the associated technologies. The nationality aspect is not very important but no doubt had I worked in a government statistical office the issues to be discussed would be different, although surely the changes there have been great too.

'When I first started statistical work in 1944 there were very few people with a specialized academic training in statistics. Most statistical work was done either by those who like me were trained in mathematics, a few of these with some minimal formal training in statistical ideas, or by natural scientists, especially biologists, or by engineers, who had acquired some knowledge of statistics out of necessity. There were very few books, and those that did exist were mostly accounts of analysis of variance, regression and so on in a form for use by nonspecialists. They gave little or no theory and hence the intellectual basis for tackling new problems was often very unclear. In my second job I was constantly faced with sets of data from carefully designed experiments of not quite standard form and I fear I must have done many analyses that were technically based more on hope obtained by extrapolation from simpler cases than on proper understanding.

'Yet paradoxically by that period, i.e. 1945–1950, many of what one now regards as the standard methods of statistics had been developed, but I suppose were not

Plate 7. Between A. M. Herzberg and M. S. Bartlett, Oxford, 1990.

widely known outside specialized groups. Thus Fisher, and very particularly Yates, had developed design of experiments and analysis of variance into a highly sophisticated art form, least squares and virtually all of its properties having even by then a long history. Principal components dates back at least to the early 1900's and canonical correlation to 1935. Time series methods, both in time and frequency domain had been developed and Neyman had set out definitive ideas on sampling.

'Why were these methods not more widely used? The answer is not so much that they were not known, although this may have been a contributing feature, but much more that some were computationally prohibitive except on a very modest scale. A typical statistician of that period would have the help of at least two or three computers. In those days a computer was a person. Such a person could use an electrically operated calculating machine to do standard balanced analyses of variance, simple regression calculations with up to two or three explanatory variables, simple chi-squared tests and maybe some other more specialized calculations. There was also a rich variety of special types of graph paper for various graphical procedures. Of course other calculations could be done but would need special planning.

'The possibility of numerical errors was ever present. Important calculations had to be checked preferably by a distinct route. I had the enormous good fortune to work for a period with Henry Daniels. He was a very powerful theoretician and I learned pretty well all I knew about theory from him. He was also meticulous. Numerical work had to be done with a sharp pencil entering the answers two (never three) digits per square

PLATE 8. With O. Kempthone and R. Hogg, Ames, Iowa 1992.

on quadrille paper (ruled in 1/4 inch squares). Errors had to be rubbed out (never crossed out) and corrected. Tables etc. had to have verbal headings.

'Probably the most distressing calculation was for time series analysis. The data were rounded to two digits and entered twice into a comptometer, an old-style cash register. From this one printed out a paper strip containing the data in columns, with each value occurring twice. This was cut down the middle to produce two strips. Paper clips were then used to attach these together at displacement h, the desired lag for a serial correlation. Corresponding values were subtracted mentally and then squared mentally and the answer entered, usually by a second person, into an adding machine. This produced $\sum(y_{i+h} - y_i)^2$, from which the lag h serial correlation can be obtained. This would be repeated for as many values of h as were required, constrained by the need to preserve the sanity of the investigators. Of course I take for granted a knowledge of the squares of the integers up to say 50 or so.

'Matrix inversion takes a time roughly proportional to the cube of the size. The largest I ever did myself was, I think, 7×7 which took, with the obligatory checks, most of one day.

'Now the consequence of all this for the analysis of data was: look very carefully at the data, to assess its broad features, possible defects and so on. Then think. Then compute.

PLATE 9. With C. R. Rao and P. J. Huber during the Karl Pearson meeting in Paris, May 2000.

'The idea of an electronic computer was known and I recall in particular the visionary statements of I. J. Good and A. D. Booth, who foresaw the impact computers would have on daily life. It is interesting that it has taken almost 50 years for what they anticipated to come to pass. I suspect that one if not both of them assumed a gentler future in which the miracles made possible by scientific progress would enhance the general quality of life, not be used as the driving force behind an ever more frantic and in many ways absurd competitive scramble.

'In statistics, as a sweeping oversimplification, the period up to 1975–1980 was taken with getting the classical procedures, and their direct developments, into a form in which they could be relatively painlessly used. Since then the new horizons opened by massive computer power, including the ability to handle large amounts of data, are being explored.

'The consequence of all this for the analysis of relatively modest amounts of data, by contrast with the attitude summarized earlier, is: first compute. Then think.

'In one sense this is a liberation and I do not doubt that on balance it is a good thing, but there is a downside on which we all need to reflect.

'How different now is statistical work from how it was when I started? First there are, of course, many more individuals with statistical training and a much deeper percolation of statistical ideas into subject-matter fields. Think for example of medical research. In many, although certainly not all, areas the specialized subject-matter medical research

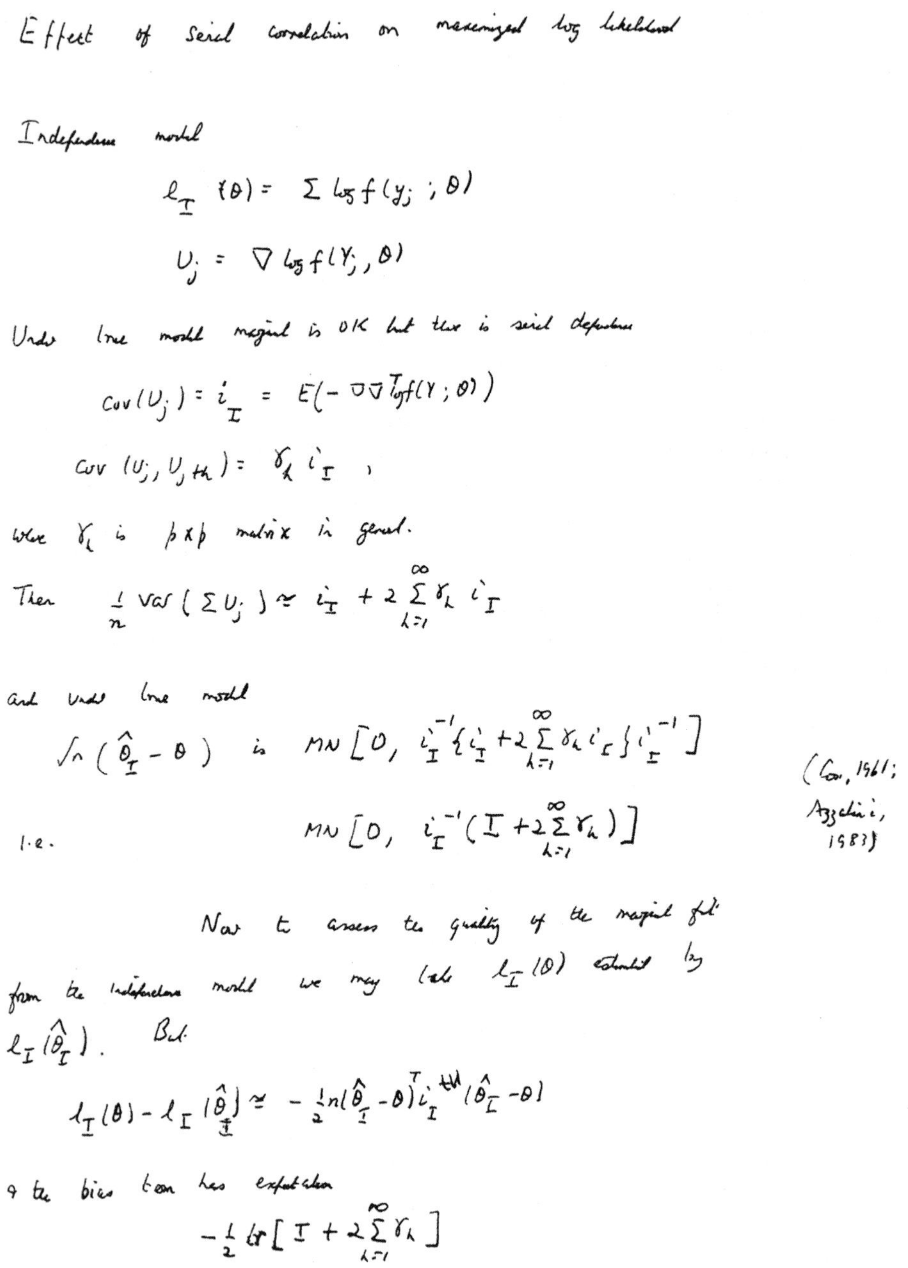

Effect of serial correlation on maximized log likelihood

Independence model

$$\ell_{\underline{I}}(\theta) = \sum \log f(y_j ; \theta)$$

$$U_j = \nabla \log f(Y_j, \theta)$$

Under true model marginal is OK but there is serial dependence

$$Cov(U_j) = i_{\underline{I}} = E(-\nabla \nabla^T \log f(Y ; \theta))$$

$$Cov(U_j, U_{j+h}) = \gamma_h \, i_{\underline{I}} \, ,$$

where γ_h is $p \times p$ matrix in general.

Then

$$\frac{1}{n} var\left(\sum U_j \right) \simeq i_{\underline{I}} + 2 \sum_{h=1}^{\infty} \gamma_h \, i_{\underline{I}}$$

and under true model

$$\sqrt{n}\left(\hat{\theta}_{\underline{I}} - \theta \right) \text{ is } MN\left[0, \; i_{\underline{I}}^{-1}\{ i_{\underline{I}} + 2 \sum_{h=1}^{\infty} \gamma_h \, i_{\underline{I}} \} i_{\underline{I}}^{-1} \right]$$

i.e.

$$MN\left[0, \; i_{\underline{I}}^{-1}(I + 2 \sum_{h=1}^{\infty} \gamma_h) \right]$$

(Cox, 1961;
Azzalini,
1983)

Now to assess the quality of the marginal fit from the independence model we may take $\ell_{\underline{I}}(\theta)$ estimated by $\ell_{\underline{I}}(\hat{\theta}_{\underline{I}})$. But

$$\ell_{\underline{I}}(\theta) - \ell_{\underline{I}}(\hat{\theta}_{\underline{I}}) \simeq - \frac{1}{2} n(\hat{\theta}_{\underline{I}} - \theta)^T i_{\underline{I}}^{+M}(\hat{\theta}_{\underline{I}} - \theta)$$

& the bias term has expectation

$$- \frac{1}{2} tr\left[I + 2 \sum_{h=1}^{\infty} \gamma_h \right]$$

PLATE 10. Sir David Cox's handwriting.

journals hinge on the use of quite advanced statistical procedures in a way that was unthinkable even 20 years ago.

'The scope for discussion with and help and advice from knowledgeable colleagues is far greater than it was; few statisticians work totally on their own. I remarked above on the virtual nonexistence of books when I started. Now the issue is more the almost overwhelming extent of the literature and not least the appearance each year of so many books, the great majority of a very good standard. Of course the technical advance of the subject leads to statisticians becoming more and more specialized into particular areas.

'Yet in some sense I wonder if things have changed all that much at a deeper level. Really serious subject-matter research problems virtually always have unique features which demand to be tackled, to some extent at least, from first principles and for this a sound knowledge of the central ideas of our subject and an appropriate attitude to scientific research are required more than very specialist information. My own work has been held back more often by insufficient scientific background in the subject-matter and, especially in my early years, by an inadequate theoretical base in statistics and insufficient mathematical depth, rather than by unfamiliarity with particular specialized techniques.

'Although most research nowadays is done in groups rather than by isolated individuals, research, whether theoretical or applied, remains an intensely personal activity and the best mode of working has to be found by trial and error by each one of us. The subject is at an exciting and demanding phase and I envy those starting on a career in statistical work.'

And finally, here is the humble side of a real scientist, provoked by two questions asked by Nancy Reid in her conversation with David Cox (Reid, 1994):

'The only other thing I wanted to ask you about is something that you mentioned to me in a letter a while ago which I think was connected with your knighthood. Your words were roughly that, after feeling when you were younger that you didn't get very much recognition for your work, you now felt you were receiving a "bizarre excess".'

David Cox: 'Yes, I think that sums it up adequately. Well, everybody needs encouragement, and of course as you get older you still need encouragement. But the time you most need it is when you starting. It would be quite wrong to think that people were ever discouraging, they weren't. It was all very low key, typically British understatement sort of thing. You never really knew what people thought, despite the relative frankness of Royal Statistical Society discussions. And I'd published a few papers with very little notion of whether anybody had paid any attention to them until I first went to the United States, where people would come and say, "Oh, I read that paper, I think we could do so and so." That sort of thing is very encouraging. And then, more recently, I've been absurdly lucky with all these pieces of recognition. Of course the system is a bad one in the sense that if you get one piece of recognition it's more likely you'll get another. It ought to be the other way around.'

'Was there a time when you suddenly felt you were a Very Important Person?'

David Cox: 'Well, I hope I've never thought in those terms. In a sense, the only thing that matters is if you can look back when you reach a vast, vast, vast age and say, "Have I done something reasonably in accord with my capability?" If you can say yes, okay. My feeling is in one sense, I've done that: in the tangible sense of books and papers, I've done more than I would have expected. In another sense I feel very dissatisfied: there are all sorts of problems that I nearly solved and gave up, or errors of judgement in doing a little something and not taking it far enough. That I nearly did something you see, this is the irritating thing. You know, if you'd no idea at all, well it doesn't matter, it's irrelevant, but if you feel you were within an inch of doing something and didn't quite do it ...'

Publications of D. R. Cox up to 2004

Cox, D. R. and Townsend, M. W. (1947). The analysis of yarn irregularity. *Proc. Internat. Wool Textile Org.*, **1**, 20–28.

Cox, D. R. and Townsend, M. W. (1948). The use of the correlogram in measuring yarn irregularity. *Proc. Internat. Wool Textile Org.*, **2**, 28–34.

Cox, D. R. (1948). A note on the asymptotic distribution of range. *Biometrika*, **35**, 310–315.

Cox, D. R. (1948). Fibre movement in drafting (with discussion). *J. Textile Inst.*, **39**, 230–240.

Cox, D. R. and Brearley, A. (1948). *An Outline of Statistical Methods for Use in the Textile Industry*, Leeds: Wool Industries Res. Assn. (Six editions from 1948 to 1960).

Cox, D. R. (1949). Theory of drafting of wool slivers. I. *Proc. Roy. Soc. A*, **197**, 28–51.

Cox, D. R. (1949). Use of the range in sequential analysis. *J. R. Statist. Soc. B*, **11**, 101–114.

Anderson, S. L. and Cox, D. R. (1950). The relation between strength and diameter of wool fibres. *J. Textile Inst.*, **41**, 481–491.

Cox, D. R. and Ingham, J. (1950). Some causes of irregularity in worsted drawing and spinning (with discussion). *J. Textile Inst.*, **41**, 376–420.

Cox, D. R. and Raper, G. F. (1950). Improved doubling methods in worsted drawing. *W. I. R. A. Bulletin*, **13**, 57–61.

Benson, F. and Cox, D. R. (1951). The productivity of machines requiring attention at random intervals. *J. R. Statist. Soc. B*, **13**, 65–82.

Brearley, A. and Cox, D. R. (1951). Designing an industrial experiment. *Textile Recorder*, **69**, 87.

Cox, D. R. (1951). Some systematic experimental designs. *Biometrika*, **38**, 312–323.

Anderson, S. L., Cox, D. R. and Hardy, L. D. (1952). Some rheological properties of twistless combed wool slivers. *J. Textile Inst.*, **43**, 362–379.

Brearley, A. and Cox, D. R. (1952). Experimentation in the textile industry. *Wool Review*, **23**, 39.

Cox, D. R. (1952). A note on the sequential estimation of means. *Proc. Camb. Phil. Soc.*, **48**, 447–450.

Cox, D. R. (1952). Estimation by double sampling. *Biometrika*, **39**, 217–227.

Cox, D. R. (1952). Sequential tests for composite hypotheses. *Proc. Camb. Phil. Soc.*, **48**, 290–299.

Cox, D. R. (1952). Some recent work on systematic experimental designs. *J. R. Statist. Soc. B*, **14**, 211–219.

Cox, D. R. (1952). The effect of skewness of the frequency distribution on the relationship between required test factors and strength variation. *Selected Government Research Reports*, **6**, 39–49.

Cox, D. R. (1952). The inter-fibre pressure in slivers confined in rectangular grooves. *J. Textile Inst.*, **43**, 87.

Cox, D. R. (1952). The relationship between required test factors and strength variation in the case of two types of failure. *Selected Government Research Reports*, **6**, 31–38.

Cox, D. R., Montagnon, P. E. and Starkey, R. D. (1952). Strength requirements for spot welds. *Selected Government Research Reports*, **6**, 189–200.

Cox, D. R. and Starkey, R. D. (1952). The relationship between required test factors and strength variation. *Selected Government Research Reports*, **6**, 19–30.

Cox, D. R. (1953). Some simple approximate tests for Poisson variates. *Biometrika*, **40**, 354–360.

Cox, D. R. and Smith, W. L. (1953). A direct proof of a fundamental theorem of renewal theory. *Scandinavisk Aktuartidskrift*, **36**, 139–150.

Cox, D. R. and Smith, W. L. (1953). The superimposition of several strictly periodic sequences of events. *Biometrika*, **40**, 1–11.

Cox, D. R. (1954). A note on the formal use of complex probabilities in the theory of stochastic processes. *Proceedings of the International Congress of Mathematics*, Amsterdam: North Holland, 284–285.

Cox, D. R. (1954). A table for predicting the production from a group of machines under the care of one operative. *J. R. Statist. Soc. B*, **16**, 285–287.

Cox, D. R. (1954). Some statistical aspects of mixing and blending. *J. Textile Inst.*, **45**, 113–122.

Cox, D. R. (1954). The design of an experiment in which certain treatment arrangements are inadmissible. *Biometrika*, 41, 287–295.

Cox, D. R. (1954). The mean and coefficient of variation of range in small samples from non-normal populations. *Biometrika*, **41**, 469–481.

Cox, D. R. and Smith, W. L. (1954). On the superposition of renewal processes. *Biometrika*, **41**, 91–99.

Cox, D. R. (1955). A use of complex probabilities in the theory of stochastic processes. *Proc. Camb. Phil. Soc.*, **51**, 313–319.

Cox, D. R. (1955). Prévision de la production d'un groupe de machines surveillées par an exécutant (with discussion). *Les Cahiers du Bureau des Temps Elementaires*, **503-02**, 3–20.

Cox, D. R. (1955). Some statistical methods connected with series of events (with discussion). *J. R. Statist. Soc. B*, **17**, 129–164.

Cox, D. R. (1955). The analysis of non-Markovian stochastic processes by the inclusion of supplementary variables. *Proc. Camb. Phil. Soc.*, **51**, 433–441.

Cox, D. R. (1955). The statistical analysis of congestion. *J. R. Statist. Soc. A*, **118**, 324–335.

Cox, D. R. and Stuart, A. (1955). Some quick sign tests for trend in location and dispersion. *Biometrika*, **42**, 80–95.

Cox, D. R. (1956). A note on the theory of quick tests. *Biometrika*, **43**, 478–480.

Cox, D. R. (1956). A note on weighted randomization. *Ann. Math. Statist.*, **27**, 1144–1151.

Cox, D. R. (1957). Note on grouping. *J. Amer. Statist. Assoc.*, **52**, 543–547.

Cox, D. R. (1957). The use of a concomitant variable in selecting an experimental design. *Biometrika*, **44**, 150–158.

Cox, D. R. and Smith, W. L. (1957). On the distribution of *Tribolium Confusum* in a container. *Biometrika*, **44**, 328–335.

Cox, D. R. (1958). *Planning of Experiments.* New York: Wiley.

Cox, D. R. (1958). Some problems connected with statistical inference. *Ann. Math. Statist.*, **29**, 357–372.

Cox, D. R. (1958). The interpretation of the effects of non-additivity in the Latin square. *Biometrika*, **45**, 69–73.

Cox, D. R. (1958). The regression analysis of binary sequences (with discussion). *J. R. Statist. Soc. B*, **20**, 215–242.

Cox, D. R. (1958). Two further applications of a model for binary regression. *Biometrika*, **45**, 562–565.

Cox, D. R. (1959). A renewal problem with bulk ordering of components. *J. R. Statist. Soc. B*, **21**, 180–189.

Cox, D. R. (1959). The analysis of exponentially distributed life-times with two types of failure. *J. R. Statist. Soc. B*, **21**, 411–421.

Cox, D. R. and Brandwood, L. (1959). On a discriminatory problem connected with the works of Plato. *J. R. Statist. Soc. B*, **21**, 195–200.

Cox, D. R. (1960). A note on tests of homogeneity applied after sequential sampling. *J. R. Statist. Soc. B*, **22**, 368–371.

Cox, D. R. (1960). On the number of renewals in a random interval. *Biometrika*, **47**, 449–452.

Cox, D. R. (1960). Regression analysis when there is prior information about supplementary variables. *J. R. Statist. Soc. B*, **22**, 172–176.

Cox, D. R. (1960). Serial sampling acceptance schemes derived from Bayes's theorem (with discussion). *Technometrics*, **2**, 353–368.

Cox, D. R. (1960). The economy of planned investigations. *Proceedings of the 4th International Conference of European Organization for Quality Control.*

Cox, D. R. (1961). A simple congestion system with incomplete service. *J. R. Statist. Soc. B*, **23**, 215–222.

Cox, D. R. (1961). Design of experiments: The control of error. *J. R. Statist. Soc. A*, **124**, 44–48.

Cox, D. R. (1961). Prediction by exponentially weighted moving averages and related methods. *J. R. Statist. Soc. B*, **23**, 414–422.

Cox, D. R. (1961). Tests of separate families of hypotheses. In *Proceedings of the Fourth Berkeley Symposium on Mathematical Statistics and Probability*, **1**, L. M. LeCam, J. Neyman and E. L. Scott (ed.). Berkeley: University of California Press, 105–123.

Cox, D. R. (1961). The role of statistical methods in science and technology. Inaugural Lecture, Birkbeck College, November 21.

Cox, D. R. and Smith, W. L. (1961). *Queues.* London: Methuen.

Atiqullah, M. and Cox. D. R. (1962). The use of control observations as an alternative to incomplete block designs. *J. R. Statist. Soc. B*, **24**, 464–471.

Barnard, G. A. and Cox, D. R. (ed.) (1962). *The Foundations of Statistical Inference.* London: Methuen.

Booth, K. H. V. and Cox, D. R. (1962). Some systematic supersaturated designs. *Technometrics*, **4**, 489–495.

Cox, D. R. (1962). Further results on tests of separate families of hypotheses. *J. R. Statist. Soc. B*, **24**, 406–424.

Cox, D. R. (1962). *Renewal Theory*. London: Methuen/New York: Wiley.

Cox, D. R. and Jessop, W. N. (1962). The theory of a method of production scheduling when there are many products. *Operat. Res. Quart.*, **13**, 309–328.

Cox, D. R. (1963). Large sample sequential tests for composite hypotheses. *Sankhya A*, **25**, 5–12.

Cox, D. R. (1963). Some models for series of events. *Bulletin of the International Statistical Institute*, **2**, 737–746.

Box, G. E. P. and Cox, D. R. (1964). An analysis of transformations (with discussion). *J. R. Statist. Soc. B*, **26**, 211–252.

Cox, D. R. (1964). Some applications of exponential ordered scores. *J. R. Statist. Soc. B*, **26**, 103–110.

Cox, D. R. (1965). A remark on multiple comparison methods. *Technometrics*, **7**, 223–224.

Cox, D. R. (1965). On the estimation of the intensity function of a stationary point process. *J. R. Statist. Soc. B*, **27**, 332–337.

Cox, D. R. and Miller, H. D. (1965). *The Theory of Stochastic Processes*. London: Methuen.

Cox, D. R. (1966). A note on the analysis of a type of reliability trial. *J. SIAM Appl. Math.*, **14**, 1133–1142.

Cox, D. R. (1966). A simple example of a comparison involving quantal data. *Biometrika*, **53**, 215–220.

Cox, D. R. (1966). Notes on the analysis of mixed frequency distributions. *Brit. J. Math. Statist. Psych.*, **19**, 39–47.

Cox, D. R. (1966). Some problems of statistical analysis connected with congestion (with discussion). In *Proceedings of the Symposium on Congestion Theory*, W. L. Smith and W. E. Wilkinson (ed.). Chapel Hill, North Carolina: University of North Carolina Press, 289–316.

Cox, D. R. (1966). Some procedures connected with the logistic qualitative response curve. In *Research Papers in Probability and Statistics (Festschrift for J. Neyman)*, F. N. David (ed.)(1966). Wiley: London, 55–71.

Cox, D. R. (1966). The null distribution of the first serial correlation coefficient. *Biometrika*, **53**, 623–626.

Cox, D. R. and Lewis, P. A. W. (1966). *The Statistical Analysis of Series of Events*. London: Methuen.

Lewis, P. A. W. and Cox, D. R. (1966). A statistical analysis of telephone circuit error data. *IEEE Trans. Commun. Technology*, **14**, 382–389.

Chambers, E. A. and Cox, D. R. (1967). Discrimination between alternative binary response models. *Biometrika*, **54**, 573–578.

Cox, D. R. (1967). Fieller's theorem and a generalization. *Biometrika*, **54**, 567–572.

Cox, D. R. (1967). *Some statistical concepts and their application*. Inaugural lecture, Imperial College of Science and Technology.

Cox, D. R. and Herzberg, A. M. (1967). A note on the analysis of a partially confounded 2^3 factorial experiment. *Technometrics*, **9**, 170.

Cox, D. R. and Lauh, E. (1967). A note on the graphical analysis of multidimensional contingency tables. *Technometrics*, **9**, 481–488.

Cox, D. R. (1968). Notes on some aspects of regression analysis (with discussion). *J. R. Statist. Soc. A*, **131**, 265–279.

Cox, D. R. (1968). Queues. In *International Encyclopaedia of Social Sciences*, D. L. Sills (ed.). New York: Collier and McMillan, 257–261.

Cox, D. R. and Hinkley, D. V. (1968). A note on the efficiency of least-squares estimates. *J. R. Statist. Soc. B*, **30**, 284–289.

Cox, D. R. and Snell, E. J. (1968). A general definition of residuals (with discussion). *J. R. Statist. Soc. B*, **30**, 248–275.

Cox, D. R. (1969). Some results connected with the logistic binary response model. *Bulletin of the International Statistical Institute*, **42**, 311–312.

Cox, D. R. (1969). Some sampling problems in technology. In *New Developments in Survey Sampling. A Symposium on the Foundations of Survey Sampling Held at the University of North Carolina, Chapel Hill*, N. L. Johnson and H. Smith (ed.). New York: Wiley, 506–527.

Draper, N. R. and Cox, D. R. (1969). On distributions and their transformations to normality. *J. R. Statist. Soc. B*, **31**, 472–476.

Herzberg, A. M. and Cox, D. R. (1969). Recent work on the design of experiments: A bibliography and a review. *J. R. Statist. Soc. A*, **132**, 29–67.

Cox, D. R. (1970). *The Analysis of Binary Data*. London: Methuen.

Cox, D. R. (1970). The continuity correction. *Biometrika*, **57**, 217–218.

Cox, D. R. (1971). A note on polynomial response functions for mixtures. *Biometrika*, **58**, 155–159.

Cox, D. R. (1971). Some aspects of life tables. *Bulletin of the International Statistical Institute, 38th Session*, 90–93.

Cox, D. R. (1971). The choice between alternative ancillary statistics. *J. R. Statist. Soc. B*, **33**, 251–255.

Cox, D. R. and Hinkley, D. V. (1971). Some properties of multiserver queues with appointments (with discussion). *J. R. Statist. Soc. A*, **133**, 1–11.

Cox, D. R. and Snell, E. J. (1971). On test statistics calculated from residuals. *Biometrika*, **58**, 589–594.

Bloomfield, P. and Cox, D. R. (1972). A low traffic approximation for queues. *J. Appl. Prob.*, **9**, 832–840.

Cox, D. R. (1972). Regression models and life-tables (with discussion). *J. R. Statist. Soc. B*, **34**, 187–220.

Cox, D. R. (1972). The analysis of multivariate binary data. *J. R. Statist. Soc. C*, **21**, 113–120.

Cox, D. R. (1972). The statistical analysis of dependencies in point processes. In *Stochastic Point Processes: Statistical Analysis, Theory, and Applications*, P. A. W. Lewis (ed.). New York: Wiley, 55–66.

Cox, D. R. and Herzberg, A. M. (1972). On a statistical problem of A. E. Milne. *Proc. Roy. Soc. A*, **331**, 273–283. [With an historical introduction by S. Chandrasekhar.]

Cox, D. R. and Lewis, P. A. W. (1972). Multivariate point processes. In *Proceedings of the Sixth Berkeley Symposium on Mathematical Statistics and Probability Theory*, **3**, L. M. LeCam, J. Neyman and E. L. Scott (ed.)(1972). Berkeley: University of California Press, 401–448.

Herzberg, A. M. and Cox, D. R. (1972). Some optimal designs for interpolation and extrapolation. *Biometrika*, **59**, 551–561.

Atkinson, A. C. and Cox, D. R. (1974). Planning experiments for discriminating between models (with discussion). *J. R. Statist. Soc. B*, **36**, 321–348.

Cox, D. R. and Hinkley, D. V. (1974). *Theoretical Statistics*. London: Chapman and Hall.

Cox, D. R. and Snell, E. J. (1974). The choice of variables in observational studies. *J. R. Statist. Soc. C*, **23**, 51–59.

Cox, D. R. (1975). A note on data-splitting for the evaluation of significance levels. *Biometrika*, **62**, 441–444.

Cox, D. R. (1975). A note on empirical Bayes inference in a finite Poisson process. *Biometrika*, **62**, 709–711.

Cox, D. R. (1975). A note on partially Bayes inference and the linear model. *Biometrika*, **62**, 651–654.

Cox, D. R. (1975). Partial likelihood. *Biometrika*, **62**, 269–276.

Cox, D. R. (1975). Prediction intervals and empirical Bayes confidence intervals. In *Perspectives in Probability and Statistics*, J. Gani (ed.), London: Academic Press, 47–55.

Peto, R., Pike, M. C., Armitage, P., Breslow, N. E., Cox, D. R., Howard, S. V., Mantel, N., McPherson, K., Peto, J. and Smith, P. G. (1976). Design and analysis of randomized clinical trials requiring prolonged observation of each patient. I Introduction and Design. *British J. Cancer*, **34**, 585–612.

Atkinson, A. C. and Cox, D. R. (1977). Robust regression via discriminant analysis. *Biometrika*, **64**, 15–19.

Cox, D. R. (1977). Nonlinear models, residuals and transformations. *Math. Operationsforch. Statist.*, **8**, 3–22.

Cox, D. R. (1977) The role of significance tests (with discussion). *Scand. J. Statist.*, **4**, 49–70.

Cox, D. R. (1977). The teaching of the strategy of statistics. *Proceedings of the 41st session. Bulletin of the International Statistical Institute*, **47**, 552–558.

Cox, D. R. and Isham, V. (1977). A bivariate point process connected with electronic counters. *Proc. Roy. Soc. A*, **356**, 149–160.

Peto, R., Pike, M. C., Armitage, P., Breslow, N. E., Cox, D. R., Howard, S. V., Mantel, N., McPherson, K., Peto, J. and Smith, P. G. (1977). Design and analysis of randomized clinical trials requiring prolonged observation of each patient. II Analysis and Examples. *British J. Cancer*, **35**, 1–39.

Cox, D. R. (1978). Foundations of statistical inference: The case for eclecticism (with discussion). *J. Australian Statist. Soc.*, **20**, 43–59.

Cox, D. R. (1978). Queues. *International Encyclopaedia of Statistics*, W. H. Kruskal and J. W. Tanur (ed.), Volume 2, New York: Free Press/London: Collier Macmillan, 834–838.

Cox, D. R. (1978). Some remarks on the role in statistics of graphical methods. *J. R. Statist. Soc. C*, **27**, 4–9.

Cox, D. R. and Hinkley, D. V. (1978). *Problems and Solutions in Theoretical Statistics*. London: Chapman and Hall.

Cox, D. R. and Isham, V. (1978). Series expansions for the properties of a birth process of controlled variability. *J. Appl. Prob.*, **15**, 610–616.

Cox, D. R. and Small, N. J. H. (1978). Testing multivariate normality. *Biometrika*, **65**, 263–272.

Barndorff-Nielsen, O. E. and Cox, D. R. (1979). Edgeworth and saddle-point approximations with statistical applications (with discussion). *J. R. Statist. Soc. B*, **41**, 279–312.

Cox, D. R. (1979). A note on the graphical analysis of survival data. *Biometrika*, **66**, 188–190.

Cox, D. R. (1979). A remark on systematic Latin squares *J. R. Statist. Soc. B*, **41**, 388–389.

Cox, D. R. and Isham, V. (1979). *Point Processes*. London: Chapman and Hall.

Cox, D. R. and Snell, E. J. (1979). On sampling and the estimation of rare errors. *Biometrika*, **66**, 125–132.

Farewell, V. T. and Cox, D. R. (1979). A note on multiple time scales in life testing. *J. R. Statist. Soc. C*, **28**, 73–75.

Cox, D. R. (1980). Local ancillarity. *Biometrika*, **67**, 279–286.

Cox, D. R. (1980). Summary views: A statistician's perspective. *Cancer Treatment Reports*, **64**, 533–535.

Cox, D. R. (1981). Statistical analysis of time series: Some recent developments (with discussion). *Scand. J. Statist.*, **8**, 93–115.

Cox, D. R. (1981). Statistical significance tests. *British J. Clin. Pharmac.*, **14**, 325–331.

Cox, D. R. (1981). Theory and general principle in statistics. *J. R. Statist. Soc. A*, **144**, 289–297.

Cox, D. R. and Snell, E. J (1981). *Applied Statistics. Principles and Examples*. London: Chapman and Hall.

Box, G. E. P. and Cox, D. R. (1982). Transformations revisited, rebutted. *J. Am. Statist. Assoc.*, **77**, 209–210.

Cox, D. R. (1982). A remark on randomization in clinical trials. *Utilitas Mathematics*, **21A**, 245–252.

Cox, D. R. (1982). Combination of data. In *Encyclopaedia of Statistical Sciences*, **2**, S. Kotz and N. L. Johnson (ed.). New York: Wiley, 45–53.

Cox, D. R. (1982). On the role of data of possibly lowered sensitivity. *Biometrika*, **69**, 215–219.

Cox, D. R. (1982). Randomization and concomitant variables in the design of experiments. In *Statistics and Probability: Essays in honor of C. R. Rao*, G. Kallianpur, P. R. Krishnaiah and J. K. Ghosh. Amsterdam: North-Holland, 197–202.

Cox, D. R. and McCullagh, P. (1982). Some aspects of analysis of covariance (with discussion). *Biometrics*, **38**, 541–561.

Cox, D. R. and Spjøtvoll, E. (1982). On partitioning means into groups. *Scand. J. Statist.*, **9**, 147–152.

Cox, D. R. (1983). A remark on censoring and surrogate response variables. *J. R. Statist. Soc. B*, **45**, 391–393.

Cox, D. R. (1983). Asymptotic theory: Some recent developments. *Questio*, **7**, 527–529.

Cox, D. R. (1983). Remarks on the analysis of additional response variables associated with survival data. *Bulletin of the International Statistical Institute, 44th Session*, **1**, 223–224.

Cox, D. R. (1983). Some remarks on overdispersion. *Biometrika*, **70**, 269–274.

Azzalini, A. and Cox, D. R.(1984). Two new tests associated with analysis of variance. *J. R. Statist. Soc. B*, **46**, 335–343.

Barndorff-Nielsen, O. E. and Cox, D. R. (1984). Bartlett adjustments to the likelihood ratio statistic and the distribution of the maximum likelihood estimator. *J. R. Statist. Soc. B*, **46**, 483–495.

Barndorff-Nielsen, O. E., and Cox, D. R. (1984). The effect of sampling rules on likelihood statistics. *International Statistical Review*, **52**, 309–326.

Barndorff-Nielsen, O. E. and Cox, D. R. (1984). The role of mathematics in theoretical statistics. In *Perspectives in Mathematics. Anniversary of Oberwolfach*, Birkhauser Verlag, Basel, 61–81.

Cox, D. R. (1984). Effective degrees of freedom and the likelihood ratio test. *Biometrika*, **71**, 487–493.

Cox, D. R. (1984). Interaction. *International Statistical Review*, **52**, 1–31.

Cox, D. R. (1984). Long-range dependence: A review. In *Statistics: An Appraisal*. H. A. David and H. T. David (ed.). Ames, Iowa: The Iowa State University Press, 55–74.

Cox, D. R. (1984). Present position and potential developments: Some personal views. Design of experiments and regression (with discussion). *J. R. Statist. Soc. A*, **147**, 306–315.

Cox, D. R. (1984). Some remarks on semi-Markov processes in medical statistics. In *International Symposium of Semi-Markov Processes and Their Applications*, Brussels, 411–421.

Cox, D. R. and Oakes, D. (1984). *Analysis of Survival Data*. London: Chapman and Hall.

Cox, D. R. (1985). Some remarks on asymptotic theory. *Bulletin of the International Statistical Institute, 45th Session*, **2**, 363–364.

Cox, D. R. (1985). Theory of statistics: Some current themes. *Bulletin of the International Statistical Institute, 45th Session*, **1**, 6.3-1–6.3-9.

Barndorff-Nielsen, O. E., Cox, D. R. and Reid, N. (1986). The role of differential geometry in statistical theory. *International Statistical Review*, **54**, 83–96.

Cox, D. R. (1986). Some general aspects of the theory of statistics. *International Statistical Review*, **54**, 117–126.

Cox, D. R. and Isham, V. (1986). The virtual waiting-time and related processes. *Adv. Appl. Prob.*, **18**, 558–573.

Cox, D. R. and Solomon, P. J. (1986). Analysis of variability with large numbers of small samples. *Biometrika*, **73**, 543–554.

McCullagh, P. and Cox, D. R. (1986). Invariants and likelihood ratio statistics. *Ann. Statist.*, **14**, 1419–1430.

Rodriguez-Iturbe, I., Cox, D. R. and Eagleson, P. S. (1986). Spatial modelling of total storm rainfall. *Proc. Roy. Soc. A*, **403**, 27–50.

Cox, D. R. (1987). Some aspects of the design of experiments. In *New Perspectives in Theoretical and Applied Statistics*, M. L. Puri, J. P. Vilaplana and W. Wertz (ed.). New York: Wiley, 3–12.

Cox, D. R. and Reid, N. (1987). Approximations to noncentral distributions. *Can. J. Statist.*, **15**, 105–114.

Cox, D. R. and Reid, N. (1987). Parameter orthogonality and approximate conditional inference (with discussion). *J. R. Statist. Soc. B*, **49**, 1–39.

Medley, G. F., Anderson, R. M., Cox, D. R. and Billard, L. (1987). Incubation period of AIDS in patients infected via blood transfusion. *Nature*, **328**, 719–721.

Rodriguez-Iturbe, I., Cox, D. R. and Isham, V. (1987). Some models for rainfall based on stochastic point processes. *Proc. Roy. Soc. A*, **410**, 269–288.

Atkinson, A. C. and Cox, D. R. (1988). Transformations. In *Encyclopaedia of Statistical Science*, **9**, S. Kotz and N. L. Johnson (ed.). New York: Wiley, 312–318.

Cox, D. R. (1988). A note on design when response has an exponential family distribution. *Biometrika*, **75**, 161–164.

Cox, D. R. (with members of Working Party) (1988). *Short-term prediction of HIV infection and AIDS in England and Wales*. Her Majesty's Stationery Office.

Cox, D. R. (1988). Some aspects of conditional and asymptotic inference: A review. *Sankhyā A*, **50**, 314–337.

Cox, D. R. and Isham, V. (1988). A simple spatial-temporal model of rainfall. *Proc. Roy. Soc. A*, **415**, 317–328.

Cox, D. R. and Solomon, P. J. (1988). A note on selective intervention. *J. R. Statist. Soc. A*, **151**, 310–315.

Cox, D. R. and Solomon, P. J. (1988). On testing for serial correlation in large numbers of small samples. *Biometrika*, **75**, 145–148.

Medley, G. F., Billard, L., Cox, D. R. and Anderson, R. M. (1988). The distribution of the incubation period for the acquired immunodeficiency syndrome (AIDS). *Proc. Roy. Soc. B*, **233**, 367–377.

Rodriguez-Iturbe, I., Cox, D. R. and Isham, V. (1988). A point process model for rainfall: further developments. *Proc. Roy. Soc. A*, **417**, 283–298.

Barndorff-Nielsen, O. E. and Cox, D. R. (1989). *Asymptotic Techniques for Use in Statistics*. London: Chapman and Hall.

Cox, D. R. (1989). The relation between the statistician and the substantive research worker. *Societa Italiana di Statistica; Statistica e Societa*, 377–382.

Cox, D. R. and Davison, A. C. (1989). Prediction for small subgroups. *Phil. Trans. Roy. Soc. B*, **325**, 185–187.

Cox, D. R. and Medley, G. F. (1989). A process of events with notification delay and the forecasting of AIDS. *Phil. Trans. Roy. Soc. B*, **325**,135–145.

Cox, D. R. and Reid, N. (1989). On the stability of maximum-likelihood estimators of orthogonal parameters. *Can. J. Statist.*, **17**, 229–233.

Cox, D. R. and Snell, E. J. (1989). *Analysis of Binary Data*, 2nd edn. London: Chapman and Hall.

Cruddas, A. M., Reid, N. and Cox, D. R.(1989). A time series illustration of approximate conditional likelihood. *Biometrika*, **76**, 231–237.

Davison, A. C. and Cox, D. R. (1989). Some simple properties of sums of random variables having long-range dependence. *Proc. Roy. Soc. A*, **424**, 255–262.

Cox, D. R. (1990). Quality and reliability: Some recent developments and a historical perspective. *J. Operational Res. Soc.*, **41**, 95–101.

Cox, D. R. (1990). Role of models in statistical analysis. *Statistical Science*, **5**, 169–174.

Cox, D. R. and Wermuth, N. (1990). An approximation to maximum likelihood estimates in reduced models. *Biometrika*, **77**, 747–761.

Darby, S. C., Doll, R., Thakrar, B., Rizza, C. R. and Cox, D. R. (1990). Time from infection with HIV to onset of AIDS in patients with haemophilia in the UK. *Statistics in Medicine*, **9**, 681–689.

Cox, D. R. (1991). Long-range dependence, non-linearity and time irreversibility. *Journal of Time Series Analysis*, **12**, 329–335.

Cox, D. R. (1991). The contribution of statistical methods to cancer research. *Cancer*, **67**, 2428–2430.

Cox, D. R. and Wermuth, N. (1991). A simple approximation for bivariate and trivariate normal integrals. *International Statistical Review*, **59**, 263–269.

Ferguson, H., Reid, N. and Cox, D. R. (1991). Estimating equations from modified profile likelihood. In *Estimating Functions*, V. P. Godambe (ed.) Oxford University Press, 279–293.

Cox, D. R. (1992). Causality: Some statistical aspects. *J. R. Statist. Soc. A*, **155**, 291–301.

Cox, D. R. (1992). Statistical analysis and survival data: An introduction. *Annual of Cardiac Surgery*, 97–100.

Cox, D. R. and Darby, S. C. (1992). Statistical and epidemiological aspects of cancer research. *Accomplishments in Cancer Research*, J. G. Fortner and J. E. Rhoads (ed.). Philadelphia: Lippincott, 344–345.

Cox, D. R., Fitzpatrick, R., Fletcher, A. E., Gore, S. M., Spiegelhalter, D. J. and Jones, D. R. (1992). Quality-of-life assessment: Can we keep it simple (with discussion)? *J. R. Statist. Soc. A*, **155**, 353–393.

Cox, D. R. and Reid, N. (1992). A note on the difference between profile and modified profile likelihood. *Biometrika*, **79**, 408–411.

Cox, D. R. and Wermuth, N. (1992). A comment on the coefficient of determination for binary responses. *American Statistician*, **46**, 1–4.

Cox, D. R. and Wermuth, N. (1992). Graphical models for dependencies and associations. *In Computational Statistics*, Vol. 1, Y. Dodge and J. Whittaker (ed.). Heidelberg: Physica-Verlag, 235–249.

Cox, D. R. and Wermuth, N. (1992). On the calculation of derived variables in the analysis of multivariate responses. *Journal of Multivariate Analysis*, **42**, 162–170.

Cox, D. R. and Wermuth, N. (1992). On the relation between interactions obtained with alternative codings of discrete variables. *Methodika*, **6**, 76–85.

Cox, D. R. and Wermuth, N. (1992). Response models for mixed binary and quantitative variables. *Biometrika*, **79**, 441–461.

Fitzpatrick, R., Fletcher, A. E., Gore, S. M., Jones, D. R., Spiegelhalter, D. J. and Cox, D. R. (1992). Quality of life measures in health care. I: Applications and issues in assessment. *British Medical Journal*, **305**, 1074–1077.

Fletcher, A. E., Gore, S. M., Jones, D. R., Spiegelhalter, D. J., Fitzpatrick, R. and Cox, D. R. (1992). Quality of life measures in health care. II Design analysis and interpretation. *British Medical Journal*, **305**, 1145–1148.

Gunby, J. A., Darby, S. C., Miles, J. H. C., Green, B. M. R. and Cox, D. R. (1992). Factors affecting indoor radon concentrations in the United Kingdom. *Health Physics*, **64**, 2–12.

Solomon, P. J. and Cox, D. R. (1992). Nonlinear component of variance models. *Biometrika*, **79**, 1–11.

Spiegelhalter, D. J., Gore, S. M., Fitzpatrick, R., Fletcher, A. E., Jones, D. R. and Cox, D. R. (1992). Quality of life measures in health care. III: resource allocation. *British Medical Journal*, **305**, 1205–1209.

Cox, D. R. (1993). Causality and graphical models. *Bulletin of the International Statistical Institute, 49th Session*, **1**, 363–372.

Cox, D. R. (1993). Some aspects of statistical models. In *Newer Thinking in Theory and Methods for Population Health Research*, K. Dean (ed.). London: Sage, 145–159.

Cox, D. R. (1993). Statistics that are functions of parameters. *Bulletin of the International Statistical Institute, 49th Session*, 293–294.

Cox, D. R. (1993). Unbiased estimating equations derived from statistics that are functions of a parameter. *Biometrika*, **80**, 905–909.

Cox, D. R. and Reid, N. (1993). A note on the calculation of adjusted profile likelihood. *J. R. Statist. Soc. B*, **55**, 467–471.

Cox, D. R. and Wermuth, N. (1993). Linear dependencies represented by chain graphs (with discussion). *Statistical Science*, **8**, 204–283.

Cox, D. R. and Wermuth, N. (1993). Some recent work on methods for the analysis of multivariate observational data in the social sciences. In *Multivariate Analysis: Future Directions*, C. R. Rao (ed.). Elsevier, 95–114.

Barndorff-Nielsen, O. E. and Cox, D. R. (1994). *Inference and Asymptotics*. London: Chapman and Hall.

Cox, D. R. and Davison, A. C. (1994). Some comments on the teaching of stochastic processes to engineers. *Int J. Cont. Engineering Educ.*, **4**, 24–30.

Cox, D. R. and Isham, V. (1994). Stochastic models of precipitation. In *Statistics for the Environment*, Volume 2, Water Related Issues, V. Barnett and K. F. Turkman (ed.). Chichester, U. K.: Wiley. 3–19.

Cox, D. R. and Wermuth, N. (1994). A note on the quadratic exponential binary distribution. *Biometrika*, **81**, 403–408.

Cox, D. R. and Wermuth, N. (1994). Tests of linearity, multivariate normality and the adequacy of linear scores. *J. R. Statist. Soc. C*, **43**, 347–355.

Cox, D. R. (1995). Some recent developments in statistical theory. *Scand Actuarial J., (Harald Cramér Symposium)*, **1**, 29–34.

Cox, D. R. (1995). The relation between theory and application in statistics (with discussion). *Test*, **4**, 207–261.

Cox, D. R. and Wermuth, N. (1995). Derived variables calculated from similar responses: Some characteristics and examples. *Computational Statistics and Data Analysis*, **19**, 223–234.

Barndorff-Nielsen, O. E. and Cox, D. R. (1996). Prediction and asymptotics. *Bernoulli*, **2**, 319–340.

Cox, D. R. and Wermuth, N. (1996). *Multivariate Dependencies*. London: Chapman and Hall.

Carpenter, L. M., Machonochie, N. E. S., Roman, E. and Cox, D. R. (1997). Examining associations between occupation and health using routinely collected data. *J. R. Statist. Soc. A*, **160**, 507–521.

Cox, D. R. (1997). Bartlett adjustment. In *Encyclopedia of Statistical Science*. Update Vol. 1, S. Kotz, C. B. Read and D. L. Banks (ed.). New York: Wiley, 43–45.

Cox, D. R. (1997). Factorisations for mixed discrete and continuous variables. *Bulletin of the International Statistical Institute, 51st session*, **57**, 2, 345–346.

Cox, D. R. (1997). The nature of statistical inference: Johann Bernoulli Lecture 1997. *Nieuw Archief voor Wiskunde*, **15**, 233–242.

Cox, D. R. (1997). Some remarks on the analysis of survival data. In *Proceedings of the First Seattle Symposium in Biostatistics: Survival Analysis*, D. Lin and T. R. Fleming (ed.) (1997). New York: Springer-Verlag, 1–9.

Cox, D. R. (1997). The current position of statistics: A personal view. *International Statistical Review*, **65**, 261–290.

Cox, D. R. and Farewell, V. T. (1997). Qualitative and quantitative aspects should not be confused. *British Medical Journal*, **314**, 73.

Fitzmaurice, G. M., Heath, A. H. and Cox, D. R. (1997). Detecting overdispersion in large-scale surveys: Application to a study of education and social class in Britain (with discussion). *J. R. Statist. Soc. C*, **46**, 415–432.

Cox, D. R. (1998). Components of variance: a miscellany. *Statistical Methods in Medical Research*, **7**, 3–12.

Cox, D. R. (1998). Some remarks on statistical education. *J. R. Statist. Soc. D*, **47**, 211–213.

Cox, D. R. (1998). Statistics, overview. In *Encyclopedia in Biostatistics*, 6, P. Armitage and T. Colton (ed.). Chichester, U. K.: Wiley, 4294–4302.

Cox, D. R. (1998). Yoke. In *Encyclopedia of Statistical Sciences*. Update Vol. 2, S. Kotz, C. B. Read and D. L. Banks (ed.). New York: Wiley, 711–712.

Cox, D. R. and Isham, V. (1998). Stochastic spatial-temporal models for rain. In *Stochastic Models in Hydrology*, O. E. Barndorff-Nielsen, S. K. Gupta and E. Waymire (ed.) (1998). Singapore: World Scientific, 1–24.

Nemunaitis, J., Poole, C., Primrose, J., Rosemurgy, A., Malfetano, J., Brown, P., Berrington, A., Cornish, A., Lynch, K., Rasmussen, H., Kerr, D., Cox, D. R. and Millar, A. (1998). Combined analysis of the studies of the effects of the matrix metalloproteinase inhibitor marimastat on serum tumour markers in advanced cancer: Selection of a biologically active and tolerable dose for longer- term studies. *Clinical Cancer Research*, **4**, 1101–1109.

Reeves, G. K., Cox, D. R., Darby, S. C. and Whitley, E. (1998). Some aspects of measurement error in explanatory variables for continuous and binary regression models. *Statistics in Medicine*, **17**, 2157–2177.

Wermuth, N. and Cox, D. R. (1998). On association models defined over independence graphs. *Bernoulli* **4**, 477–494.

Wermuth, N. and Cox, D. R. (1998). On the application of conditional independence to ordinal data. *International Statistical Review*, **66**, 181–199.

Wermuth, N. and Cox, D. R. (1998). Statistical dependence and independence. In *Encyclopedia in Biostatistics*, 6, P. Armitage and T. Colton (ed.). Chichester, U. K.: Wiley, 4260–4264.

Wong, M. Y. and Cox, D. R. (1998). A note on the robust interpretation of regression coefficients. *Test*, **7**, 287–294.

Cox, D. R. (1999). Some remarks on failure-times, surrogate markers, degradation, wear and the quality of life. *Lifetime Data Analysis*, **5**, 307–314.

Cox, D. R. (1999). The generation of some multivariate structures. *Bulletin of the International Statistical Institute, 52nd Session*, **1**, 227–228.

Cox, D. R. (1999). Variables, types of. In *Encylopedia of Statistical Sciences*, Update Vol. 3, S. Kotz, C. B. Read and D. L. Banks (ed.). New York: Wiley, 772–775.

Cox, D. R., Darby, S. C., Reeves, G. K. and Whitley, E. (1999). The effects of measurement errors with particular reference to a study of exposure to residential radon. In *Radiation Dosimetry*, E. Ron (ed.). National Cancer Institute, 139–151.

Cox, D. R. and Wermuth, N. (1999). Derived variables for longitudinal studies. *Proc. Nat. Acad. Sci. (USA)*, **96**, 12273–12274.

Cox, D. R. and Wermuth, N. (1999). Likelihood factorizations for mixed discrete and continuous variables. *Scand. J. Statist.*, **25**, 209–220.

Rodriguez-Iturbe, I., Porporato, A., Ridolfi, L., Isham, V. and Cox, D. R. (1999). Probabilistic modelling of water balance at a point: the role of climate, soil and vegetation. *Proc. Roy. Soc. A*, **455**, 3789–3805.

Cox, D. R. (2000). Some remarks on likelihood factorization. In *IMS Lecture Note Series 36, Papers in honor of W. van Zwet*, A. van der Vaart *et al.* (ed.), 165–172.

Cox, D. R. (2000). The five faces of Bayesian statistics. *Calcutta Statistical Association Bulletin*, **50**, 127–136.

Cox, D. R., Isham, V. and Northrop, P. (2000). Statistical modelling and analysis of spatial patterns. In *The Geometry of Ecological Systems*, U. Dieckman, R. Law and J. A. J. Metz (ed.). Cambridge University Press, 65–88.

Cox, D. R. and Reid, N. (2000). *The Theory of the Design of Experiments*. London: Chapman and Hall/CRC Press.

Cox, D. R. and Wermuth, N. (2000). On the generation of the chordless four-cycle. *Biometrika*, **87**, 206–212.

Gravenor, M. B., Cox, D. R., Hoinville, L. J., Hoek, A. and McLean, A. R. (2000). Scrapie in Britain during the BSE years. *Nature*, **40**, 584–585.

Moore, R. J., Jones, D. A., Cox, D. R. and Isham, V. S. (2000). Design of the HYREX raingauge network. *Hydrology and Earth Systems Sciences*, **4**, 523–530.

Rotnitzky, A., Cox, D. R., Bottai, M. and Robins, J. (2000). Likelihood-based inference with singular information matrix. *Bernoulli*, **6**, 243–284.

Wheater, H. S., Isham, V. S., Cox, D. R., Chandler, R. E., Kakou, A., Northrop, P. J., Oh, L., Onof, C. and Rodriguez-Iturbe, I. (2000). Spatial-temporal rainfall fields: modeling and statistical aspects. *Hydrology and Earth Systems Sciences*, **4**, 581–601.

Cox, D. R. (2001). Biometrika: The first 100 years. *Biometrika*, **88**, 3–11.

Cox, D. R. and Donnelly, C. A. (2001). Mathematical biology and medical statistics: contributions to the understanding of AIDS epidemiology. *Statistical Methods in Medical Research*, **10**, 141–154.

Cox, D. R. and Wermuth, N. (2001). Causal inference and statistical fallacies. In *The International Encyclopedia of Social and Behavioural Sciences*, Vol. 3, N. J. Smelsep and P. B Baltes (ed.). Oxford: Elsevier, 1554–1561.

Cox, D. R. and Wermuth, N. (2001). Some statistical aspects of causality. *European Sociological Review*, **17**, 65–74.

Gravenor, M. B., Cox, D. R., Hoinville, L. J., Hoek, A. and McLean, A. R. (2001). The flock-to-flock force of infection for scrapie in Britain. *Proc. Roy. Soc. London B*, **268**, 587–592.

Law, G. R., Cox, D. R., Machonochie, N. E. S., Simpson, E., Roman, E. and Carpenter, L. M. (2001). Large tables. *Biostatistics*, **2**, 163–171.

Wermuth, N. and Cox, D. R. (2001). Graphical models: Overview. In *International Encyclopedia of Social and Behavioural Sciences*, Vol. 9, N. J. Smelsep and P. B Baltes (ed.) (2001). Oxford: Elsevier, 6379–6386.

Wong, M. Y. and Cox, D. R. (2001). A test of multivariate independence based on a single factor model. *J. Multivariate Analysis*, **79**, 219–225.

Cox, D. R. (2002). Karl Pearson and the chi-squared test. In *Goodness of Fit Test and Model Validity*, C. Huber-Carol *et al.* (ed.) Boston: Birkhauser, 3–8.

Cox, D. R. (2002). Stochasticity. In *Encyclopedia of Environmetrics*, **4**. A. H. El-Shaarawi and W. W. Piegorsch (ed.). New York: Wiley, 2142–2143.

Cox, D. R. and Hall, P. (2002). Estimation in a simple random effects model with nonnormal distributions. *Biometrika*, **89**, 831–840.

Cox, D. R., Isham, V. S. and Northrop, P. J. (2002). Floods: some probabilistic and statistical approaches. *Phil. Trans. Roy. Soc. A*, **360**, 1389–1408.

Cox, D. R. and Solomon, P. J. (2002). *Components of Variance.* London: Chapman and Hall/CRC Press.

Cox, D. R. and Wermuth, N. (2002). On some models for multivariate binary variables parallel in complexity with the multivariate Gaussian distribution. *Biometrika*, **89**, 462–469.

Berrington, A. and Cox, D. R. (2003). Generalized least squares for the synthesis of correlated information. *Biostatistics*, **4**, 423–431.

Bird, S. M., Cox, D. R., Farewell, V. T., Goldstein, H., Holt, T. and Smith, P. C. (2003). Performance indicators: Good, bad, and ugly. *Royal Statistical Society.* Working Party on Performance Monitoring in the Public Services.

Carpenter, L. M., Laws, G., Roman, E., Cox, D. R., Machonochie, N. E. S. and Simpson, E. (2003). Occupation and cancer: the application of a novel graphical approach to routinely collected registration data. *Health Statistics Quarterly*, **17**, 5–12.

Cox, D. R. (2003). Communication of risk: health hazards from mobile phones. *J. R. Statist. Soc. A*, **166**, 241–246.

Cox, D. R. (2003). Conditional and marginal association for binary random variables. *Biometrika*, **90**, 982–984.

Cox, D. R. (2003). Henry Ellis Daniels. *Biographical Memoirs of Fellows of the Royal Society*, **49**, 133–146.

Cox, D. R. (2003). Some remarks on statistical aspects of econometrics. In *Stochastic Musings*, J. Panaretos (ed.). Mahwah, New Jersey: Lawrence Erlbaum, 20–28.

Cox, D. R. and Wermuth, N. (2003). A general condition for avoiding effect reversal after marginalization. *J. R. Statist. Soc. B*, **65**, 937–941.

Cox, D. R. and Wermuth, N. (2003). Conditional and marginal association and the preservation of monotonicity. *Bulletin of the International Statistical Institute, 54th Session*, **60**, Book 1, 221–227.

Donnelly, C. A., Cox, D. R., Bourne, J., Le Fevre, A. Woodroffe, R. Gettinby, G. and Morrison, I. (2003). Impact of localized badger culling on tuberculosis in British cattle. *Nature*, **426**, 934–937.

Kardoun, O. J. W. F., Salom, D., Schaafsma, W., Steerneman, A. J. M., Willens, J. C. and Cox, D. R. (2003). Reflections on fourteen cryptic issues concerning the nature of statistical inference (with discussion). *International Statistical Review*, **71**, 277–318.

Cox, D. R. (2004). Some remarks on model criticism. In *Methods and Models in Statistics*, N. Adams, M. Crowder, D. J. Hand and D. Stephens (ed.). London: Imperial College Press, 13–22.

Cox, D. R. and Reid, N. (2004). A note on pseudo-likelihood constructed from marginal densities. *Biometrika*, **91**, 729–737.

Cox, D. R. and Wermuth, N. (2004). Causality: a statistical view. *International Statistical Review*, **72**, 285–305.

Cox, D. R. and Wong, M. Y. (2004). A simple procedure for the selection of significant efforts. *J. R. Statist. Soc. B*, **66**, 395–400.

Wermuth, N. and Cox, D. R. (2004). Joint response graphs and separation induced by triangular systems. *J. R. Statist. Soc. B*, **66**, 687–718.

1
Stochastic models for epidemics

Valerie Isham

1.1 Introduction

Early in 1987, the AIDS epidemic was still in its relatively early stages in the United Kingdom, although well underway in the United States. In preparation for a one-day Royal Statistical Society meeting on the statistical aspects of AIDS (Royal Statistical Society 1988) later that year, an informal workshop was organised by David Cox to discuss the various mathematical modelling approaches then being employed. At David's instigation, I prepared a review paper (Isham 1988) on models of the transmission dynamics of HIV and AIDS for the one-day meeting, based in part on the workshop, and thus began my involvement with epidemic modelling. It has proved to be an area full of theoretical challenges and complex applied problems. Understanding the dynamics of the spread of an infectious disease brings possibilities for its control. Human infections such as influenza, malaria and HIV are still major worldwide causes of morbidity and mortality. In 2001, the UK experienced the severe economic and social effects of an epidemic of foot and mouth disease, while early in 2003 the world became aware of a new infection, Severe Acute Respiratory Syndrome (SARS), spreading rapidly across the globe. In both cases, the ease and speed of modern communications meant that considerable transmission had already taken place before the threat of an epidemic was recognized. In such situations, strict control strategies must be imposed rapidly if they are to be effective in curtailing the spread of infection and preventing a large-scale epidemic. A clear understanding of the basic theory of epidemic models is a prerequisite for this and, equally, for informing the development of the successful vaccination policies against a wide range of infections (Anderson and May 1991) routinely used around the world.

In this chapter, I will concentrate on a review of some of the topics that have pre-occupied researchers working on models for the transmission of infection over the last fifteen years or so, looking at progress that has been made and identifying continuing challenges. The topics discussed will be a personal selection and the references given will be indicative rather than complete. I will concentrate on model structure and general modelling issues rather than on models for specific infections, although some of these will be used for illustration, and I will focus almost entirely on models of infections for animal/human hosts and mention processes on regular spatial lattices, more relevant for plant diseases, only in passing. I will also touch only briefly on statistical aspects. It is interesting to look back a decade to the issues identified in the discussion paper of Mollison *et al.* (1994): heterogeneity, thresholds and persistence, non-stationarity, control. In spite of considerable advances, many of these are equally topical today and are destined to be the focus of research effort for some while to come. As is appropriate for this volume, my focus is on *stochastic* models. Those contributing to that Royal Statistical Society workshop in 1987 included Norman Bailey and Maurice Bartlett, both of whom had already played pivotal roles in developing the theory of stochastic epidemic models (Bailey 1975, Bartlett 1960). Nevertheless, at that stage, most applied work on specific infections relied on deterministic models, as described in the key text of Anderson and May (1991) and the later review of Hethcote (2000). One of the most noticeable changes of the last fifteen years has been the increased acceptance by very many biologists, not only of the role that mathematical modelling has to play in the solution of many of their most pressing problems, but also that such models will often need to incorporate *intrinsic* stochasticity. Thus, not only is there the long-acknowledged need to incorporate uncertainty in parameter values—generally achieved via a numerical sensitivity analysis—but, more importantly, the fact that the size of a population of infected hosts or of infecting parasites within a single host is random and integer-valued is recognised to have many important consequences. These include effects on such issues as the possible fade-out and re-emergence of an infection, the prediction of the course of an individual realization of an epidemic, and the determination of the appropriate period of application of a control treatment.

The purposes for which models are developed are many and varied. The suitability of a model and the appropriateness of the assumptions on which it is based depend entirely on its purpose. Models may be used for a careful exposition of issues and general understanding of transmission dynamics, to reach general qualitative conclusions, or for real-time use in a particular epidemic. They can also help in formulating what data should be collected in order to answer particular questions of interest. Simple models can lead to qualitative understanding and provide robust approximations. More complex models inevitably require more detailed assumptions about the nature of the underlying process and tend to have correspondingly more unknown parameters. Such models often require numerical solution or simulation, but can still allow those factors and sources of heterogeneity largely responsible for driving the dynamics to be distinguished from those having little influence. A particular strength of models is their ability to answer 'what if' questions. Thus they have an important role to play in determining control strategies, formulating ways to detect a future outbreak and developing contingency plans to deal with such an outbreak—see, for example Keeling *et al.* (2003) in the context of foot and mouth disease. In recent years, drug resistance has become an important issue, and the evolution of such resistance, and means of its control, are major concerns (Austin *et al.* 1999). Epidemic models have an important role to play in underpinning work in this area.

In some applications, however, recognition of the complex structure of the contacts that may result in transmission of infection has resulted in numerical simulation of individual-based models, in which each host is represented separately. Such models are often wholly, or largely, deterministic and seek explicitly to represent the detailed complexity of real-world contacts, rather than subsuming less-important details in random variation. In a few cases, individual-based models are founded on very detailed observation of a particular population, although even then their construction may only depend on marginal rather than joint properties of the population. Nevertheless, in the context of a specific infection, such models may give useful information about the potential spread and control of the infection in that population, although that information is unlikely to be applicable in any other situation. More often, though, individual-based models require vast numbers of poorly substantiated assumptions, especially concerning parameter values, to which the robustness of the results obtained is uncertain. In addition, such models generally convey little insight into which mechanisms are the most important in driving the transmission of infection, and those to which the dynamics are relatively insensitive. The claim is that such models are needed because of their 'realistic' assumptions, in contrast to the simple population structure and within-subgroup homogeneous mixing that underlie most population models. It is therefore vitally important to explore how model properties are affected by population heterogeneity and structure, and in recent years much effort has been directed at addressing such questions.

1.2 Historical background

Mathematical modelling of infectious diseases has a long history; see, in particular, Bailey (1975). The starting point is generally taken to be a paper by Daniel Bernoulli (Bernoulli 1760) on the prevention of smallpox by inoculation; an account of Bernoulli's model-based analysis of data can be found in Daley and Gani (1999). However, as Bailey (1975) points out, it was another hundred years or so before the physical basis for the cause of infectious diseases became well established. Thus, the pace of progress only really picked up early in the twentieth century, with the work of people such as Hamer (1906), Ross (Ross 1911, 1916, Ross and Hudson 1917a,b), and Kermack and McKendrick (1927), which established the principle of *mass action* or *homogeneous mixing*—by which the rate of new infections is proportional to the current numbers of susceptibles and infectives in the population—and the now-familiar deterministic equations for the general epidemic model. Homogeneously mixing models are also referred to as *mean field* models. Very many of the epidemic models in use today have this general epidemic (SIR) model, or its stochastic counterpart, as their basis and thus it can be used to illustrate many of the main issues. In the SIR model, if $x(t), y(t)$, and $z(t)$ represent, respectively, the numbers of susceptible, infective and removed individuals (i.e. those recovered but immune, isolated, or playing no further part in the transmission of infection for whatever reason) in the population at time t, then

$$\mathrm{d}x(t)/\mathrm{d}t = -\alpha x(t)y(t)/n, \qquad (1.1)$$
$$\mathrm{d}y(t)/\mathrm{d}t = \{\alpha x(t)/n - \gamma\}y(t), \qquad (1.2)$$
$$\mathrm{d}z(t)/\mathrm{d}t = \gamma y(t). \qquad (1.3)$$

The variables x, y and z are not restricted to integer values and thus it is only appropriate to use this deterministic model for large populations. In some versions of the model, the parameter α above is replaced by $n\beta$, but the formulation given here is more appropriate when an infection is spread by direct contact and an individual makes potentially infectious contacts at a fixed rate α regardless of the population size. When n is fixed, the distinction is not important but confusion may arise if β is regarded as fixed when the limiting situation as $n \to \infty$ is considered. Note, however, that the latter parameterization is appropriate if the variables x, y and z represent population proportions. The parameter γ can be interpreted as the reciprocal of the mean infectious period. In the natural stochastic formulation of this model, infectious periods are independently and exponentially distributed with parameter γ.

In the deterministic SIR model, the number of infectives grows as long as the proportion of susceptibles in the population exceeds $\gamma/\alpha = 1/R_0$, where the reproduction ratio, $R_0 = \alpha/\gamma$, represents the mean number of effective contacts made by an infective during an infectious period. If $R_0 < 1$, then this condition cannot be met and only a minor outbreak can result. Thus the target, in controlling a particular outbreak of infection, is to take steps to bring R_0 below one by reducing the rate of infectious contacts or the duration of the infectious period. With an infection for which $R_0 > 1$, a population will be protected from epidemic outbreaks as long as the proportion of susceptibles is kept below the threshold by vaccination; this effect is known as *herd immunity*. For an endemic infection, a population of fixed size, but subject to demographic changes and without an effective vaccination strategy, will be subject to recurrent epidemics when the proportion of susceptibles in the population builds up sufficiently. Estimates of R_0 are sometimes based on the *final size* of an epidemic, that is, the total number infected, $\lim_{t\to\infty} z(t)$. It is easy to show that this limit is the solution of the equation $n - z - x(0)\mathrm{e}^{-zR_0/n} = 0$. Various approximations are then possible depending on assumptions about R_0, $x(0)$, and n. For example, if $R_0 > 1$ and $x(0) \simeq n$ exceeds the threshold value (n/R_0) by an amount ν then, approximately, as $t \to \infty$, the limiting value of $x(t)$ will be ν below the threshold giving a final size of 2ν. However, this approximation requires ν to be small, that is, R_0 close to 1; for large R_0, the final size behaves as $n(1 - \mathrm{e}^{-R_0})$.

Stochastic epidemic models were also being developed early in the twentieth century, alongside the deterministic ones, and McKendrick (1926) discussed a stochastic version of the general epidemic model. However, at that time, there was rather more interest in discrete-time models, and chain binomial models in particular, and further discussion of continuous time stochastic SIR models will be postponed to Section 1.3. The best-known chain binomial model was proposed by Reed and Frost in lectures given in 1928 (Wilson and Burke 1942,

1943). However, their work had been preceded some 40 years earlier by En'ko in 1889 (Dietz 1988, 1989). In the standard Reed–Frost model, given the numbers S_t, I_t of susceptibles and infectives at time t, S_{t+1} has a binomial distribution with index S_t and mean $S_t(1-p)^{I_t}$, and $I_{t+1} = S_t - S_{t+1}$. In effect, individuals are assumed to be infective for a single time unit and in that time can infect (independently, with probability p) any member of the population who is susceptible. This means that the number of potentially infectious contacts scales with the population size, although the Reed–Frost model is usually only applied to small populations, e.g. households, where this is not a problem. In contrast, in the En'ko formulation it is assumed that each individual makes a fixed number, K, of contacts at random with replacement from among the $n-1$ remaining individuals in the population, infection being transmitted if the person contacted is infectious, whereby $\mathrm{E}(S_{t+1} \mid S_t, I_t) = S_t\{1 - I_t/(n-1)\}^K$. Dietz and Schenzle (1985) discuss generalizations where K is a random variable. For example, if K has a Poisson distribution with mean $-(n-1)\ln(1-p)$, then the expected value of S_{t+1} given (S_t, I_t) is the same as for the Reed–Frost model, although the corresponding distributional relationship does not hold. More generally, if appropriate assumptions are made about the distribution of K—essentially, that the number of contacts per unit time is suitably small—then

$$\mathrm{E}(I_{t+1} \mid S_t, I_t) = S_t\mathrm{E}\left[1 - \{1 - I_t/(n-1)\}^K\right] \simeq S_tI_t\mathrm{E}(K)/(n-1),$$

which corresponds to the homogeneous mixing assumption of the continuous-time SIR models described above.

An alternative modification of the Reed–Frost model is to assume that the probability that a susceptible escapes infection by a single, specified infective is $(1-p)^{1/n}$, rather than $(1-p)$. In this case, given S_t, I_t, the mean number of infectives at time $t+1$ is given by

$$\mathrm{E}(I_{t+1} \mid S_t, I_t) = S_t\{1 - (1-p)^{I_t/n}\} \simeq S_tI_tp/n$$

for small p. Thus this modification is in a sense equivalent to the use of the factor $1/n$ in eqn. (1.2) for the continuous time deterministic SIR model, and allows comparison of epidemics in host populations of different sizes. In particular, if two populations have identical *proportions* of infectives and susceptibles at time t, then the expected proportions at time $t+1$ will also be the same. If n is large and a single infective is introduced into a population of $n-1$ susceptibles then, to a first approximation, p can be interpreted as the probability that there will be some spread of infection.

In the last 80 years, deterministic and stochastic epidemic models, in both discrete and continuous time, have been much studied. Together with the texts already cited (Bartlett 1960, Bailey 1975, Anderson and May 1991, Daley and Gani 1999), the following books between them contain a wealth of information on epidemic models, as well as providing an invaluable source of further references, covering many more topics than there will be scope to touch upon in this chapter: Becker (1989), Renshaw (1991), Mollison (1995), Grenfell and Dobson (1995), Isham and Medley (1996), Andersson and Britton (2000), Diekmann and Heesterbeek (2000).

1.3   Stochastic models for homogeneous populations

Any stochastic epidemic model has a deterministic counterpart, obtained by setting the deterministic population increments equal to the expected values of the conditional increments in the stochastic model. Thus, many stochastic models will have the same deterministic analogue. Useful insight into the stochastic model can often be gained from its corresponding deterministic model, as long as the population size is large. For example, in the stochastic SIR model, given the current state $(X(t), Y(t), Z(t))$ specifying the numbers of susceptible, infective and removed individuals, respectively, we have

$$\begin{aligned} &\mathrm{E}\{Y(t + \mathrm{d}t) - Y(t) \mid X(t) = x(t), Y(t) = y(t), Z(t) = z(t)\} \\ &\quad = \{\alpha x(t)/n - \gamma\} y(t) \mathrm{d}t + o(\mathrm{d}t) \end{aligned} \tag{1.4}$$

corresponding to the deterministic equation (1.2) for $y(t)$. However, it follows from (1.4) that

$$\frac{\mathrm{d}\mathrm{E}\{Y(t)\}}{\mathrm{d}t} = \left\{ \frac{\alpha \mathrm{E}\{X(t)\}}{n} - \gamma \right\} \mathrm{E}\{Y(t)\} + \frac{\alpha}{n} \mathrm{cov}\{X(t), Y(t)\}, \tag{1.5}$$

from which it is clear that the solution of the deterministic equations is not simply the mean of the stochastic process. This is an inevitable consequence of the non-linearity of the transition rates of the stochastic model. Nevertheless, for a broad class of processes, the deterministic solution is a good approximation to the stochastic mean of a major outbreak when n is large.

A thorough discussion of the properties of deterministic and stochastic versions of the SIR model is given by Bailey (1975); valuable accounts and updates for many model variants can be found in Lefèvre (1990), Andersson and Britton (2000) and Diekmann and Heesterbeek (2000). Even for these simple models, explicit results are often difficult to obtain, and much effort and ingenuity has been expended in deriving methods of solution and approximations. In particular, the threshold behaviour of the stochastic SIR model for large populations (Whittle 1955, Bailey 1975, Andersson and Britton 2000) is broadly analogous to that of the deterministic model described in Section 1.2. If the initial number of infectives is small, essentially all contacts of infectives are with susceptibles and a branching process approximation is appropriate. Thus, if $R_0 \leq 1$, there will be only a small outbreak of infection and the distribution of the final size of the epidemic will be *J-shaped*, i.e. with a mode at or close to zero and decreasing monotonically thereafter. The time to extinction of the infection will be $O(1)$ as $n \to \infty$. If $R_0 > 1$ and I_0 denotes the initial number of infectives, then with probability $1 - (1/R_0)^{I_0}$ the branching process explodes, corresponding to a major outbreak of infection. As the number of infectives builds up, the approximation ceases to be appropriate and a central limit effect takes over. In this case, the final size distribution will be *U-shaped*, i.e. bimodal, and the time to extinction will be $O(\log n)$ as $n \to \infty$ (Barbour 1975). The central limit effect can be examined formally by looking at a sequence of models where the initial proportions of susceptibles and infectives are kept fixed, so that the process is kept well away from the boundary of the state space, as the population size n increases to infinity. In this case, the SIR process tends to a Gaussian diffusion about the deterministic solution. Both the means and the variances of the numbers in the three states scale with the population size, and thus the *proportions* tend to the deterministic solution with probability one. These results were first obtained by Whittle (1957) and later extended (Daley and Kendall 1965, Kurtz 1970, 1971, 1981, Barbour 1972, 1974, Daniels 1991) to a very general class of density-dependent Markov jump processes. This threshold behaviour at $R_0 = 1$ is, in fact, an asymptotic result for large populations, with a major outbreak of infection being essentially one in which the size of the epidemic can be arbitrarily large. Many authors (e.g., Bailey (1975) using model simulations) have observed that threshold values strictly larger than one are appropriate for finite populations. In a discussion of this issue, Nåsell (1995) suggests defining the threshold as the value of R_0 for which the distribution of the total size of the epidemic switches from J-shape to U-shape. Based on numerical work, he conjectures that this threshold has the form $R_0 = 1 + O(n^{-1/3})$.

In the past, the view seems to have been widely held that a deterministic model gives the mean behaviour of the corresponding stochastic system, at least asymptotically, and that for large populations there is therefore little to be gained from using a stochastic model that will generally be more difficult to analyse. However, it is now widely accepted that both deterministic and stochastic models have their strengths and can provide insight and understanding. Even with large populations, chance fluctuations do not always average out to have rather little overall effect. But even when they do, it may be important to take the variability of individual realisations into account, for example in predicting the course of an individual outbreak (Isham 1991, 1993*b*). The latter paper discusses variants of the SIR model used in connection with the AIDS epidemic: as both means and variances scale with the population size n, the variation about the deterministic limit is of order $\sqrt{n}$, so that there is considerable uncertainty about predicted values, even apart from the effects of parameter uncertainty. Similarly, stochastic effects can be important in determining a control strategy where some measure is to be implemented 'until the epidemic is over'. Once numbers of cases are small, there is considerable uncertainty in the residual time to extinction that cannot be conveyed by a deterministic treatment. Stochastic effects also play an essential role in questions of recurrence and extinction (or *fade-out*) of infections. As already noted, with a deterministic epidemic model governed by a set of differential equations, an open population with recruitment of susceptibles will be subject to recurrent epidemics if initially exposed to a source of infection for which the reproduction ratio R_0 exceeds unity. The infection never wholly dies out and can regenerate from arbitrarily small amounts of residual infection. Thus, for example, tiny fractions of an infected fox—the 'attofox' of Mollison (1991)—can generate repeated waves of rabies. In contrast, a stochastic epidemic that reaches a state in which only a single infective remains may well soon die out completely, and is in any case certain to do so eventually, unless there is a continuing external source of infection.

The attention on fade-out has focused mainly on childhood infections, especially measles. Measles is a seasonal infection driven largely by the school year, and time series of pre-vaccination data typically show an annual cycle together with major outbreaks recurring at rather regular intervals of somewhere between two and five years. In small populations the infection often goes extinct. Much of the early work seeking to explain these recurrent epidemics is due to the seminal work of Bartlett (Bartlett 1949, 1956, 1957, 1960), who introduced the idea of a *critical community size* below which an infection will rapidly go extinct unless it is reintroduced from outside; in a data-analytic investigation he describes the critical size as that for which there is a 50:50 chance of fade-out following a major outbreak. A more recent discussion of the role of community size in measles fade-out is given by Keeling and Grenfell (1997). In theoretical studies, terms such as 'fade-out' and 'major outbreak' need formal definition. For example, Nåsell (1999) considers an open SIR model (i.e. with demography) and looks at the population size for which the expected time to extinction from the quasi-stationary distribution (see below) takes a fixed value. In comparing theoretical results with empirical data, however, allowance must be made for complicating features of real infections, such as population structure, vaccination coverage and the durations of disease stages—the assumed exponential distributions of simple models are seldom appropriate.

In a basic measles model, an extra *latent* stage is usually included in the SIR model to give an SEIR model. Those in the latent (i.e. exposed) class are infected but not yet infectious. Demography is also included, and a steady-state population is generally assumed, achieved via equal per capita birth and death rates applying to all members of the population. In a deterministic version of the model, using reasonably realistic parameter values, the number infected oscillates about a stable point but, in contrast to the data, these oscillations are damped. In addition, Schenzle (1984) commented that it is difficult to find a set of parameters that gives an acceptable fit simultaneously to all aspects of an epidemic. Damped oscillations were first noted by Soper (1929) in a model variant with no latent class and an exponentially growing population since no deaths were included in the model. Although this model is in one sense rather unrealistic, it gives useful qualitative results, and the corresponding stochastic formulation has been used by Bartlett and many subsequent authors. In particular, Bartlett (1956) showed by simulation that the stochastic formulation of the model does lead to undamped oscillations, on a time scale that depends on population size, and to the possibility of fade-out. An alternative resolution within the deterministic framework is to assume that the contact rate α varies seasonally, see for example Dietz (1976), which can also generate the required undamped oscillations. Nevertheless, it is generally accepted that grouping children into age bands has an important effect over and above the timing effects of the school year, and thus Schenzle (1984) used an age-structured model, explicitly incorporating such effects, successfully to reproduce the stable biennial cycle observed in data for England and Wales.

Measles data for large populations show an apparently chaotic behaviour (Schaffer 1985), due to interaction between the natural periodicity of the non-linear dynamics and the seasonal periodicity of the contact rate (Bartlett 1990). In this respect, it is important to note that the level of spatial and/or temporal aggregation of the data will have a considerable effect on the observed properties. For a while, the chaotic properties of deterministic measles models were the focus of considerable activity; for a review of some of this work see Isham (1993a). However, as already noted, demographic stochasticity plays an essential role in the recurrence and extinction properties of epidemic models. Recently, Nåsell (2002a) has carefully investigated deterministic and stochastic versions of an SEIR model, with a periodic contact rate and allowing reintroduction of infection from an external source, using a mixture of analytic approximation and simulations. He concludes that stochasticity is necessary for acceptable representation of recurrent epidemics. In particular, as is already well known (Grenfell 1992) the high level of seasonal forcing needed if the SEIR model is to be chaotic gives an unrealistically low incidence of infections in the troughs between the epidemics. In this work Nåsell also shows how the amount of damping in the oscillations of the non-seasonal deterministic model reflects the variability in the times between outbreaks in the corresponding stochastic model. In an alternative approach (Finkenstädt *et al.* 2002) measles data for England and Wales are explored by use of a model that is a combination of a mechanistic stochastic SIR epidemic model and a statistical time series. The model runs in discrete time and uses an algorithm to reconstruct the size of the susceptible class from information on births and cases. The number of infectives at time $t+1$ then has a negative binomial distribution whose expectation is a function of the numbers of susceptibles and infectives at time t, including an additional Poisson-distributed number of immigrant infectives, and is proportional to $X^\theta Y^\psi$. The constant of proportionality is allowed to vary seasonally and the parameters θ and ψ allow possible departures from homogeneous mixing, (for which $\theta = \psi = 1$). It is found that the model, when fitted to data, can successfully reproduce the dynamics of measles outbreaks over a range of population sizes from small towns to large cities, with a seasonal transmission rate that closely corresponds to the effects of the school year.

Many of the issues touched upon above in the context of stochastic models for measles present substantial theoretical challenges more generally. In particular, current research continues to be concerned with important questions on thresholds for *invasion*—under what conditions does a small amount of initial infection invade an almost entirely susceptible population?—and *persistence*—if it does so, what is the necessary condition for the infection to persist and become endemic? These questions have been addressed with reference to a number of different models—see Nåsell (2002b), for example—and various properties have been used to examine threshold behaviour. In a population with no immigration of infectives after the initial infection, an endemic situation can result for long time periods, even when eventual extinction is certain, as long as there is replacement of susceptibles. This can be achieved through births of susceptibles, as in the open SIR model discussed in the context of measles above, or in a closed population if individuals can be infected more than once, as for example in an SIS model, where infectives return to the susceptible state on recovery instead of becoming immune. The latter model is mathematically more tractable although generally less obviously realistic. In models such as these where eventual extinction is certain, the distributions of the time to extinction from a variety of initial conditions have been studied. From any particular initial state, essentially there are two qualitatively different possibilities, either the infection will die away rather quickly, or it will reach a state of quasi-stationary equilibrium—that is, an equilibrium distribution conditional on extinction not having occurred (Darroch and Seneta 1967)—about the endemic equilibrium of the deterministic model, in which it will remain for some random time, before eventually a sufficiently large random fluctuation will lead to extinction.

The certainty of eventual extinction for the stochastic SIS model contrasts with the deterministic version of the model where, if $R_0 > 1$, any initial infection, however small, will invade a susceptible population and reach an endemic equililibrium. Nevertheless, various other properties of this model do exhibit threshold behaviours. For example, starting from a single infective in an otherwise wholly susceptible population, it is easy to see that there will be an initial growth in the expected number of infectives if $R_0 > 1 + 1/(n-1)$, which could be thus be defined as a threshold value (Jacquez and Simon 1993). Alternatively, the time to that extinction also exhibits a threshold behaviour. Andersson and Djehiche (1998) consider an initial condition in which I_0 goes to infinity with n, and show that the time to extinction is asymptotically exponentially distributed with a mean that grows exponentially with n if $R_0 > 1$. While this same behaviour may result if $R_0 > 1$ but $I_0 = O(1)$, it is no longer certain. When $R_0 \leq 1$, the time to extinction stays small or behaves as $\log n$ as $n \to \infty$ depending on the initial number of infectives. Nåsell (1996, 1999) discusses approximations for the mean time to extinction when the initial distribution is the quasi-stationary distribution and also shows that the quasi-stationary distribution itself exhibits a threshold behaviour, being approximately geometric for $R_0 < 1$ and approximately normal when $R_0 > 1$; see also Clancy and Pollett (2003).

The properties of the SIR model with demography have similarly been much studied—see Bailey (1975) and Andersson and Britton (2000), for example—although results are rather more difficult to obtain than for the SIS model discussed above; this is especially true of the time to extinction, although approximations are available (Nåsell 1999). As expected, in the limiting $n \to \infty$ case and assuming that the initial proportion of infectives scales with n, there is a threshold at $R_0 = 1$ below which an infection dies out rather rapidly and above which it increases to an endemic level with Gaussian fluctuations about that level.

In the last few paragraphs, some recent results to do with invasions and persistence, times to extinction, thresholds and critical community sizes have been outlined. There are many other questions that could be asked involving different models or different initial conditions. For example, one might be interested in the probability that an infection becomes extinct, or the further time in which it does so, conditionally upon a major outbreak having already resulted from the introduction of a single infective. A particular challenge for the future is to extend results such as those described above, to allow seasonally varying parameters or structured populations where assumptions of homogeneous mixing do not apply (see Section 1.4). If individuals are of several types depending, for example, on behavioural or spatial characteristics, with different rates of contact within and between groups, how does this affect the time to extinction? As mentioned earlier, this is particularly relevant for the case of the recurrent epidemics such as measles.

The models described in this section have all been variations on an SIR theme, with a simple and relatively tractable mathematical structure. In particular, they assume that the hosts are identical and homogeneously mixing; models for heterogeneous populations will be discussed in the next section. They also assume that the times that an individual spends in distinct states—latent, infectious, and so forth—are exponentially distributed. In order to make models more nearly reflect the realities of disease progression it is important to allow alternative distributions with a non-zero mode; for example, Keeling and Grenfell (1998), Lloyd (2001) and Keeling and Grenfell (2002) discuss the effect on the persistence of measles. In some cases the variation of the distribution may be very small, as for the incubation period for childhood diseases such as measles or chicken pox, while in others, such as the period between HIV infection and AIDS diagnosis, a very wide range of values is possible. The usual 'forward equation' approach for Markov models can be used to cover variable rates of disease progression if the states of the system are extended to include information about the elapsed time from infection; the gamma distribution—via Erlang's 'method of stages', see Cox (1955b), for example—and the Weibull distribution are especially tractable possibilities algebraically. In addition, explicitly incorporating elapsed time from infection into the models allows the rate of transmission to depend on this time, as is appropriate for HIV for example, where infectives are thought to be at their most infectious soon after infection and then again later when developing the symptoms of AIDS. However, an alternative and very elegant approach is due to Sellke (1983); see also Andersson and Britton (2000). In this, an individual becomes infected when the total *infection pressure* $(\alpha/n)\int_0^t Y(t)\mathrm{d}t$—in which infectives are weighted by their infectious periods, which can have an arbitrary distribution—reaches an exponentially distributed level. More generally,

Clancy (1999a,b) uses martingale methods to extend results on the SIR model, allowing a broad class of infection and removal rate functions, and discusses optimal strategies for intervention via isolation/or and immunization.

1.4 Heterogeneity

1.4.1 Sources of heterogeneity

So far, we have concentrated almost entirely on models for populations of identical, homogeneously mixing hosts. However, for applied purposes, it is often vital to incorporate sources of heterogeneity, even though the aim is a simple, parsimonious model. It is helpful, then, to distinguish intrinsic heterogeneity of the hosts, for example a variation in susceptibility due to genetics or immune status, from heterogeneity of mixing whereby the rate of contact between a pair of hosts depends on their spatial separation or relative positions in a social network. In the next subsection, we concentrate on intrinsic host heterogeneity, postponing discussion of heterogeneous mixing to Section 1.4.3 that concentrates on structured populations in which mixing within local groups takes place at a higher rate than with the rest of the population, and Section 1.4.4 where mixing determined by contact networks is discussed. Models are almost always temporally homogeneous, although for endemic diseases and epidemics over very long time periods it may be necessary to allow for changing behaviour and contact patterns in the host population, the introduction or alteration of control policies, and so forth. Although it is usually assumed that changes to the host population occur independently of the infection process, there is also the possibility that changes happen in response to the infection process itself. An additional source of heterogeneity that will not be discussed in any detail here, comes from the infecting parasites themselves. There are many different coexisting parasite strains causing infections such as influenza, malaria and dengue, and it is necessary to take account of the interactions between these and the host's immunological system, as infection with one strain may confer subsequent immunity to other closely related strains. It has been shown, for example, that treating the strains of a malaria parasite species as indistinguishable can have misleading consequences for the transmissibility of the infection (Gupta *et al.* 1991, 1994). From a modelling perspective, parasite heterogeneity means that the state space of the process has to be substantially increased to include the past infection history of the hosts, and in any particular case the aspects of this history that are most important in determining the dynamics of the infection will need to be established. The increased complexity of these models means that most work in this area so far has been deterministic; an exception is the recent simulation study of influenza evolution by Ferguson *et al.* (2003).

1.4.2 Heterogeneity of hosts

Age is one obvious source of variation between hosts that may be relevant to transmission of infection and/or disease progression. Thus, for example, the period from infection with HIV to diagnosis of full AIDS is known to vary with host age (Billard *et al.* 1990). In such cases, it may be important to model the age structure of the population. Of course, it may also be that individuals mix differentially according to their age, as in the case of measles discussed earlier. For sexually transmitted diseases, hosts will need to be divided into male and female, and in some cases it will be necessary to distinguish couples in monogamous relationships and to model the changes in these partnerships over time. In general, such population structure is easily incorporated, by dividing the population first on the basis of relevant explanatory variables such as age and gender, and then subdividing these categories by infection status: susceptible, infective, and so forth. The infection states themselves may be further divided. For example, the degree of susceptibility may vary with the immunological status of the host, and a further important area for future research is to link susceptibility to infection in epidemic models with detailed models of the workings of the host's immune system. It is simplest to assume that the host heterogeneity can be described by discrete variables, although continuously varying characteristics can also be considered if necessary. The overall framework is often essentially that of the SIR model, but with a much larger state space. The large number of variables and parameters may mean that model properties have to be determined numerically or via simulation.

Most of the models allowing host heterogeneity focus on the complexities of a particular infection. However, Ball and Clancy (1995) discuss a multitype model in which infectives are randomly allocated a type on infection, and obtain asymptotic results for the final size and *severity*, that is, the area under the trajectory of infectives, of the epidemic, as the initial population size increases. A special case of this model, with just two types, allows the infection to be carrier-borne—that is, a proportion of infectives develop symptoms immediately and are withdrawn from circulation, while the others remain symptom-free and able to transmit infection. Host heterogeneity is also a feature of a class of models for contagion proposed by Dodds and Watts (2005). In these models, which have a homogeneously mixing SIR basis, contacts take place at regular time points and each infectious contact between an infective and susceptible results in a random 'dose' of infection being transmitted, with the susceptible only being infected when the accumulated dose over the last T time points reaches some randomly assigned threshold. A more natural alternative might be to include all past doses with an exponentially decaying weight. Individuals are able to recover once this accumulated dose falls below the threshold. In an essentially deterministic analysis, Dodds and Watts (2005) show that for a range of model specifications the resulting dynamics of the system falls into one of three universal classes depending on the values of two basic parameters: the probabilities that an individual will be infected as a result of one, and two, infectious contacts.

A vast number of models have been constructed to describe the dynamics of specific infections, both for human and animal infections, although until recently their analysis has mainly focused on deterministic approximations. For the most part, the simple homogeneous SIR model has to be extended to allow for host heterogeneity and for extra infection states. An enormous amount of work has gone into modelling sexually transmitted diseases, such as gonorrhea, hepatitis B, and especially HIV and AIDS in various homosexual, heterosexual, and intravenous drug-using communities. As well as measles, there are many other childhood diseases, such as mumps, chicken pox (varicella), rubella, and pertussis, and infections such as influenza and tuberculosis that cause considerable morbidity and mortality, particularly in developing countries. Models for smallpox have received recent attention because of bioterrorism concerns. In the case of tropical diseases such as malaria, schistosomiasis and onchocerciasis, there is no direct host–host transmission of the parasite, and models must include complicated interactions between parasites, human hosts, secondary hosts and intermediate vectors. For malaria, it is the process of biting by the mosquito vector that must be modelled; for onchocerciasis, the vector is a blackfly; for schistosomiasis, the schistosome fluke spends part of its life-cycle in a second host (a snail) with infection transmitted between the humans and snails through water contact.

Both schistosomiasis and onchocerciasis are macroparasitic diseases, which adds an extra complication to the models. All the infections that we have discussed so far have been of microparasites (viruses and bacteria are familiar examples), which replicate within the host and rapidly reach an equilibrium level, so that it is appropriate to describe the state of the host population by the numbers of susceptibles, infectives, and so forth. This equilibrium level may depend on properties of the hosts such as the competence of their immune systems, in which case it will be appropriate to include such sources of host heterogeneity in the model. Macroparasites such as nematodes, on the other hand, have part of their life-cycle outside the host, so that the parasite load only builds up through reinfection. This reinfection process, the individual parasite burdens of the hosts, and the external parasite population must all be incorporated in a full stochastic model, which will involve many complex dependencies and non-linearities. Thus, most models for macroparasite infections concentrate on a particular part of the process, while making drastically simplifying assumptions about other aspects. One approach is to use a *hybrid* model (Nåsell 1985), in which some part of the process, for example the dynamics of the host population (Kretzschmar 1989*a*,*b*), is replaced by a deterministic approximation while allowing for random variability of the hosts' parasite loads. An alternative approach is not to model the parasite population external to the hosts, but to assume that infections of hosts occur in a non-homogeneous Poisson process that is independent of the parasite loads of the hosts. This may be appropriate, together with an additional simplification—to ignore the demography of the host population—for a managed animal population early in the season. In this case (Herbert and Isham 2000, Cornell *et al.* 2003) many other features of the physical process such as clumped infections, immune responses affecting parasite establishment, fertility and mortality, and non-exponentially distributed durations of parasite stages, can all be incorporated in an analytically tractable model. Other approximations have also been discussed; for example, infection may be assumed to be transmitted directly between hosts, using a branching process to approximate the initial stages of the spread of infection between identical (and immortal) hosts (Barbour and Kafetzaki 1993, Barbour *et al.* 1996). In models of specific diseases, interest generally focuses on control issues, the effects of prospective vaccination strategies or treatment interventions, and sometimes on short-term prediction. This pre-supposes that the mechanisms controlling the dynamics of the particular infection are already well understood, so that the model can be used as a firm basis for such investigations, but this may not be the case; for example, the age-intensity profiles of parasite loads from cross-sectional data on schistosomiasis (Chan and Isham 1998) show that the load builds up during the childhood to reach a maximum before decreasing with age. To what extent is this due to a protective acquired immunity rather than to behaviour changes? For example, children may spend more time in/near the snail-infested water than adults. In onchocerciasis, the parasite loads increase with age and there is a debate over the opposing roles of protective acquired immunity and parasite-induced immunosuppression, with experimental investigations giving contradictory findings. Duerr *et al.* (2002) use a model-based analysis to support the importance of immunosuppressive mechanisms. More generally, Duerr *et al.* (2003) compare the effects of a range of different mechanisms—including age-dependent exposure, parasite-induced host mortality, clumped infections and density-dependent effects on the parasites—on age-intensity profiles and dispersion patterns. The latter are of interest because macroparasite loads are generally overdispersed with most hosts having few parasites, and a few hosts having large numbers of parasites. An additional concern and challenge for the future is that almost all epidemic models are concerned with a single infection, and while this may be appropriate for many short-lived microparasitic infections, in other cases between-species immunological interactions may be important. Thus, the susceptibility of a particular host to infection or reinfection by a specific parasite may depend not only on the host's immune response to their infection history with the same parasite species, but also to that with other species. Cross-species immune responses may be immunosuppressive or protective: Behnke *et al.* (2001) discuss both types of response for multispecies helminth infections, while Bruce *et al.* (2000) provide evidence for cross-protective interactions between different malaria species. As yet there is little stochastic modelling of such mechanisms, but see Bottomley *et al.* (2005).

While analysis of the deterministic version corresponding to any stochastic model is a useful starting point, less stringent approximations can be helpful in providing information about variability and are often straightforward even for relatively complicated processes. For simplicity, consider the closed SIR model, which is a bivariate process. The non-linearity of the model means that the forward equations for the expected numbers of infectives and susceptibles (see eqn(1.5) for the former) involve second-order terms; setting the covariance terms in these equations to zero recovers the equations for the deterministic approximation. Similarly, the forward equations for second-order moments involve third-order terms, and so on. In the method of *moment closure* this set of equations is closed by imposing a suitable relationship between these moments. For example, for the SIR model one might choose to assume that the joint distribution of susceptibles and infectives is bivariate normal, thus obtaining a set of five moment equations to be solved (Isham 1991, 1993*b*); these equations are precisely those that apply in the limiting Gaussian diffusion discussed in Section 1.3. Moment closure has been found to give acceptable approximations in other cases too; for example Nåsell (2003) considers the quasi-stationary distribution of the stochastic logistic model, a finite state birth–death process. The success of the moment closure method can sometimes be attributed to central limit effects, as in the case of the density-dependent Markov processes. However, the method can often work surprisingly well for small populations, and close to the boundary of the state space (Isham 1995); Stark *et al.* (2001) explore the connection with the existence of inertial manifolds as a possible explanation for this. Distributions other than the multivariate normal may also be used to suggest suitable relations between the moments: for example, in macroparasite infections, the parasite load is often well approximated by a negative binomial distribution. A problem is that once one moves away from the multivariate normal distribution, it may be difficult to find a joint distribution with the right marginals, but where the covariances are not inappropriately constrained; see Herbert and Isham (2000) for a discussion of a negative binomial approximation. Alternatively, Keeling (2000) suggests an approximation based on standardized so-called *multiplicative* moments such as V^*, T^* where $E(X^2)=[E(X)]^2V^*$, $E(X^3)=[E(X)]^3V^{*3}T^*$ and so forth, which he finds gives good results in cases where the multivariate normal approximation is unacceptable—for example, it can give negative mean population sizes. For a univariate model, this approach is equivalent to assuming a lognomal distribution.

Space constraints mean that it has not been possible to refer to even a tiny proportion of the models on specific single-strain, multistrain and multispecies infections here; Anderson and May (1991) and collections of workshop papers such as Gabriel *et al.* (1990) and Isham and Medley (1996) provide a convenient general starting point for models of many infections, but a quick scan through recent journals or an Internet search will generate a wealth of fascinating examples. Historically, the complexities of specific infections have led to a focus on the use of deterministic models, but a greater understanding of the importance of intrinsic stochasticity together with considerable advances in computing power means that the use of stochastic models is becoming increasingly feasible. However, this does not necessarily mean resorting to detailed stochastic simulations. There is also a role for toy models to illuminate particular aspects of a complex process as well as for a combination of analytic approximations and numerical evaluation to investigate properties of interest. A single example will have to suffice. A simple model (Herbert and Isham 2000) for the early-season dynamics of a macroparasite infection in a managed age-cohort of hosts was mentioned earlier. The stochastic feedback of the current parasite population via the reinfection process of the hosts can be simulated (Cornell *et al.* 2004), and shows complicated interactions between stochastic effects, including the need for the parasites to mate within hosts, and spatial effects governed by the size of the host population. A toy model—a bisexual Galton–Watson metapopulation process (Cornell and Isham 2004), whose properties can be explored using a mixture of algebraic approximations and numerical evaluation—sheds light on these interactions.

1.4.3 Heterogeneity of mixing

In all the models described so far, apart from the age-structured measles models mentioned in Section 1.3, homogeneous mixing has been assumed, with all members of the host population equally likely to be in potentially infectious contact. For the pre-vaccination measles data discussed earlier, the inclusion of age structure in the models plays an important role. However, most measles cases are in young children, with transmission taking place at pre-school, school or in the family, and at school children mix predominantly within their own age group. Thus age may be a surrogate for social structure in the population and an alternative strategy is to model this structure directly. A general approach is then to divide the host population into groups (or *types*), where it is assumed that hosts mix homogeneously within each group. Contacts between groups are modelled by use of a *mixing matrix* whose (i, j)th element p_{ij} specifies the probability that a host in group i will have an potentially infectious contact, i.e. one in which transmission will take place if one host is infected and the other is susceptible, with a host in group j. This group j host is chosen at random from members of the group. Then, for example, if there are $X_i(t)$ susceptibles and $Y_i(t)$ infectives in group i at time t, and all individuals make such contacts at rate α, the total rate at which infections take place in group i is $\alpha X_i \sum_j p_{ij} Y_j / n_j$, where n_j is the size of group j. As usual in such models, the chance of a within-group infection is taken to be $p_{ii} Y_i / n_i$ rather than using the more realistic factor $n_i - 1$ in the denominator—a negligible approximation in most cases. Clearly the numbers of i–j contacts must balance, so the elements of the mixing matrix must satisfy the constraints $n_i p_{ij} = n_j p_{ji}$. Various structures for the mixing matrix have been proposed, including *proportional mixing* where $p_{ij} = n_j / \sum_l n_l$, which is just homogeneous mixing of the whole population; *restricted mixing* where $p_{ij} = \delta_{ij}$ ($\delta_{ij} = 1$ if $i = j$, $\delta_{ij} = 0$ otherwise) so that individuals only mix within their subgroup; *preferred mixing* where $p_{ij} = \epsilon \delta_{ij} + (1 - \epsilon) n_j / \sum_l n_l$, where individuals have a reserved proportion of contacts with their own subgroup and the rest are spread proportionately through the whole population; and various more complicated possibilities such as mixing schemes that model the social or geographical locations in which contacts take place (Jacquez *et al.* 1989,

Koopman *et al.* 1989). In some cases the mixing matrix is estimated from empirical data. It is then relevant to note that if demography is included in the model so that the group sizes, n_j, are not fixed, the constraints on the mixing matrix mentioned above mean that its elements will vary with time; one possibility is then to model $n_i p_{ij}$, using a log-linear model for example (Morris 1991, 1996).

It is straightforward to generalize the above mixing structure to allow the distinct subgroups to have different contact rates α_i ($\alpha_i \neq \alpha_j$). The constraints on the elements of the mixing matrix are then $n_i \alpha_i p_{ij} = n_j \alpha_j p_{ji}$, with corresponding changes to the various special cases, e.g., proportional mixing corresponds to $p_{ij} = \alpha_j n_j / \sum_l \alpha_l n_l$, so that it is the 'contact' that is chosen at random rather than the person contacted. The variability of contact rates is particularly important for sexually transmitted diseases where the host population may consist of large numbers of individuals with low contact rates and a rather smaller number with very high rates. In this case, the spread of infection is determined not just by the average contact rate over the population but by its variability (Anderson *et al.* 1986). For example, Jacquez and Simon (1990) show that, for a deterministic model with proportional mixing, the rate of growth of infectives at the start of the epidemic is determined by $\mu_\alpha + \sigma_\alpha^2 / \mu_\alpha$ where μ_α and σ_α^2 denote, respectively, the mean and variance of the contact rate over the population.

An important concern of both theoretical and practical interest is how to generalize the concept of the reproduction ratio, R_0, for general heterogeneous and structured populations. The definition of R_0 was first put on a firm footing by Heesterbeek (1992), see also Diekmann and Heesterbeek (2000). Essentially, R_0 is the spectral radius (dominant eigenvalue) of the first-generation matrix, which specifies the expected numbers of type j infections directly caused by a type i infective; the formulation allows hosts to move between types during their lifetimes and applies equally with heterogeneity of hosts due to disease status or immune level, and with heterogeneity of host mixing. As for a homogeneous population (see Section 1.3), threshold results for a stochastic model are less straightforward than for its deterministic counterpart. The condition for invasion in a deterministic model is that $R_0 > 1$, but for stochastic models the threshold results are rather more complicated. Keeling and Grenfell (2000) discuss the effect of population structure, and in particular the number of neighbours of each individual, in the context of a simple stochastic SIR-type model in a closed population. As expected, the smaller the size of the neighbourhood, the higher the threshold for invasion.

Much of the theoretical work on epidemic models for metapopulations is due to F. Ball and his collaborators. The emphasis is on models for small groups (households) within a large population, where the assumed mixing structure is relatively simple; properties such as the final size and severity of an epidemic, and asymptotic results when the number of households becomes large, are of particular interest. One important distinction is whether infection spreads between groups through the movement of infectives, infection then usually being restricted to susceptibles within the same group, or whether the group membership is fixed but infectious contacts occur between groups. For example, in a series of papers (Ball and Clancy 1993, Clancy 1994, 1996), the model includes movement of infectives between groups, with the rate of contact of the infective with a susceptible depending on the infective's original and current groups and the group of the susceptible. A simplifying special case is then to assume that infection is only transmitted within groups. An important example of the alternative approach is provided by Ball *et al.* (1997) who consider epidemics with two levels of mixing: local and global. Individuals mix mainly within their own (local) group, but have occasional contacts with the rest of the population. The key result is that the reproduction ratio for the epidemic is then simply a product of two factors: the reproduction ratio for the homogeneously mixing epidemic when the local contact rate is set to zero, and an 'amplification factor' A, say, which is the mean size of an epidemic started by a single randomly chosen infective when the global contact rate is set to zero. Thus, if A—and hence necessarily the group sizes—are sufficiently large, a substantial outbreak can result even if the level of global contacts is extremely low. This result is not restricted to when the groups form a partition of the population, and applies equally when individuals have overlapping sets of local neighbours, a scenario explored in more detail in Ball and Neal (2002). It can be seen that A needs to be reduced to control the spread of infection, and a vaccination strategy that aims to reduce the number of each individual's unvaccinated neighbours to be as near as possible equal is intuitively optimal. In more recent work, Ball and Lyne (2002) look at optimal vaccination strategies for a household model, where the costs of the vaccination programme are included. In a further level of generality, Ball and Lyne (2001) consider a multitype household model where the infection rate depends on the types of the infected and susceptible individuals as well as on whether the contact is a local or global one. Most of the work on household models considers SIR-type epidemics, but Ball (1999) investigates the spread of an SIS model in a population with a large number of small households, determining an appropriate threshold parameter for invasion in this case, while Arrigoni and Pugliese (2002) obtain limiting results when both the household size and their number go to infinity.

1.4.4 Spatial models

A particular challenge in epidemic modelling is the appropriate way to allow for spatial population structure, whereby the rate of contact between hosts depends on their spatial separation. The hierarchical framework of the metapopulation models described above can be used as a surrogate for spatial structure. In the case of measles, outbreaks seldom take place in isolated communities, and it is often important to look jointly at spatially coupled epidemics (Grenfell 1992). In general, the chance of extinction of infection from the whole population decreases as the mixing between the subpopulations is increased, due to the so-called 'rescue' effect. However, this effect will be lessened if the epidemics in the subpopulations are nearly in phase due to a strong degree of seasonal forcing. One of the earliest explicitly spatial models for the spread of infection was a metapopulation model for the spread of influenza, first in the USSR and later globally (Rvachev and Longini 1985), where data on the transportation networks were used to provide information about migration routes. The more general setting of Ball *et al.* (1997) in which there is no partition of the population into fixed groups, but in which each individual has their own set of neighbours, is particularly appropriate in a spatial context. In *small worlds* models (Watts and Strogatz 1998, Newman *et al.* 2002), individual hosts have some simple spatial structure, with most contacts reserved for spatially neighbouring hosts and just a few long-distance contacts. Such models are highly relevant to epidemics characterized by local spread plus a few large jumps, as occurred with the recent outbreaks of foot and mouth disease and SARS. The *great circle model*, in which hosts are equally spaced round a circle, is used as a simple illustrative example by Ball *et al.* (1997), and further explored by Ball and Neal (2003) who derive asymptotic results including a threshold theorem for invasion to occur and a central limit theorem for the final epidemic conditionally upon its occurrence.

In the metapopulation models, it is normally the case that any individual host may be directly infected by any other, even though the chance of this may reflect spatial separation and could be extremely small. Contact networks provide a way of representing a more complex population structure, where only hosts connected by edges of the network have any direct contact. These networks can reflect structure in geographical or social space and, traditionally, their study has been the province of social scientists who have generally been concerned with the collection of data and characterisation of empirical networks in relatively small populations. Social networks are particularly relevant to the spread of infections transmitted by direct contact—for example, sexually transmitted diseases (Johnson 1996)—and study has focused especially on HIV. However, in recent years mathematicians have become increasingly involved in network models. Two sorts of networks are discussed: directed graphs showing 'who is infected by whom' and undirected graphs representing the underlying contact structure of the population. Much theoretical work is devoted to the latter, and particularly to investigating the properties of simple random graphs (Bollobás 1985) and of Markov random graphs (Frank and Strauss 1986). In a simple random graph, an edge exists independently between each pair of vertices with probability p. In order that the *degree* of a vertex (i.e. the number of edges emanating from it) is well behaved as the size of the network (n) increases, it is usual to assume that $p = \beta/n$, so that for large n the degree distribution is approximately Poisson. Alternatively, the degree distribution may be specified and/or estimated from empirical data; a graph can then be constructed by a random pairing of the arms from each vertex (Newman *et al.* 2002). The problem when the sum of all the degrees is odd can be removed by resampling the degree of one node. If the degree distribution has a power-law tail then the graph is said to be *scale-free* (Barabasi and Albert 1999, Barabasi *et al.* 1999). Scale-free networks will be characterized by rather few individuals having very large numbers of contacts and acting as 'hubs'. The degree distributions of empirical networks are frequently far from Poisson and power-law tails with exponents of between -2 and -3 are typical (Newman 2003), in which case the degree has an infinite variance. If such a graph were used to model an epidemic, this would indicate an infinite R_0 (see e.g., Jones and Handcock (2003)). However, this is not a practical concern since homogeneity assumptions are almost certainly inappropriate in an arbitrarily large population. In any case, simply specifying the local properties of a network by an empirically reasonable degree distribution is not sufficient to guarantee that the most probable global networks resemble observed data. For example, empirical networks typically have many more short loops than is the case for a random graph. Since it is much easier to collect data, estimate parameters and apply control policies at a local level, this raises the general question of how local properties determine global structure. Even for a homogeneously mixing population, where all pairs of individuals are in potential contact, a 'who is infected by whom' random graph framework can be useful in investigating epidemic properties that are not time dependent (Ball *et al.* 1997), such as the probability of a large outbreak, or the final size of an epidemic. In particular, a simple random graph structure is appropriate for an SIR epidemic with a constant infectious period and Barbour and Mollison (1990) use this representation to obtain asymptotic results for the Reed–Frost model.

A Markov random field, defined on a set of sites equipped with a neighbourhood structure, has the property that the values of the field on disjoint sets of sites are conditionally independent, given the values on a neighbourhood region (Dobrushin 1968). In its most familiar specification, the sites are the vertices of a graph (often a regular lattice) whose edges define the neighbourhood structure. Thus the graph is pre-specified and the field—which might be binary in a special case—has a Markov property. In contrast, for a Markov random graph the sites on which a neighbour relation is defined are the edges (or *ties*) between vertices. For example, a common assumption is that two edges are neighbours if they share a common vertex. A binary Markov random field is then defined on these edges, the resulting random graph being determined by those edges for which the field takes the value 1. Thus, in contrast to a simple random graph, a Markov random graph has dependent edges. The Hammersley–Clifford theorem (Preston 1973) shows that for Markov random graphs the global properties are completely determined by the local conditional independence properties, and that the joint distribution of the graph has a simple exponential form, where the exponent is a sum of contributions from cliques of neighbouring edges. This exponential form for the joint distribution makes Markov random graph models particularly tractable from a statistical perspective and likelihood-based methods of inference using Markov chain Monte Carlo algorithms are being developed (Snijders 2002). It also means that many simple special cases can be considered as possible models for networks with varying amounts of local dependence, and these can be easily simulated and explored. However, the homogeneity assumptions involved are usually unrealistically strong and lead to realizations that are unrepresentative of empirical networks. In particular, most of the probability may be concentrated on small parts of the state space with either very few or very large numbers of edges (Handcock 2003). Such models are therefore inappropriate except for very small populations. For larger networks, empirically plausible ways to construct neighbourhoods that might lead to processes more closely resembling observed networks are proposed by Pattison and Robins (2002). One possibility is that network ties take place in a social setting, while another is that the neighbourhood structure depends on the realization of the process—so that, for example, two edges might be neighbours if they are separated by a single edge. The use and interpretation of fitted random graph models needs care, even supposing that completely observed network data are available for model fitting. In particular, there is the question of how appropriate the fitted model is to other scenarios, and whether it provides more than a description of a specific social structure. Even if it does have some general applicability, it will be important to ascertain the robustness of the network properties to uncertainty in the values of the fitted parameters.

Most attention has concentrated on the network structure of the population *per se*, for example the size of the largest component, or the degree of assortative mixing (in which a high-degree node is likely to be connected to another such node) exhibited, and relatively little to considering realistic disease dynamics. The most intuitively natural approach is to view a contact network as the underlying structure on which an infection evolves. In this case, it may be assumed simply that if an arbitrary node in the graph is infected, the size of the epidemic will be that of the connected component containing the index case. Alternatively, the edges of the graph can be given associated probabilities specifying the chances that node B will become infected if A is, and vice versa. In either case, the timing of contacts, duration of infectivity and concurrency of partnerships are all ignored, which is certainly not always appropriate (Morris and Kretzschmar 1989). Whether an epidemic takes off will depend not on the simple mean number of neighbours but, rather, on a size-biased mean because those having more contacts will themselves be more likely to become infected; for example, Andersson (1998) obtains asymptotic results for a simple discrete time random graph model. An alternative approach is to model the directed 'who infects whom' graph as a single entity (Andersson and Britton 2000). Care is needed in this approach because of the dependencies due to the fact that an infective with a long infectious period will tend to infect many neighbours. When the contact network is regarded as the framework on which an infection spreads, the stationary of that network may be an issue. Except over short time periods, at the very least it will be necessary to include births of susceptibles and deaths of all individuals together with their associated network edges. Very little attention has been paid to the question of the evolution of the underlying contact networks—but see Read and Keeling (2003) who use simulation to look at the effects of different network structures on disease transmission—and even less to the idea that the network may evolve in response to the disease dynamics. One of the obvious difficulties here is that it is hard enough to collect complete data on a network of contacts at a (more or less) single time instant, and collecting longitudinal data to support a model of network evolution is generally infeasible, except in the smallest population. However, in recent work, Snijders (2001) has been developing continuous-time Markov chain models of network dynamics, albeit in relatively small closed populations of perhaps a few hundred individuals. The idea is to start from a fixed initial configuration, and then to model the evolution of the network from this starting point. The creation of links between individuals in the network is allowed to depend on explanatory variables, which themselves evolve, that can represent individual characteristics such as behaviour or infection status. Each individual seeks to create or dissolve links so as to optimize an objective function.

None of the models described so far is a fully spatial model for the spread of infection. Much of the work most closely meeting this description is that on regular spatial lattices (Durrett and Levin 1994, Durrett 1995), which arises from work in statistical physics, but unfortunately there is no room to discuss this here. These lattice processes are most obviously applicable for models of plant diseases and, although they are sometimes applied in other contexts, the underlying homogeneous space makes them generally inappropriate for human/animal diseases. The challenge of producing satisfactory stochastic spatial epidemic models has been around a long time, and Bartlett (1960) notes their intractability. Thus, work allowing for general spatial transmission has traditionally been deterministic. There is a need for simple spatial stochastic models that allow a non-uniform population density in continuous space and, in recent years, plausible stochastic models have often been explored by simulation. For example, Bolker (2003) uses a mixture of moment approximation and simulation in a model that allows both underlying spatial heterogeneity and population density to affect transmission. For spatial processes, the velocity of the spread of infection and the diameter of the infected region are added to the usual concerns relating to thresholds for invasion and persistence, duration and final size of the epidemic.

In building models of spatial-temporal processes, there is the whole question of which spatial, temporal and population scales to use. A good example of this issue is provided by measles data that are available, for England and Wales, in the form of weekly reports of numbers of cases on a town-by-town basis, the sizes of which are highly variable. The data can be analysed as they stand or aggregated into larger spatial units. In a very small town, a homogeneous mixing assumption may not be too unrealistic, but the data for a large town or city will inevitably already consist of the superposition of coupled epidemics in many subgroups. Properties such as the persistence of infection (or the chance of fade-out) will depend on the level of aggregation of the data (Keeling and Grenfell 1997). In general, the appropriate scale or scales to use will depend on the purposes of the modelling, but it may be sensible to think in terms of a hierarchy of models, so that local spread over short time scales is represented by an individual-based model, and then these fine-scale models are used as the building blocks in a larger-scale model, and so on. This general approach is particularly suitable for metapopulation models as, for example, in Ball *et al.* (1997).

Spatial models are complicated to analyse, and techniques such as pair approximation (Keeling 1999*a*,*b*) that attempt to encapsulate complicated spatial information within a deterministic framework are sometimes used. The basic idea is as follows. In a homogeneously mixing model, the rate of new infections is proportional to the product $X(t)Y(t)$ of current numbers of susceptibles and infectives. In a spatial model, the chance that a particular susceptible-infective pair will generate a new infection needs to be weighted appropriately by a function of their spatial separation. With an underlying contact network, these weights are binary and $X(t)Y(t)$ needs to be replaced by the number of susceptible-infective pairs of neighbours $[XY](t)$, say. Thus, a forward equation for $E(X(t))$ will involve $E([XY](t))$, that for the expected number, $E([XX](t))$, of susceptible–susceptible pairs will involve the expected number of triples $E([XXY](t))$, while that for $E([XY](t))$ will involve $E([XXY](t))$ and $E([YXY](t))$, and so on. By considering increasingly long strings of susceptible and infective nodes from the network, where adjacent elements of the string are neighbours, an arbitrary spatial structure can be represented. However, the pair approximation method simplifies this process by replacing the expected number of triples of the form $[ABC]$ by a simple formula only involving pairs, thus closing the set of simultaneous equations. The simplest version is to use $E([ABC]) \simeq E[AB]E[BC]/E[B]$, where we drop the dependence on t for notational simplicity. However, this approximation will give an overestimate, essentially because it is based on a conditional independence assumption and does not allow for the fact that if node j is a neighbour of both i and k, then there is a good chance that i and k are neighbours. A better approximation attempts to allow for this by using a correction factor of the form $(1 - 1/m)\{1 - \phi + \phi(n/m)(E[AC]/E[A]E[C])\}$, where ϕ is the probability that i and k are neighbours given that they are both neighbours of j (for example, $\phi = 0.4$ for a hexagonal lattice), m is the average number of neighbours per node, and n is the size of the network. There is a similarity between the methods of pair approximation and moment closure. For moment closure, there is no spatial structure and the transition rates are functions of the total numbers in various subpopulations. The set of moment equations characterizing the joint distribution of these numbers can be closed by an approximation whereby higher-order moments such as $E\{X^2(t)Y(t)\}$ are specified in terms of lower-order moments; these relationships may be suggested by distributional properties. Thus, moment closure gives a set of approximations to the moments of the joint distribution of the state of the system. With pair approximation, spatial information is included, but the allowance for spatial structure is at the expense of stochasticity. Only mean properties are considered, so that the effect of the spatial structure of, say, the infectives on the mean number of suscepti-bles is approximated but there is no attempt to look at other moments of the distribution, although this would be possible, at least in principle. Thus the ap-proximation is two-fold, involving both the use of a deterministic model and the approximation of expected numbers of spatially neighbouring nodes with partic-ular infection states. In models for populations in continuous spaces, a method similar to pair approximation—rather confusingly sometimes called the method of moments (Law and Dieckmann 2000, Dieckmann and Law 2000)—that looks at the correlation structure of the spatial field used. There, an equation for the evolution of the expected number of pairs of particular types at a specific sepa-ration, ζ say, will involve the number of triples $[ABC; \zeta, \xi]$ at separations ζ and ξ. This is then approximated by $[AB; \zeta][BC; \xi]/[B]$.

Pair approximation methods were used (Ferguson *et al.* 2001*a,b*) to model the recent UK foot and mouth epidemic, using information on farm positions and the network of contacts between farms to give numbers of pairs of neigh-bouring farms in the various infection states, as a means of exploring possible intervention strategies such as 'ring' culling around infected farms, and vacci-nation. This approach contrasts with a more explicitly spatial and stochastic approach (Keeling *et al.* 2001) involving the full spatial structure of the farm network. In this approach, a matrix of probabilities that each farm will infect each other farm, if infected itself, is constructed that evolves as an infection spreads. Realizations of an epidemic can therefore be simulated to predict its future course, using appropriate parameter values estimated externally. Again, the model was used as a means of investigating possible control strategies; see Kao (2002) for comparative discussion of these issues.

1.5 Statistical issues and further discussion

I have concentrated here on the development of stochastic models for the trans-mission of infection. If these models are to be used to draw conclusions about specific infections, rather than to gain mathematical understanding of general classes of infection processes, they will usually have to be fitted and validated against data. This section will be confined to a few general remarks; for further discussion and a more detailed introduction to the statistical issues touched on briefly below see Becker (1989), Becker and Britton (1999) and Andersson and Britton (2000). Some of the simplest models, such as the SIR model, are broadly applicable in many situations. In that case, one or both easily inter-pretable parameters—the reproduction ratio, R_0, and the mean infectious pe-riod, $1/\gamma$—can often be assumed known or can be determined experimentally, although if necessary they can be easily estimated from data on the course of an epidemic; see, e.g., Becker (1989). In general, if parameter values are to be estimated from experiments then these need careful statistical design. If such estimates *are* available externally—i.e. independently of the model—then model properties can be compared with those of empirical data to validate the model.

With relatively complicated models of specific infections, there will generally be a need not only to estimate at least some of the parameters of the model from epidemic data, and to test the sensitivity of the inferences made from the model to uncertainties in their values, but also to validate the assumed structure of the model. It is particularly helpful if an inferential framework can be constructed within which a hierarchy of increasingly complex models can be compared. It is important to ensure that any conclusions drawn are robust to model uncertainty. In this regard, and especially when strong simplifying assumptions have been made, it can be reassuring if similar conclusions result from a range of different modelling approaches; for example, this was the case in the recent UK foot and mouth outbreak.

In an ideal world, one would have complete observation of a single epidemic, or better still, multiple replications thereof, or a series of linked outbreaks of an endemic infection. More often, the information is incomplete and data are aggregated over time periods—even the whole course of the epidemic as with final size data—or over large heterogeneous populations. Alternatively, perhaps only cross-sectional data are available. In such cases, a complete epidemic model specification may be unnecessary. In this regard, it is interesting to ask to what extent population structure can be inferred from incomplete data, especially from a single realization of an epidemic process. For example, if there is very little mixing between the two homogenously mixing subgroups in a simple metapopulation model, and a single infective is seeded in one subgroup, there may be a clear time-lag between the epidemics in the two groups. As the between-subgroup mixing increases the time-lag will tend to disappear. To what extent can such population heterogeneity be inferred from data for a single epidemic realization, especially when only partial information—perhaps the total numbers of new cases in the population in discrete time intervals—is recorded?

In predicting the future course of a partially observed epidemic, a range of predictions can be given incorporating such uncertainties and these should, additionally, include the effects of random variation between realizations. If, as is generally the case, parameters are estimated by model fitting, then the more parameters are fitted, the more the capability to validate the model by comparison of observed and fitted properties is reduced, as is the predictive power of the fitted model with regard to future epidemics. As with any spatial and/or temporal stochastic process, the properties to be used for fitting and for validation need to be carefully considered and will depend on the purposes for which the modelling is being carried out. When there is complete observation of the times of infection and removal (recovery) for an epidemic outbreak, likelihood methods can be used to obtain parameter estimates. Maximum likelihood estimation is also sometimes possible with incomplete observation, for example with final size data. More generally though, likelihood methods tend to be unwieldy with anything other than the smallest populations. Martingale estimating equations or the use of the EM algorithm to estimate the missing data then provide useful approaches. However, in recent years, the very rapid growth of computer-intensive methods in general and of Markov chain Monte Carlo methods in particular has led increasingly to their application in fitting epidemic models (Gibson and Renshaw 2001, O'Neill 2002), one of the earliest uses being parameter estimation for a partially observed SIR model in a Bayesian framework (O'Neill and Roberts 1999). Nevertheless, in particular cases, the need for rapid answers during the course of an outbreak has led to much simpler approaches, such as choosing parameter values to maximize the likelihood over a fixed grid (as in Riley *et al.* (2003) for SARS).

One of the most important purposes in fitting models to epidemic data is the investigation of potential control strategies. Control may be achieved by reducing the rate of contacts between susceptibles and infectives, the probability that a contact results in transmission, or the duration of infectivity. As, for example, in the case of foot and mouth disease and SARS, these reductions might be achieved by travel restrictions, biosecurity measures, or quarantine arrangements. Another possibility is to reduce the pool of susceptibles by vaccination. In this case, models can be used to determine the impact of a range of vaccination strategies, by making various assumptions about the effects of the vaccine—does it confer indefinite total protection, total protection for a proportion of the population and none for the rest, partial protection for each individual, does the immunity wane over time, or perhaps the effect is not to protect the vaccinated individual from being infected but to make an infective less infectious to others, and so on? Allowing for vaccination status in the models adds an extra layer to the heterogeneity of the population. Various measures of vaccine efficacy have been proposed, and the design of studies and methods of their estimation are important statistical issues; see, e.g., Datta *et al.* (1999) and Becker *et al.* (2003). Heterogeneity of mixing is an added complication in assessing possible vaccination strategies. Even for a perfect vaccine, if there is a limited supply, or limited resources with which to administer it, how should the vaccine be allocated? In the case of the simple metapopulation models described in Section 1.4.3, aiming to reduce the numbers of susceptibles in each subgroup to the same level makes intuitive sense, but assumes that numbers of susceptibles are known and that it is socially and operationally feasible to carry out such a strategy.

For almost all of the research described above, it has been assumed that a small number of infectives—perhaps just one—is introduced into a wholly susceptible population. Then the interest is in the probability that the infection will spread and lead to a major outbreak, the properties of that outbreak conditionally on it doing so, and how best to control transmission. But models also have a role to play in answering another important problem: how best to detect an emerging new infection, such as SARS; or a new outbreak of a pre-existing one, such as foot and mouth disease. This topic that has become especially topical with the current emphasis on detection of bioterrorism attacks, and has spurred renewed interest in models for smallpox, which was successfully eradicated worldwide by 1980. Returning closer to home, however, and with the reason for this volume in mind, those who have followed David Cox's interests over the last few years will know of his pre-occupation with badgers and the spread of tuberculosis in cattle (Donnelly *et al.* 2003). Potential models for this exemplify many of the complications discussed in this chapter including the presence of more than one host species, uncertain transmission routes, complex spatial considerations and incomplete observations. This work has recently led David to think about the question of how to detect a new outbreak of infection, and to know when, and how, to instigate control measures. Perhaps he will soon provide the answer to this question in yet another seminal contribution to applied probability!

Acknowledgements

My thanks are due to all those, far too numerous to list, who have shared with me their considerable insights on epidemic models. In particular, I am grateful to the participants of many epidemic modelling workshops, and their organisers and sponsors, including Sydney (1990), Skokloster (1990), Oberwolfach (1991, 1994, 1999), INI (1993), Stockholm (1995), Woudschoten (1996), Skye (1997), Canberra (1998), PIMS (1999), Trieste (1999), DIMACS (2002, 2003), Copenhagen (2003), Mariefred (2003) and IMA (2003), from whom I have learned much. But above all, my thanks go to David Cox, without whose gentle spur I might never have discovered the fascination of epidemic models and without whose friendship and collaboration my life would have been immeasurably the poorer.

2
Stochastic soil moisture dynamics and vegetation response

Amilcare Porporato and Ignacio Rodríguez-Iturbe

2.1   Introduction

In this chapter we present an overview of a stochastic soil moisture model that was originally proposed in Rodríguez-Iturbe *et al.* (1999) following previous analysis by Cox and Isham (1986). The model has been used by the authors and co-workers to investigate analytically the relationship between the hydrologic and vegetation dynamics—ecohydrology—in water-controlled ecosystems (Rodríguez-Iturbe and Porporato 2005).

Water-controlled ecosystems are complex evolving structures whose characteristics and dynamic properties depend on many interrelated links between climate, soil, and vegetation (Rodríguez-Iturbe 2000, Porporato and Rodríguez-Iturbe 2002, Rodríguez-Iturbe and Porporato 2005). On the one hand, climate and soil control vegetation dynamics, while on the other hand vegetation exerts important control on the entire water balance and is responsible for feedback to the atmosphere.

Two characteristics make quantitative analysis of the problem especially daunting: the very large number of different processes and phenomena that make up the dynamics; and the extremely large degree of variability in time and space that the phenomena present. The first obviously demands simplifying assumptions in the modelling scheme while preserving the most important features of the dynamics, while the second demands a stochastic description of some of the processes controlling the overall dynamics, especially precipitation.

The dynamics of the soil–plant–atmosphere system are interpreted on the daily time scale. Moreover, for the sake of simplicity, only cases with negligible seasonal and interannual rainfall components are considered. The growing season is thus assumed to be statistically homogeneous: the effects of the initial soil moisture condition are considered to last for a relatively short time and the probability distribution of the soil water content to settle in a state independent of time. Possible extensions to transient conditions are discussed in Rodríguez-Iturbe and Porporato (2005).

2.2   Soil water balance at a point

2.2.1   Basic model

Soil moisture is the key variable synthesizing the interrelationship between climate, soil, and vegetation. Its dynamics are described by the soil water balance, i.e. the mass conservation of soil water as a function of time. We will consider the water balance vertically averaged over the root zone, focusing on the most important components as sketched in Fig. 2.1.

The state variable regulating the water balance is the relative soil moisture, s, which represents the fraction of pore volume containing water. The total volume of soil is given by the sum of the volumes of air, water, and mineral components, i.e. $V_s = V_a + V_w + V_m$. The porosity is defined as

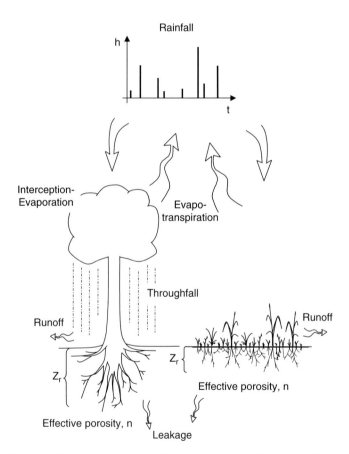

Fig. 2.1. Schematic representation of the various mechanisms of the soil water balance with emphasis on the role of different functional vegetation types. After Laio *et al.* (2001*b*).

$$n = \frac{V_a + V_w}{V_s},$$

and the volumetric water content, θ, is the ratio of water volume to soil volume, so that the relative soil moisture is

$$s = \frac{V_w}{V_a + V_w} = \frac{\theta}{n}, \qquad 0 \leq s \leq 1.$$

If we denote the rooting depth by Z_r, then snZ_r represents the volume of water contained in the root zone per unit area of ground.

With the above definitions and under the simplifying assumption that the lateral contributions can be neglected—that is, there are negligible topographic effects over the area under consideration—the vertically averaged soil moisture balance at a point may be expressed as

$$nZ_r \frac{ds(t)}{dt} = \varphi\{s(t), t\} - \chi\{s(t)\}, \qquad (2.1)$$

where t is time, $\varphi\{s(t), t\}$ is the rate of infiltration from rainfall, and $\chi\{s(t)\}$ is the rate of soil moisture losses from the root zone. The terms on the right-hand side of eqn (2.1) represent water fluxes, i.e. volumes of water per unit area of ground and per unit of time (e.g. mm/day).

The infiltration from rainfall, $\varphi\{s(t), t\}$, is the stochastic component of the balance. It represents the part of rainfall that actually reaches the soil column,

$$\varphi\{s(t), t\} = R(t) - I(t) - Q\{s(t), t\},$$

where $R(t)$ is the rainfall rate, $I(t)$ is the amount of rainfall lost through canopy interception, and $Q\{s(t), t\}$ is the rate of runoff.

The water losses from the soil are from two different mechanisms,

$$\chi\{s(t)\} = E\{s(t)\} + L\{s(t)\}, \qquad (2.2)$$

where $E\{s(t)\}$ and $L\{s(t)\}$ are the rates of evapotranspiration and drainage, respectively. The normalized form of the losses will be denoted $\rho(s) = \chi(s)/(nZ_r)$.

2.2.2 Rainfall modelling

On small spatial scales, where the contribution of local soil-moisture recycling to rainfall is negligible, the rainfall input can be treated as an external random forcing, independent of the soil moisture state. Since both the occurrence and amount of rainfall can be considered to be stochastic, the occurrence of rainfall will be idealized as a series of point events in continuous time, arising according to a Poisson process of rate λ and each carrying a random amount of rainfall extracted from a given distribution. The temporal structure within each rain event is ignored and the marked Poisson process representing precipitation is physically interpreted on a daily time scale, where the pulses of rainfall correspond to daily precipitation assumed to be concentrated at an instant in time.

The depth of rainfall events is assumed to be an independent random variable h, described by an exponential probability density function

$$f_H(h) = \alpha^{-1} \exp(-h/\alpha), \qquad h \geq 0, \tag{2.3}$$

where α is the mean depth of rainfall events. In the following, we will often refer to the value of the mean rainfall depth normalized by the active soil depth,

$$\frac{1}{\gamma} = \frac{\alpha}{nZ_r}. \tag{2.4}$$

Both the Poisson process and the exponential distribution are commonly used in simplified models of rainfall on a daily time scale.

Canopy interception is incorporated in the stochastic model by simply assuming that, depending on the kind of vegetation, a given amount of water can potentially be intercepted from each rainfall event (Rodríguez-Iturbe *et al.* 1999). The rainfall process is thus transformed into a new marked-Poisson process, called a censored process, where the frequency of rainfall events is now

$$\lambda' = \lambda \int_\Delta^\infty f_H(h)\mathrm{d}h = \lambda \mathrm{e}^{-\Delta/\alpha},$$

and the depths have the same distribution as before.

2.2.3   Infiltration and runoff

It is assumed that, when the soil has enough available storage to accommodate the totality of the incoming rainfall event, the increment in water storage is equal to the rainfall depth of the event; whenever the rainfall depth exceeds the available storage, the excess is converted in surface runoff. Since it depends on both rainfall and soil moisture content, infiltration from rainfall is a stochastic, state-dependent component, whose magnitude and temporal occurrence are controlled by the entire soil moisture dynamics. The probability distribution of the infiltration component may be easily written in terms of the exponential rainfall–depth distribution, eqn (2.3), and the soil moisture state s. Referring to its dimensionless counterpart y, that is, the infiltrated depth of water normalized by nZ_r, one can write

$$f_Y(y, s) = \gamma \mathrm{e}^{-\gamma y} + \delta(y - 1 + s) \int_{1-s}^\infty \gamma \mathrm{e}^{-\gamma u}\mathrm{d}u, \qquad 0 \leq y \leq 1 - s, \tag{2.5}$$

where γ is defined in eqn (2.4). Equation (2.5) is thus the probability distribution of having a jump in soil moisture equal to y, starting from a level s. The mass at $(1 - s)$ represents the probability that a storm will produce saturation when the soil has moisture s. This sets the upper bound of the process at $s = 1$, making the soil moisture balance evolution a bounded shot-noise process.

2.2.4   Evapotranspiration and drainage

The term $E\{s(t)\}$ in eqn (2.2) represents the sum of the losses resulting from plant transpiration and evaporation from the soil. Although these are governed by different mechanisms, we will consider them together.

When soil moisture is high, the evapotranspiration rate depends mainly on the type of plant and climatic conditions such as leaf-area index, wind speed, air temperature, and humidity. Providing soil moisture content is sufficient to permit the normal course of the plants' physiological processes, evapotranspiration is assumed to occur at a maximum rate $E_{\max}$, which is independent of s. When soil moisture content falls below a given point s^*, which depends on both vegetation and soil characteristics, plant transpiration on the daily time scale is reduced by stomatal closure to prevent internal water losses and soil water availability becomes a key factor in determining the actual evapotranspiration rate. Transpiration and root water uptake continue at a reduced rate until soil moisture reaches the so-called wilting point s_{w}. Below the wilting point, s_{w}, soil water is further depleted only by evaporation at a very low rate up to the so-called hygroscopic point, s_{h}.

From the above arguments, daily evapotranspiration losses are assumed to occur at a constant rate $E_{\max}$ for $s^* < s < 1$, and then to decrease linearly with s, from $E_{\max}$ to a value E_{w} at s_{w}. Below s_{w}, only evaporation from the soil occurs and the loss rate is assumed to decrease linearly from E_{w} to zero at s_{h}.

Drainage or deep infiltration is modelled as starting from zero at the so-called field capacity s_{fc} and then increasing exponentially to K_{s} at saturation. The total soil water losses are shown in Fig. 2.2.

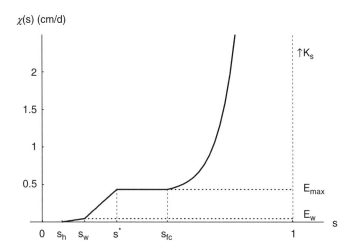

FIG. 2.2. Soil water losses (evapotranspiration and drainage), $\chi(s)$, as function of relative soil moisture for typical climate, soil, and vegetation characteristics in semi-arid ecosystems. After Laio *et al.* (2001*b*).

2.3   Probabilistic evolution of the soil moisture process

The probability density function of soil moisture, $p(s,t)$, can be derived from the Chapman–Kolmogorov forward equation for the process under analysis (Rodríguez-Iturbe *et al.* 1999, Cox and Miller 1965, Cox and Isham 1986)

$$\frac{\partial}{\partial t}p(s,t) = \frac{\partial}{\partial s}\{p(s,t)\rho(s)\} - \lambda'p(s,t) + \lambda'\int_{s_\mathrm{h}}^{s} p(u,t)\,f_Y(s-u,u)\mathrm{d}u. \qquad (2.6)$$

The complete solution of eqn (2.6) presents serious mathematical difficulties. Only formal solutions in terms of Laplace transforms have been obtained for simple cases when the process is not bounded at $s = 1$ (Cox and Isham 1986, and references therein). Here we focus on the steady-state case. Under such conditions, the bound at 1 acts as a reflecting boundary. Accordingly, the solutions for the bounded and unbounded case only differ by an arbitrary constant of integration. The general form of the steady-state solution can be shown to be (Rodríguez-Iturbe *et al.* 1999)

$$p(s) = \frac{C}{\rho(s)}\exp\left\{-\gamma s + \lambda'\int\frac{\mathrm{d}u}{\rho(u)}\right\}, \qquad s_\mathrm{h} < s \le 1, \qquad (2.7)$$

where C is the normalization constant. The limits of the integral in the exponential term of eqn (2.7) must be chosen to assure the continuity of $p(s)$ at the end points of the four different components of the loss function (Cox and Isham 1986, Rodríguez-Iturbe *et al.* 1999).

Figure 2.3 shows some examples of the soil moisture probability density function. The two different types of soil are loamy sand and loam with two different values of active soil depth. These are chosen in order to emphasize the role of soil in the soil moisture dynamics. The role of climate is studied only in relation to changes in the frequency of storm events λ, keeping the mean rainfall depth α and the maximum evapotranspiration rate E_max fixed. A coarser soil texture corresponds to a consistent shift of the probability density function toward drier conditions, which in the most extreme case can reach a difference of 0.2 in the location of the mode. The shape of the probability density function also undergoes marked changes, with the broadest probability density functions for shallower soils.

2.4   Long-term water balance

As described in eqns (2.1) and (2.2), rainfall is first partitioned in interception, runoff, and infiltration, the last of which is itself divided into drainage and evapotranspiration. For the purposes of describing vegetation conditions, the amount of water transpired can be further divided into water transpired under stressed conditions and water transpired under unstressed conditions.

Figure 2.4 presents examples of the behaviour of the various components of the water balance normalized by the mean rainfall rate, for some specific rainfall, soil, and vegetation characteristics. The influence of the frequency of rainfall

Shallow Soil ------------------> Deep Soil

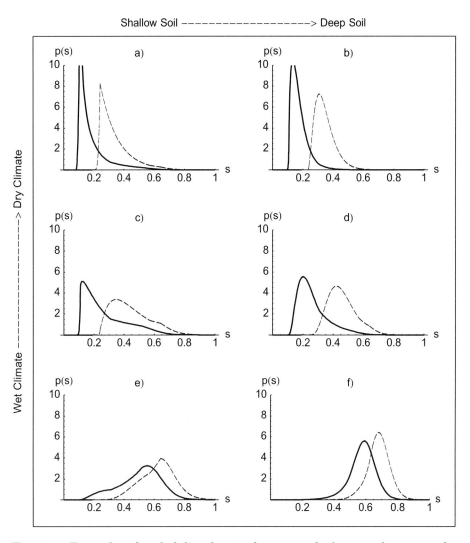

FIG. 2.3. Examples of probability density functions of relative soil moisture for different types of soil, soil depth, and mean rainfall rate. Continuous lines refer to loamy sand, dashed lines to loam. Left column corresponds to $Z_r = 30$ cm, right column to 90 cm. Top, centre, and bottom graphs have a mean rainfall rate λ of 0.1, 0.2, and 0.5 d^{-1} respectively. Common parameters for all graphs are $\alpha = 1.5$ cm, $\Delta = 0$ cm, $E_w = 0.01$ cm/d, and $E_{max} = 0.45$ cm/d. After Laio *et al.* (2001*b*).

events, λ, is shown in part a) of Fig. 2.4 for a shallow loam. Since the amount of interception changes proportionally to the rainfall rate, it is not surprising that the fraction of water intercepted remains constant when normalized by the total

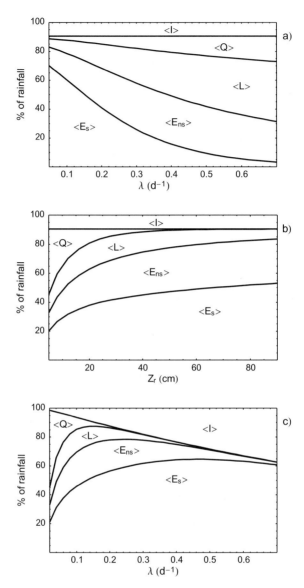

FIG. 2.4. Components of the water balance normalized by the total rainfall. a) Water balance as a function of the mean rainfall rate λ, for a shallow loamy soil ($Z_r = 30$ cm, $\alpha = 2$ cm). b) Water balance as a function of the soil depth Z_r, for a loamy sand, with $\alpha = 2$ cm, $\lambda = 0.2$ d^{-1}. c) Water balance for a loamy sand as a function of the frequency of rainfall events for a constant mean total rainfall during a growing season, $\Theta = 60$ cm. Other common parameters are $E_w = 0.01$ cm/d, $E_{\max} = 0.45$ cm/d, and $\Delta = 0.2$ cm. After Laio *et al.* (2001*b*).

rainfall, $\alpha\lambda$. The percentage of runoff increases almost linearly. More interesting is the interplay between drainage and the two components of evapotranspiration. The fraction of water transpired under stressed conditions rapidly decreases from $\lambda = 0.1$ to about $\lambda = 0.4$, while the evapotranspiration under unstressed conditions evolves much more gently. This last aspect has interesting implications for vegetation productivity.

Part b) of Fig. 2.4 shows the role of the active soil depth in the water balance. For relatively shallow soils there is a strongly non-linear dependence on soil depth of all the components of the water balance, with the obvious exception of interception, which is constant because the rainfall is constant. For example, changing $nZ_r = 5$ to $nZ_r = 20$ practically doubles the amount of water transpired in this particular case.

Part c) of Fig. 2.4 shows the impact on water balance when the frequency and amount of rainfall are varied while keeping constant the total amount of rainfall in a growing season. The result is interesting, due to the existence of two opposite mechanisms regulating the water balance. On one hand runoff production, for a given mean rainfall input, strongly depends on the ratio between soil depth and the mean depth of rainfall events. The rapid decrease of runoff is thus somewhat analogous to that in part b) of Fig. 2.4, where a similar behaviour was produced by an increase in soil depth. On the other hand, interception increases almost linearly with λ. The interplay between these two mechanisms determines a maximum of both drainage and evapotranspiration at moderate values of λ— of course the position of the maxima changes according to the parameters used. This is particularly important from the vegetation point of view, since the mean transpiration rate is linked to productivity of ecosystems. The role not only of the amount, but also of the timing of rainfall in soil moisture dynamics, is made clear by the existence of an optimum for transpiration/productivity, which is directly related to the climate–soil–vegetation characteristics.

2.5 Minimalist models of soil moisture dynamics

A possible simplification of the previous model is to assume that drainage and runoff losses take place instantaneously above a given value s_1, that is, at an infinite rate, $K_s \rightarrow \infty$, so that s_1 is the effective upper bound of the process. In this way, one can introduce a new variable

$$x = \frac{s - s_w}{s_1 - s_w},$$

which will be called the effective relative soil moisture, and define the available water storage as $w_0 = (s_1 - s_w)nZ_r$. In the new variable x, the only relevant parameters are w_0, λ, α, and E_{max}. Using dimensional analysis, the problem can be expressed as a function of only two dimensionless variables,

$$\gamma = \frac{w_0}{\alpha}, \qquad \frac{\lambda}{\eta} = \frac{\lambda\,w_0}{E_{max}}.$$

On the assumption that evapotranspiration losses increase linearly with soil moisture (Porporato *et al.* 2004), $\rho = \eta x$, the soil moisture probability density function turns out to be a truncated gamma density,

$$p(x) \propto x^{\lambda/\eta - 1}\, e^{-\gamma\, x}, \qquad 0 < x < 1; \tag{2.8}$$

owing to the truncation, the shape of this density depends on both the scale parameter γ and the shape parameter λ/γ.

The corresponding behaviour of the probability density function of the effective relative soil moisture as a function of the governing parameters may be used for a general classification of soil moisture regimes. Accordingly, the boundaries between different shapes of the probability density function may be used to define an 'arid' regime, corresponding to probability density functions with zero mode, an 'intermediate' regime, corresponding to soil moisture probability density functions with a central maximum, and a 'wet' regime, with the mode at saturation, as indicated in Fig. 2.5. A further distinction within the intermediate regime can be made on the basis of plant response to soil moisture dynamics. Using the effective relative soil moisture value x^* as a threshold marking the onset of plant water stress, the dashed line of slope $1/x^*$ in Fig. 2.5 becomes the place where the mode of the effective relative soil moisture probability density

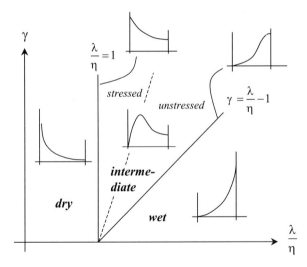

FIG. 2.5. Classification of soil water balance, based on the shape of the effective relative soil moisture probability density function, as a function of the two governing parameters, λ/η and γ, that synthesize the role of climate, soil, and vegetation. The dashed line, showing $\gamma = 1/x^*(\lambda/\eta - 1)$, is the locus of points where the mode of the soil moisture probability density function is equal to the threshold x^* marking the onset of plant water stress. After Porporato *et al.* (2004).

function is equal to x^* and thus where plants are more likely to be at the boundary between stressed and unstressed conditions. Accordingly, it may be used to divide water-stressed or semi-arid types of water balance on the left side from unstressed ones on the right side; see Fig. 2.5.

2.6 Crossing properties and mean first passage times of soil moisture dynamics

2.6.1 General framework

The frequency and duration of excursions of the soil moisture process above and below some levels directly related to the physiological dynamics of plants are of crucial importance for ecohydrology. The analysis of the crossing properties is also important to gain insights on the transient dynamics of soil moisture, either at the beginning of the growing season, to evaluate the time to reach the stress levels after the soil winter storage is depleted, or after a drought, to estimate the time to recover from a situation of intense water stress.

The derivation of exact expressions for the mean first passage times for systems driven by white shot noise, as is the soil moisture dynamics described above, has received considerable attention, both because of its analytical tractability and because of the large number of its possible applications. We follow here Laio *et al.* (2001a), where simple interpretable results for the mean first passage times of stochastic processes driven by white shot noise are derived for cases relevant to soil moisture dynamics.

In order to analyse the crossing properties of the Markovian process just described, one obtains the corresponding backward or adjoint equations from the forward equations. Using the backward equation it is then possible to write the differential equation that describes the evolution of the probability density, $g_T(s_0, t)$, that a particle starting from s_0 inside an interval $\{\xi', \xi\}$ leaves the interval for the first time at a time t. For the process under consideration we obtain

$$\frac{\partial g_T(s_0, t)}{\partial t} = -\rho(s_0)\frac{\partial g_T(s_0, t)}{\partial s_0} - \lambda\, g_T(s_0, t) + \lambda \int_{s_0}^{\xi} f_Y(z - s_0, s_0)g_T(z, t)\mathrm{d}z. \quad (2.9)$$

One does not need to solve the partial integro-differential equation (2.9) to obtain the first passage time statistics of the process. An expression involving the mean time for exiting the interval $[\xi', \xi]$, $\overline{T}_{\xi'\xi}(s_0)$, is obtained from eqn (2.9) by multiplying by t and then integrating between 0 and ∞. Using the definition of the mean time of crossing, $\overline{T}_{\xi'\xi}(s_0) = \int_0^\infty t\, g_T(s_0, t)\mathrm{d}t$ and integrating by parts the term with the time derivative, one gets

$$-1 = -\rho(s_0)\frac{\mathrm{d}\,\overline{T}_{\xi'\xi}(s_0)}{\mathrm{d}s_0} - \lambda \overline{T}_{\xi'\xi}(s_0) + \lambda \int_{s_0}^{\xi} \gamma \mathrm{e}^{-\gamma(z - s_0)}\, \overline{T}_{\xi'\xi}(z)\mathrm{d}z, \quad (2.10)$$

where the exponential part of the jumps distribution, $f_Y(z - s_0, s_0)$, has been used in the integral on the right-hand side because, with the hypothesis that

$\xi < s_{\rm b}$, the presence of the bound at $s_{\rm b}$ becomes irrelevant. The integro-differential equation (2.10) was also obtained by Masoliver (1987) in a more general way. A similar procedure may be followed to derive the equations for the higher-order moments of $g_T(s_0, t)$.

On differentiating eqn (2.10) with respect to s_0 and reorganizing the terms, the following second order differential equation is obtained:

$$\rho(s_0)\frac{\mathrm{d}^2 \overline{T}_{\xi'\xi}(s_0)}{\mathrm{d}s_0{}^2} + \left\{\lambda + \frac{\mathrm{d}\rho(s_0)}{\mathrm{d}s_0} - \gamma\rho(s_0)\right\}\frac{\mathrm{d}\overline{T}_{\xi'\xi}(s_0)}{\mathrm{d}s_0} + \gamma = 0. \qquad (2.11)$$

Equation (2.11) needs two boundary conditions, the first of which may be obtained from eqn (2.10) evaluated at $s_0 = \xi$,

$$\rho(\xi)\left.\frac{\mathrm{d}\overline{T}_{\xi'\xi}(s_0)}{\mathrm{d}s_0}\right|_{s_0=\xi} = 1 - \lambda\overline{T}_{\xi'\xi}(\xi).$$

For the second condition one must consider whether the lower limit ξ' is above or below $s_{\rm h}$. In the first case, ξ' is an absorbing barrier of the model, so that the boundary condition is $\overline{T}_{\xi'\xi}(\xi') = 0$. In contrast, when $\xi' < s_{\rm h}$, ξ' cannot be reached by the trajectory, and the average exiting time from the interval becomes the mean first passage time over the threshold ξ. In this case the second boundary condition is obtained by setting $s_0 = s_{\rm h}$ in eqn (2.10), i.e.,

$$\overline{T}_{\xi'\xi}(s_{\rm h}) = \int_{s_{\rm h}}^{\xi} \gamma e^{-\gamma(z-s_{\rm h})}\overline{T}_{\xi'\xi}(z)\mathrm{d}z + \frac{1}{\lambda}.$$

Below, we will deal with the case where only one threshold, generically indicated as ξ, is effectively present. The mean first passage time over a threshold ξ above the initial point s_0 was obtained as (Laio *et al.* 2001*a*)

$$\overline{T}_\xi(s_0) = \frac{P(\xi)}{p(\xi)\rho(\xi)} + \int_{s_0}^{\xi}\left\{\frac{\lambda P(u)}{p(u)\rho^2(u)} - \frac{1}{\rho(u)}\right\}\mathrm{d}u, \qquad (2.12)$$

where $P(x) = \int_0^x p(z)\mathrm{d}z$. When the starting point s_0 coincides with the threshold ξ, the integral on the right-hand side vanishes and the mean crossing time is

$$\overline{T}_\xi(\xi) = \frac{P(\xi)}{p(\xi)\rho(\xi)}. \qquad (2.13)$$

As a consequence, under steady-state conditions the mean rate of occurrence or frequency of the upcrossing (or downcrossing) events of the threshold ξ can be obtained from eqn (2.13) as

$$\nu(\xi) = p(\xi)\rho(\xi). \qquad (2.14)$$

Returning to the mean first passage times, some manipulation of eqn (2.12) leads to the synthetic expression

$$\overline{T}_{\xi}(s_0) = \overline{T}_{s_0}(s_0) + \gamma \int_{s_0}^{\xi} \overline{T}_u(u) du, \tag{2.15}$$

which, along with eqn (2.13), completely defines the mean first passage time from s_0 to ξ for $s_0 < \xi$.

When instead the mean first passage time over a threshold ξ' below the initial point s_0 is considered and $\xi \to \infty$, one obtains

$$\overline{T}_{\xi'}(s_0) = \int_{\xi'}^{s_0} \frac{1}{\rho^2(u)p(u)} \{\lambda - \lambda P(u) + p(u)\rho(u)\} du$$

$$\tag{2.16}$$

$$= \overline{T}_{s_0}(s_0) - \overline{T}_{\xi'}(\xi') + \frac{1}{\nu(\xi')} - \frac{1}{\nu(s_0)} + \gamma \int_{\xi'}^{s_0} \left\{\frac{1}{\nu(u)} - \overline{T}_u(u)\right\} du,$$

where $\overline{T}_{s_0}(s_0)$, $\overline{T}_{\xi'}(\xi')$, and $\overline{T}_u(u)$ are calculated from eqn (2.13) and $\nu(\xi')$, $\nu(s_0)$, and $\nu(u)$ from eqn (2.14).

Unlike $\overline{T}_{\xi}(s_0)$, $\overline{T}_{\xi'}(s_0)$ depends on the presence of the bound at $s = s_b$ and on the shape of $\rho(s)$ for $s > s_0$. This is clear from the presence in eqn (2.16) of $\nu(\xi')$, $\nu(s_0)$, and $\nu(u)$, which in turn contain the normalizing constant of the density.

In Fig. 2.6, $\overline{T}_{s^*}$ is studied as a function of the frequency of the rainfall events λ and of the mean rainfall depth α, in such a way that the product $\alpha\lambda$ remains

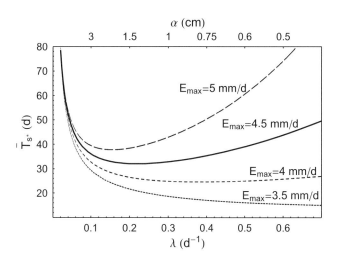

FIG. 2.6. Mean duration of excursions below s^*, $\overline{T}_{s^*}(s^*)$, as a function of the frequency of the rainfall events λ when the total rainfall during the growing season is kept fixed at 650 mm for different values of the maximum evapotranspiration rate. The root depth is $Z_r = 60$ cm, the soil is a loam. After Laio *et al.* (2001*a*).

constant for different values of the maximum evapotranspiration rate. This is to compare environments with the same total rainfall during a growing season, but with differences in the timing and average amount of the precipitation events. In the case of high maximum transpiration rates, plants may experience longer periods of stress either where the rainfall events are very rare but intense or where the events are very frequent and light. From a physical viewpoint, this is due to the relevant water losses by drainage, runoff or canopy interception, indicating possible optimal conditions for vegetation. Figure 2.6 indicates the existence of an optimum ratio between λ and α. The same behaviour was encountered in part c) of Fig. 2.3 in connection with the water balance.

2.6.2 Minimalist model with linear losses

In the case of linear losses bounded above at $x = 1$, the mean first passage time of the threshold ξ or ξ' with starting point x_0 can be calculated from eqns (2.14), (2.15), and (2.16), using the expression for the steady-state probability density function, eqns (2.8), and (2.9). For $x_0 = \xi$ one obtains

$$
\overline{T}_\xi(\xi) = \frac{1}{\lambda} + \frac{1}{\lambda} e^{\gamma\xi} (\gamma\xi)^{-\lambda\xi/\eta} \left\{ \Gamma\left(1 + \frac{\lambda\xi}{\eta}\right) - \Gamma\left(1 + \frac{\lambda\xi}{\eta}, \gamma\xi\right) \right\}
$$

$$
= \frac{1}{\lambda} \, {}_1F_1\left[1, 1 + \frac{\lambda}{\eta}, \gamma\xi\right],
$$

where ${}_1F_1$ denotes the confluent hypergeometric Kummer function (Abramowitz and Stegun 1964).

The quantities $\overline{T}_\xi(\xi)$ and $\nu(\xi)$ are plotted as functions of ξ in Fig. 2.7. Common features for all the curves are the increase of $\overline{T}_\xi(\xi)$ with ξ and the presence of a maximum in the crossing frequency, $\nu(\xi)$. In particular, as the maximum evapotranspiration increases, so does the duration of excursions below any given threshold ξ, while the value $\xi_{\max}$ for which $\nu(\xi)$ has a maximum is usually very close to the mean value of the steady-state distribution. This is because they both represent levels of x around that the trajectory preferably evolves. However, the two values only coincide in the unbounded case: in fact, one can set $\rho(x)p(x) = \nu(x)$ and obtain the equation for the threshold that experiences the maximum crossing frequency. When the loss function is linear, one obtains $\xi_{\max} = \lambda/(\gamma\eta)$. Only in the unbounded case is such a value equal to the mean steady-state value.

It also is interesting to point out that the mean time of crossing, $\overline{T}_\xi(\xi)$, does not decrease monotonically as ξ goes to zero, since there are very low soil moisture values below which the process very seldom goes. Although such levels, which are not shown in Fig. 2.7, are usually very low and thus of little practical interest in water-controlled ecosystems, they could be higher and thus more relevant when the conditions are humid, especially in the case of very low transpiration rates. An example of such a case is shown in Fig. 2.8.

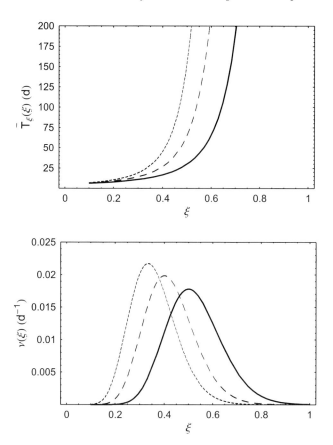

FIG. 2.7. Mean duration of an excursion below ξ, $\overline{T}_\xi(\xi)$ (top), and frequency of upcrossing of ξ, $\nu(\xi)$ (bottom), as a function of the threshold value ξ for the minimalist soil moisture process with linear losses. $\lambda = 0.2$ d^{-1} and $\alpha = 1$ cm, $Z_{\mathrm{r}} = 100$ cm. The three curves refer to different $E_{\max}$: 0.4 cm/day (solid), 0.5 cm/day (long dashes), and 0.6 cm/day (short dashes). After Rodríguez-Iturbe and Porporato (2005).

2.7 Plant water stress

2.7.1 Definition

The reduction of soil moisture content during droughts lowers the plant water potential and leads to a decrease in transpiration. This, in turn, causes a reduction of cell turgor and relative water content in plants, which brings about a sequence of damages of increasing seriousness. Using the links between soil moisture and plant conditions, the mean crossing properties of soil moisture can be used to define an index of plant water stress that combines the intensity, duration, and frequency of periods of soil water deficit. Porporato *et al.* (2001)

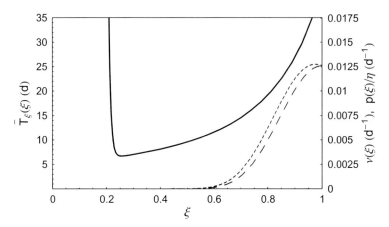

FIG. 2.8. Mean duration of an excursion below ξ, $\overline{T}_\xi(\xi)$ (continuous line), frequency of upcrossing of ξ, $\nu(\xi)$ (dashed line), and soil moisture probability density function, $p(\xi)$ (dotted line), as a function of the threshold value ξ for the minimalistic soil moisture process with linear losses. $\lambda = 0.2$ d^{-1} and $\alpha = 1$ cm, $Z_r = 100$ cm, $E_{max}=0.2$ cm/d. After Rodríguez-Iturbe and Porporato (2005).

use the points of incipient and complete stomatal closure as indicators of the starting and maximum points of water stress. They define a 'static' stress ζ that is zero when soil moisture is above the level of incipient reduction of transpiration, s^*, and grows non-linearly to a maximum value equal to one when soil moisture is at the level of complete stomatal closure or wilting, s_w, i.e.

$$\zeta(t) = \begin{cases} \left\{\frac{s^*-s(t)}{s^*-s_w}\right\}^q, & s_w \le s(t) \le s^*, \\ 0, & \text{otherwise,} \end{cases}$$

where q is a measure of the non-linearity of the effects of soil moisture deficit on plant conditions. The mean water stress given that $s < s^*$, indicated by $\overline{\zeta'}$, can be calculated from the soil moisture probability density function.

A more complete measure of water stress, $\overline{\theta}$, which combines the previously defined 'static' stress, $\overline{\zeta'}$, with the mean duration and frequency of water stress through the variables $\overline{T}_{s^*}(s^*)$ and $\overline{n}_{s^*}$; the latter is the mean number of crossings during a growing season of duration T_{seas} (Porporato *et al.* 2001). This is referred to as 'dynamic' water stress or mean total dynamic stress during the growing season and is defined as

$$\overline{\theta} = \begin{cases} \left(\frac{\overline{\zeta'}\,\overline{T}_{s^*}}{k\,T_{seas}}\right)^{\overline{n}_{s^*}^{-r}}, & \overline{\zeta'}\,\overline{T}_{s^*} < k\,T_{seas}, \\ 1, & \text{otherwise.} \end{cases}$$

The role of the parameters k and r is discussed in Porporato *et al.* (2001). The dynamic water stress well describes plant conditions and has been used with success in different ecohydrological applications (e.g. Rodríguez-Iturbe and Porporato 2005).

2.7.2 Vegetation patterns along the Kalahari precipitation gradient

As a case study of the stochastic model of soil moisture and plant water stress dynamics we briefly summarize the application to the Kalahari precipitation gradient (Porporato *et al.* 2003).

The mean annual precipitation in the Kalahari decreases from north to south going from more than 1500 mm/year in the northern tropical region to less than 300 mm/year near the Kalahari desert. Porporato *et al.* (2003) examined the historic record of daily rainfall data for a set of stations distributed along the Kalahari transect and found that the decrease in the mean rainfall amount is mostly a consequence of a reduction in the rate of storm arrivals. Thus, assuming a spatially uniform mean rainfall depth per event, $\alpha = 10$ mm, they varied the mean rate of event arrivals λ in the range between 0.5 and 0.1 d^{-1} when going from north to south along the transect. Mean typical values for tree and grass parameters were estimated from the literature. Since grass roots are typically concentrated closer to the soil surface, the parameter Z_r representing the effective rooting depth was assumed to be $Z_r = 100$ cm for trees and $Z_r = 40$ cm for grasses.

Figure 2.9 shows the dynamical water stress computed as a function of the rate of event arrivals for the mean parameter values representative of trees and grasses. As expected, the general behaviour is one of progressive increase of plant

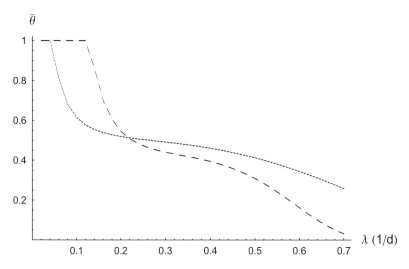

FIG. 2.9. Behaviour of the dynamical water stress as a function of the mean rate of arrival of rainfall events for trees (dashed line) and grasses (dotted line). After Porporato *et al.* (2003).

stress going from wet to dry climates. The plateau for intermediate values of λ is a consequence of the interplay between the frequency of periods of water stress, which attains its maximum in such a zone, and the duration of the stress periods, which increases with decreasing λ. The dependence of the dynamic stress on λ is more marked for trees than for grasses. In particular, for very low rainfall amounts grasses are able to lower water stress more significantly than trees are able to. This is partially due to their lower wilting point, which reduces the effect of water deficit on the plant. More importantly, grasses benefit from their shallow rooting depth, which allows increased access to light rainfall events and reduces the occurrence of long periods of water stress.

The point of equal stress, which could be interpreted as identifying a general region of tree–grass coexistence, is located near $\lambda = 0.2$ d^{-1}, which corresponds to a total rainfall of approximately 420 mm for the wet season months October to April. The fact that the slope of the two curves near the crossing point is fairly shallow may help to explain the existence of a wide region suitable for tree–grass coexistence at average rainfall rates. The pronounced interannual variability of both rainfall parameters might further enhance the possibility of coexistence by randomly driving the ecosystem from an increase in grasses during dry years to tree encroachment during wet years (Porporato *et al.* 2003).

2.8   Conclusions

A simplified stochastic model for the soil water balance that explicitly accounts for rainfall variability has been used to investigate analytically the impact on the soil water balance of climate, soil, and vegetation. The crossing analysis of such a process has been employed to define a measure of plant water stress that well describes vegetation response in water-controlled ecosystems. The model presented here demonstrates that, with careful simplifications and the inclusion of stochastic components, problems with complex dynamics that are encountered in biogeochemical sciences can be approached theoretically in a coherent and synthetic way. Further developments, comparison with data, and applications of the model may be found in Rodríguez-Iturbe and Porporato (2005).

Different aspects of theoretical ecohydrology offer further interesting areas of research. In particular, we mention the analysis of transient properties and the distribution of crossings of soil moisture dynamics, the extension to include spatial variability, especially in the presence of topography and river networks, and the analysis of second- and higher-order stochastic processes resulting from the coupling of soil moisture dynamics with plant biomass and soil nutrient dynamics.

3
Theoretical statistics and asymptotics

Nancy Reid

3.1 Introduction

Cox and Hinkley's *Theoretical Statistics* (1974) was arguably the first modern treatment of the foundations and theory of statistics, and for the past thirty years has served as a unique book for the study of what used to be called mathematical statistics. One of its strengths is the large number of examples that serve not only to illustrate key concepts, but also to illustrate limitations of many of the classical approaches to inference. Since the time of its publication there has been considerable development of the theory of inference, especially in the area of higher-order likelihood asymptotics. The paper by Barndorff-Nielsen and Cox (1979) read to the Royal Statistical Society initiated many of these developments. There have also been very dramatic advances in the implementation of Bayesian methodology, particularly with the development of sampling methods for computing posterior distributions.

A current approach to the study of theoretical statistics should be informed by these advances. Barndorff-Nielsen and Cox (1994) presents an approach to inference based on likelihood asymptotics, as does Pace and Salvan (1997). Severini (2000a) covers much of the same material at a somewhat more elementary level. Bayesian methodology is covered in books on that topic, for example Bernardo and Smith (1994), while Schervish (1995) presents an encyclopedic treatment of both classical and Bayesian approaches to inference. Davison (2003) provides a masterful and wide-ranging account of the use of statistical models in theory and practice, with a welcome emphasis on many practical aspects, and this approach is closest in spirit to the one I would like to follow. At a more advanced level there is not to my knowledge a treatment of the main themes of theoretical statistics that attempts to incorporate recent advances in a unified framework.

In this chapter I will attempt to set out some ideas for such a framework, building on topics and examples from Cox and Hinkley (1974), and very heavily influenced by Barndorff-Nielsen and Cox (1994). The emphasis is on some theoretical principles that have their basis in asymptotics based on the likelihood function. There are, of course, many important advances and developments in statistical theory and methodology that are not subsumed by asymptotic theory or Bayesian inference. These include developments in non-parametric and semi-parametric inference as well as a large number of more specialized techniques for particular applications. The explosion in statistical methods in a great many areas of science has made a comprehensive study of the field of statistics much more difficult than perhaps it was, or seems to have been, thirty years ago. In light of David Cox's view that a theory of statistics forms a basis both for understanding new problems and for developing new methods, I try here to focus on how to structure this theory for students of statistics.

In Section 3.2 I sketch a framework for the implementation of higher-order asymptotics and Bayesian inference into my ideal course or text on the theory of statistics. In Section 3.3 I note topics from a more standard treatment of theoretical statistics that have been omitted and provide some brief conclusions.

3.2 Likelihood-based asymptotics

3.2.1 Introduction

The likelihood function provides the foundation for the study of theoretical statistics, and for the development of statistical methodology in a wide range of applications. The basic ideas for likelihood are outlined in many books, including Chapter 2 of Cox and Hinkley (1974). Given a parametric model $f(y; \theta)$ on a suitably defined sample space and parameter space, the likelihood and log likelihood functions of θ are defined as

$$L(\theta) = L(\theta; y) \propto f(y; \theta) \qquad \ell(\theta) = \ell(\theta; y) = \log L(\theta; y) + a(y).$$

If θ is a scalar parameter a plot of $L(\theta)$ or $\ell(\theta)$ against θ for fixed y provides useful qualitative information about the values of θ consistent with that observed value of y. For examples see Barndorff-Nielsen and Cox (1994, pp. 96–97). Particularly useful summaries include the score function $u(\theta; y) = \ell'(\theta)$, the maximum likelihood estimate $\widehat{\theta} = \arg \sup_\theta \ell(\theta)$, the observed Fisher information $j(\theta) = -\ell''(\theta)$ and the log likelihood ratio $\ell(\widehat{\theta}) - \ell(\theta)$. Standardized versions of these are used to measure the discrepancy between the maximum likelihood estimate $\widehat{\theta}$ and the parameter θ. We define them as follows:

$$\text{score} \qquad s(\theta) = \ell'(\theta)\{j(\widehat{\theta})\}^{-1/2}, \tag{3.1}$$

$$\text{Wald} \qquad q(\theta) = (\widehat{\theta} - \theta)\{j(\widehat{\theta})\}^{1/2}, \tag{3.2}$$

$$\text{likelihood root} \qquad r(\theta) = \text{sign } (q)\sqrt{[2\{\ell(\widehat{\theta}) - \ell(\theta)\}]}. \tag{3.3}$$

With independent identically distributed observations $y = (y_1, \ldots, y_n)$, an application of the central limit theorem to the score function $u(\theta; y)$ can be used to show that each of s, q, and r has a limiting standard normal distribution, under the model $f(y; \theta)$, and subject to conditions on f that ensure consistency of $\widehat{\theta}$ and sufficient smoothness to permit the requisite Taylor series expansions.

These limiting results suggest that in finite samples we can approximate the distribution of s, q, or r by the standard normal distribution. In Figure 3.1 we illustrate these approximations for a single observation from the location model

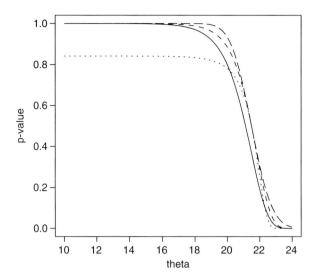

FIG. 3.1. *p*-value functions from the exact distribution of $\widehat{\theta}$ (solid) and the approximate normal distribution based on the likelihood root r (dash), the score statistic s (dotted), and the Wald statistic q (long dash). The model is $f(y; \theta) = \exp\{-(y - \theta) - e^{-(y-\theta)}\}$ and the observed value of y is 21.5. This illustrates the use of asymptotic theory to give approximate *p*-values; in this case the approximations are quite different from each other and from the exact *p*-value.

$f(y; \theta) = \exp\{-(y - \theta) - e^{-(y-\theta)}\}$. I have plotted as a function of θ the exact probability $F(y; \theta)$ and the normal approximations to this exact probability: $\Phi(s)$, $\Phi(q)$, and $\Phi(r)$, for a fixed value of y, taken here to be 21.5. This is called here the *p*-value function, and is used to calibrate the log likelihood function by identifying values of θ consistent with the data on a probability scale. For example, the value of θ that corresponds to a *p*-value of $1 - \alpha$ corresponds to a lower $100\alpha\%$ confidence bound for θ. The approximations illustrated in Figure 3.1 are called *first-order* approximations, as they are based on the limiting distribution.

A function of y and θ that has a known distribution is often called a *pivotal* statistic, or pivotal function. The *p*-value functions based on r, q, and s shown in Figure 3.1 are all pivotal functions, at least approximately, since the distribution of, for example, r, is known to be approximately standard normal, and so the distribution of $\Phi(r)$ is approximately $U(0,1)$. The exact *p*-value function constructed from the known distribution $f(y; \theta)$ is an exact pivotal function. An important aspect of the theory of higher-order asymptotics based on the likelihood function is the joint dependence of the likelihood function and derived quantities on both the parameter and the data.

TABLE 3.1. Selected exact p-values compared to the third-order approximation $\Phi(r^*)$, for the location model illustrated in Figure 3.1.

θ	14.59	16.20	16.90	17.82	18.53	20.00	20.47	20.83
Exact	0.999	0.995	0.990	0.975	0.950	0.800	0.700	0.600
$\Phi(r^*)$	0.999	0.995	0.990	0.974	0.949	0.799	0.700	0.600
θ	21.41	21.69	21.98	22.60	22.81	23.03	23.17	23.43
Exact	0.400	0.300	0.200	0.050	0.025	0.010	0.005	0.001
$\Phi(r^*)$	0.401	0.302	0.202	0.051	0.025	0.010	0.005	0.001

Somewhat remarkably, a simple combination of r and s leads to a pivotal function for which the normal approximation is much more accurate. Table 3.1 compares the exact p-value to that based on a normal approximation to the quantity

$$r^*(\theta) = r(\theta) + \frac{1}{r(\theta)} \log \frac{s(\theta)}{r(\theta)}. \qquad (3.4)$$

The derivation of r^* uses so-called higher-order asymptotics, the basis of which is an asymptotic expansion whose leading term gives the first-order normal approximation to the distribution of r. The approximation illustrated in Table 3.1 is often called a *third-order* approximation, as detailed analysis shows that the approximation has relative error $O(n^{-3/2})$, in the *moderate deviation* range $\hat{\theta} - \theta = O_p(n^{-1/2})$. However, before theoretical motivation is provided for this particular construction the numerical evidence is compelling, in this and any number of continuous one-parameter models. The following sections consider methods for obtaining similar results in somewhat more complex models. The emphasis is on constructing p-value functions for a scalar parameter, based on a scalar summary of the data. Section 3.2.2 considers reduction of the data to a scalar summary, and Section 3.2.3 discusses models with nuisance parameters.

3.2.2   Marginal and conditional distributions

With a sample of size $n = 1$ and a scalar parameter the variable is in one-to-one correspondence with the parameter, and plots and tables are easily obtained. In some classes of models we can establish such a correspondence and obtain analogous results. In full exponential families this correspondence is achieved by considering the marginal distribution of the maximum likelihood estimator or the canonical statistic. In transformation families this correspondence is achieved by considering the conditional distribution given the maximal ancillary. Thus, these two families can serve as prototypes for proceeding more generally. The discussion of exponential and transformation families in Chapter 2 of Barndorff-Nielsen and Cox (1994) seems ideal for this, and there are excellent treatments as well in Pace and Salvan (1997), Severini (2000a), and Davison (2003). An important and useful result is the derivation, in transformation models, of the exact distribution of the maximum likelihood estimator conditional on the maximal ancillary. The illustration of this is particularly simple for the location model, where we have (Cox and Hinkley, 1974, Example 4.15),

$$f(\widehat{\theta} \mid a; \theta) = \frac{\exp\{\ell(\theta; \widehat{\theta}, a)\}}{\int \exp\{\ell(\theta; \widehat{\theta}, a)\} d\theta}, \tag{3.5}$$

with

$$\ell(\theta; \widehat{\theta}, a) = \sum \log f(y_i - \theta) = \sum \log f(a_i + \widehat{\theta} - \theta), \tag{3.6}$$

where the maximal ancillary $a = (a_1, \ldots, a_n) = (y_1 - \widehat{\theta}, \ldots, y_n - \widehat{\theta})$.

Laplace approximation to the integral in the denominator gives an approximation to the conditional density of $\widehat{\theta}$, called the p^* approximation

$$p^*(\widehat{\theta} \mid a; \theta) = c |j(\widehat{\theta})|^{1/2} \exp\{\ell(\widehat{\theta}) - \ell(\theta)\}. \tag{3.7}$$

An approximation to the integral of this density can be integrated to give the normal approximation to the distribution of r^* defined in (3.4); see for example Davison (2003, Chapter 12). This argument generalizes to transformation families, of which the location model is the simplest example (Barndorff-Nielsen and Cox, 1994, §6.2).

Related calculations can be carried out in one-parameter exponential families with density $f(y_i; \theta) = \exp\{\theta y_i - c(\theta) - d(y_i)\}$. For $y = (y_1, \ldots, y_n)$ we have

$$f(y; \theta) = \exp\{\theta y_+ - nc(\theta) - \tilde{d}(y_+)\};$$

where $y_+ = \sum y_i$ is a minimal sufficient statistic for θ. A higher-order approximation to the distribution of y_+ or $\widehat{\theta}$ leads to a normal approximation to the distribution of

$$r^*(\theta) = r(\theta) + \frac{1}{r(\theta)} \log \frac{q(\theta)}{r(\theta)}, \tag{3.8}$$

where now $q(\theta)$ is the Wald statistic eqn (3.2). This approximation can be derived from a saddlepoint approximation to the density of the sufficient statistic y_+, just as eqn (3.4) is derived by integrating the p^* approximation (3.7). An example is given by the Pareto distribution

$$f(y; \theta) = \theta(1 + y)^{-(\theta+1)}, \quad y > 0, \quad \theta > 0;$$

the p-values based on a simulated sample of size 5 are illustrated in Figure 3.2. The exact distribution is computed using the fact that $\log(1 + y)$ follows an exponential distribution with rate θ.

Examples such as these are perhaps useful for motivation of the study of the r^* approximation, but they are also essentially defining the inference goal as a set of confidence bounds, or equivalently p-values for testing arbitrary values

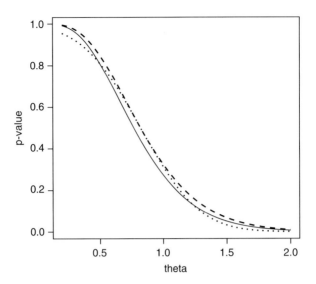

F IG. 3.2. A one-parameter exponential family illustration of the asymptotic theory. The approximate *p*-value based on r^* cannot be distinguished from the
exact value (solid). The normal approximation to the likelihood root (dashed)
and Wald statistic (dotted) are much less accurate.

of the parameter. The intervals will not be nested if the log likelihood function
is multimodal, so it is important to plot the likelihood function as well as the
p-value function.

In location families, or more generally in transformation families, conditioning serves to reduce the dimension of the data to the dimension of the parameter.
Assume we have a model with a single unknown parameter that is neither an
exponential family model nor a transformation model. The solution that arises
directly from consideration of higher-order approximations is to construct a one-
dimensional conditional model using an *approximate* ancillary statistic.

As an example, consider sampling from the bivariate normal distribution,
with only the correlation coefficient as an unknown parameter (Cox and Hinkley,
1974, Example 2.30; Barndorff-Nielsen and Cox, 1994, Example 2.38). The log
likelihood function is

$$\ell(\theta) = -\frac{n}{2}\log(1-\theta^2) - \frac{1}{2(1-\theta^2)}\sum_{i=1}^{n}(y_{1i}^2+y_{2i}^2) + \frac{\theta}{1-\theta^2}\sum_{i=1}^{n}y_{1i}y_{2i}, \quad -1 < \theta < 1.$$
$$(3.9)$$

In this model the minimal sufficient statistic $(\Sigma(y_{1i}^2 + y_{2i}^2), \Sigma y_{1i}y_{2i})$ does not
have a single component that is exactly ancillary, although either of Σy_{1i}^2 or
Σy_{2i}^2 is ancillary. It is possible to find an approximate ancillary by embed-
ding the model in a two-parameter exponential family, for example by treating

$\alpha_1 \Sigma(y_{1i} + y_{2i})^2/\{2n(1+\theta)\}$ and $\alpha_2 \Sigma(y_{1i} - y_{2i})^2/\{2n(1-\theta)\}$ as independent χ_n^2 random variables, where $\alpha_1 = \alpha/(1-\theta)$, $\alpha_2 = \alpha/(1+\theta)$, thus recovering eqn (3.9) when $\alpha = 1$. An illustration of this approach in a different $(2,1)$ family is given in Barndorff-Nielsen and Cox (1994, Example 7.1).

It is also possible to use a method based on a local location family, as described in Fraser and Reid (1995). The main idea is to define the approximately ancillary statistic indirectly, rather than explicitly, by finding vectors in the sample space tangent to an approximate ancillary. This is sufficient to provide an approximation to the p-value function. The details for this example are given in Reid (2003, §4), and the usual normal approximation to the distribution of r^* still applies, where now

$$r^*(\theta) = r(\theta) + \frac{1}{r(\theta)} \log \frac{Q(\theta)}{r(\theta)}, \tag{3.10}$$

and

$$Q(\theta) = \{\varphi(\widehat{\theta}) - \varphi(\theta)\}\{j_{\varphi\varphi}(\widehat{\theta})\}^{1/2} \tag{3.11}$$

is the Wald statistic, eqn (3.2) in a new parametrization $\varphi = \varphi(\theta; y)$. This new parametrization is the canonical parameter in an approximating exponential model and is computed by sample space differentiation of the log likelihood function:

$$\varphi(\theta) = \ell_{;V}(\theta; y), \tag{3.12}$$
$$j_{\varphi\varphi}(\widehat{\theta}) = |\ell_{\theta;V}(\widehat{\theta})|^{-2} j(\widehat{\theta}).$$

The derivation of V is outlined in Reid (2003, §3); briefly if $z_i(y_i; \theta)$ has a known distribution, then the ith element of V is

$$V_i = -\left(\frac{\partial z_i}{\partial y_i}\right)^{-1} \left(\frac{\partial z_i}{\partial \theta}\right)\Big|_{\widehat{\theta}}. \tag{3.13}$$

For model (3.9) we have

$$\ell_{;V}(\theta) = n\{\theta(t - \widehat{\theta}s) - (s - \widehat{\theta}t)\}/\{(1-\theta^2)(1-\widehat{\theta}^2)\}$$
$$\ell_{\theta;V}(\widehat{\theta}) = n(\widehat{\theta}s + t)/(1-\widehat{\theta}^2)^2,$$

with $s = \Sigma y_{1i} y_{2i}$, $t = \Sigma(y_{1i}^2 + y_{2i}^2)$. The pivotal statistic used was $(z_{1i}, z_{2i}) = [(y_{1i} + y_{2i})^2/\{2(1+\theta)\}, (y_{1i} - y_{2i})^2/\{2(1-\theta)\}]$.

The derivations of approximations like these require a fairly detailed description of asymptotic theory, both the usual first-order theory and less familiar extensions such as Edgeworth, saddlepoint, and Laplace approximations, as set out, for example, in Barndorff-Nielsen and Cox (1989). It is difficult to know how much of this is essential for a modern approach to statistical theory, but one hopes it is possible to describe the main results without being overly technical or overly vague. I have not been specific about the definition of *approximate ancillarity*, which can become rather technical. The construction using vectors based on a pivotal statistic gives a means of proceeding, and while the ancillary statistic related to this is not uniquely determined, the resulting approximation is unique to third order.

3.2.3   Models with nuisance parameters

We now consider vector parameters θ, and in order to be able to extend the construction of a p-value function we will assume that the parameter of interest ψ is a scalar, and that $\theta = (\psi, \lambda)$. The simplest way to obtain a log likelihood function for ψ is to use the profile log likelihood function $\ell_p(\psi) = \ell(\psi, \widehat{\lambda}_\psi)$ where $\widehat{\lambda}_\psi$ is the constrained maximum likelihood estimate. Properties and illustrations of the profile log likelihood function are described in Barndorff-Nielsen and Cox (1994, Chapters 3–5). Asymptotically normal statistics analogous to eqns (3.1), (3.2), and (3.3) can be obtained from the profile log likelihood.

The profile likelihood does not, in general, correspond to the density of an observable random variable, however, and as a result the first-order approximations can be very poor, especially in the case of high-dimensional nuisance parameters, such as inference for the variance parameter in a normal theory linear regression, and in more extreme examples where the dimension of the nuisance parameter increases with the sample size. These examples can be used to motivate adjustments to the profile likelihood to accommodate the presence of nuisance parameters.

As an example, consider a one-way analysis of variance (Cox and Hinkley, 1974, Examples 5.20, 9.5, 9.24), in which $y_{ij} \sim N(\lambda_i, \psi)$ for $i = 1, \ldots, n; j = 1, \ldots, k$. A comparison of the profile log likelihood and the log likelihood from the marginal distribution of the within-group sum of squares $\Sigma_{ij}(y_{ij} - \bar{y}_{i.})^2$ illustrates both the effect on the maximum point and on the curvature; see Figure 3.3. The marginal log likelihood for variance components in normal theory linear models is often called a residual log likelihood, and has been discussed in much more general settings; see, for example, Cox and Solomon (2003, Chapter 4).

I think the easiest way to motivate the adjustment that in this example produces the marginal log likelihood is to take a Bayesian approach, and derive the Laplace approximation to the marginal posterior density

$$\pi_m(\psi \mid y) = \frac{\int \exp\{\ell(\psi, \lambda; y)\pi(\psi, \lambda)d\lambda}{\int \exp\{\ell(\psi, \lambda; y)\pi(\psi, \lambda)d\psi d\lambda}$$

(3.14)

$$\doteq \frac{1}{\sqrt{(2\pi)}} \exp\{\ell_p(\psi) - \ell_p(\widehat{\psi})\}\{j_p(\widehat{\psi})\}^{1/2} \frac{|j_{\lambda\lambda}(\widehat{\psi}, \widehat{\lambda})|^{1/2}}{|j_{\lambda\lambda}(\psi, \widehat{\lambda}_\psi)|^{1/2}} \frac{\pi(\psi, \widehat{\lambda}_\psi)}{\pi(\widehat{\psi}, \widehat{\lambda})}$$

where $j_p(\psi) = -\ell_p''(\psi)$, and $j_{\lambda\lambda}(\psi, \lambda)$ is the nuisance parameter component of the observed Fisher information matrix.

This approximation is relatively easy to derive, and can be used to develop two important and closely related aspects of higher-order asymptotics in a

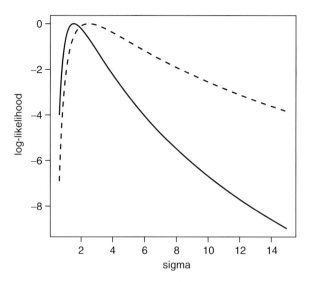

FIG. 3.3. The profile log likelihood (solid) and the marginal log likelihood (dashed) for σ in the one-way analysis of variance model, with three observations from each of five groups. Profiling does not properly account for uncertainty in the estimation of the nuisance parameter λ.

frequentist setting. First, the 'likelihood function' associated with this marginal posterior is an adjusted profile log likelihood

$$\ell_a(\psi) = \ell_p(\psi) - \frac{1}{2} \log |j_{\lambda\lambda}(\psi, \widehat{\lambda}_\psi)|.$$

It seems at least plausible that one could construct non-Bayesian inference for ψ by ignoring the prior, that is, assigning a uniform prior to both ψ and λ. This shows directly that the adjusted log likelihood is not invariant to reparameterization, and could lead into a discussion of orthogonal parameterization (Barndorff-Nielsen and Cox, 1994, §2.7; Cox and Reid, 1987).

Second, an extension of the derivation of the r^* approximation in the location model leads to an approximation to the marginal posterior survivor function of the same form

$$\int_\psi^\infty \pi_m(\psi \mid y) d\psi \doteq \Phi(r^*), \tag{3.15}$$

where

$$r^* = r^*(\psi) = r_p(\psi) + \frac{1}{r_p(\psi)} \log \frac{q_B(\psi)}{r_p(\psi)},$$

$$r_p(\psi) = \text{sign}\,\{q_B(\psi)\}\sqrt{[2\{\ell_p(\widehat{\psi}) - \ell_p(\psi)\}]},$$

$$q_B(\psi) = -\ell_p'(\psi)\{j_p(\widehat{\psi})\}^{-1/2} \frac{|j_{\lambda\lambda}(\psi, \widehat{\lambda}_\psi)|^{1/2}}{|j_{\lambda\lambda}(\widehat{\psi}, \widehat{\lambda})|^{1/2}} \frac{\pi(\widehat{\psi}, \widehat{\lambda})}{\pi(\psi, \widehat{\lambda}_\psi)}. \tag{3.16}$$

In a non-Bayesian context, higher-order inference in the presence of nuisance parameters is relatively straightforward in two models that, while specialized, nonetheless cover a fair number of examples. The first is independent sampling from the regression-scale model

$$y_i = x_i^{\mathrm{T}}\beta + \sigma e_i, \tag{3.17}$$

where the e_i are independent with mean zero and follow a known distribution. A generalization of the location family result, eqn (3.5), gives a density on $\mathbb{R}^{p+1}$, where p is the dimension of β, by conditioning on the residuals $(y_i - x_i^{\mathrm{T}}\widehat{\beta})/\widehat{\sigma}$. In this conditional distribution all nuisance parameters are exactly eliminated by marginalizing to the t-statistic for a component of β (or for $\log\sigma$);

$$t_j = (\widehat{\beta}_j - \beta)/\widehat{se}(\widehat{\beta}_j).$$

This is essentially the same as Bayesian inference for β_j using the prior $d\beta d\log\sigma$, so the approximation given in eqn (3.15) can be used. Software for this is implemented in the `marg` library of S (Brazzale, 2000, Chapter 6). Table 3.2 gives 95% confidence intervals for a regression coefficient fitting model, eqn (3.17), with a number of different error distributions. It is computed from the house price data from Sen and Srivastava (1990). In this example the confidence intervals are quite stable over a range of error distributions, although the first- and third-order confidence intervals are rather different.

In exponential family models, if the parameter of interest is a component of the canonical parameter, then the nuisance parameter can be eliminated by conditioning on the remaining components of the sufficient statistic, leading to approximations like eqn (3.15) for models such as logistic regression, Poisson regression, and inference about the shape of a Gamma distribution. The appropriate definition of r^* can be derived from the saddlepoint approximation as

$$r^* = r^*(\psi) = r_{\mathrm{p}}(\psi) + \frac{1}{r_{\mathrm{p}}(\psi)}\log\frac{q_{\mathrm{E}}(\psi)}{r_{\mathrm{p}}(\psi)},$$

$$r_{\mathrm{p}}(\psi) = \mathrm{sign}\ (\widehat{\psi} - \psi)\sqrt{[2\{\ell_{\mathrm{p}}(\widehat{\psi}) - \ell_{\mathrm{p}}(\psi)\}]},$$

$$q_{\mathrm{E}}(\psi) = (\widehat{\psi} - \psi)\{j_{\mathrm{p}}(\widehat{\psi})\}^{1/2}\frac{|j_{\lambda\lambda}(\widehat{\psi}, \widehat{\lambda})|^{1/2}}{|j_{\lambda\lambda}(\psi, \widehat{\lambda}_\psi)|^{1/2}}. \tag{3.18}$$

TABLE 3.2. Comparison of 95% confidence intervals for a regression parameter under different models for the error distribution. Fitting and inference is carried out in R using the `marg` library of Brazzale (2000). The data set has 26 observations, 4 covariates, and an unknown scale parameter. The confidence intervals are for the coefficient of the covariate `frontage`.

	First order		Third order	
Student (3)	-0.07	0.65	-0.16	0.69
Student (5)	-0.09	0.65	-0.15	0.70
Student (7)	-0.08	0.66	-0.14	0.70
Normal	-0.08	0.66	-0.13	0.71

This is implemented in `cond` in Brazzale (2000), where a number of examples are given. In regression models for discrete data, the question of continuity correction arises, and the correct interpretation of the third-order approximations, which are inherently continuous, is not completely clear. Experience seems to indicate that the normal approximation to the distribution of r^* gives good approximations to the mid-p-value, $\mathrm{pr}(y > y^0) + (1/2)\,\mathrm{pr}(y = y^0)$. The continuity correction implemented in Brazzale (2000) instead replaces y^0 by $y^0 \pm 1/2$, according as y^0 is in the right or left tail, and computes r^* at this new data value. Discussion of continuity correction is given in Davison and Wang (2002), Pierce and Peters (1992), and Severini (2000b).

Outside these two classes there is not an exact conditional or marginal density for the parameter of interest. To find an approximate solution, it turns out that the higher-order approximation is a generalization of the t-pivot result for regression-scale models. The approach developed by Barndorff-Nielsen (1986) and outlined in Barndorff-Nielsen and Cox (1994, §6.6) is based on an approximation to the conditional density of $\widehat{\theta}$ given a, which is transformed to the joint density of $(r^*(\psi), \widehat{\lambda}_\psi)$, where now

$$r^* = r^*(\psi) = r_\mathrm{p}(\psi) + \frac{1}{r_\mathrm{p}(\psi)} \log \frac{u(\psi)}{r_\mathrm{p}(\psi)}$$

and to show that this leads to the standard normal approximation to the density of r^*. The complementing statistic $u(\psi)$ is

$$u(\psi) = |\ell_{;\widehat{\theta}}(\widehat{\theta}) - \ell_{;\widehat{\theta}}(\widehat{\theta}_\psi) \quad \ell_{\lambda;\widehat{\theta}}(\widehat{\theta}_\psi)|/\{|j_{\lambda\lambda}(\widehat{\theta}_\psi)||j(\widehat{\theta})|\}^{1/2}; \tag{3.19}$$

this assumes that the log likelihood $\ell(\theta)$ can be expressed as $\ell(\theta; \widehat{\theta}, a)$. In applications it can be quite difficult to compute the sample space derivatives in u, and a method of approximating them when the model can be embedded in a full exponential family is proposed by Skovgaard (1996, 2001).

An approach developed in Fraser (1990) and Fraser and Reid (1995), and outlined in Reid (2003) uses the parametrization $\varphi(\theta)$ given at (3.12) to generalize Q of (3.11) to

$$\begin{aligned} Q(\psi) &= \{\nu(\widehat{\theta}) - \nu(\widehat{\theta}_\psi)\}/\widehat{\sigma}_\nu \\ &= \frac{|\varphi(\widehat{\theta}) - \varphi(\widehat{\theta}_\psi) \quad \varphi_{\lambda'}(\widehat{\theta}_\psi)|}{|\varphi_{\theta'}(\widehat{\theta})|} \frac{|j_{\theta\theta}(\widehat{\theta})|^{1/2}}{|j_{\lambda\lambda}(\widehat{\theta}_\psi)|^{1/2}}. \end{aligned} \tag{3.20}$$

The close connection between u and Q is discussed further by Reid (2003) and Fraser *et al.* (1999). Both ν and $\widehat{\sigma}_\nu$ are computed directly from $\varphi(\theta)$ and the matrix of derivatives $\varphi_{\theta\mathrm{T}}(\theta)$, and can be implemented numerically if necessary. This enables approximate inference for parameters in generalized linear models other than linear functions of the canonical parameters. In the two special cases discussed above this simplifies to the versions discussed there; that is, for inference in generalized linear models with canonical link function Q simplifies to q_E given in eqn (3.18), and in a regression-scale model, the general version reproduces the marginal approximation using q_B, eqn (3.16), but without the prior.

We illustrate this using the Behrens–Fisher problem (Cox and Hinkley, 1974, §5.2iv). Suppose we have independent samples of size n_1 and n_2 from normal distributions with parameters $\theta = (\mu_1, \sigma_1^2, \mu_2, \sigma_2^2)$, and the parameter of interest is $\psi = \mu_1 - \mu_2$. The canonical parameter of the full exponential model is

$$\varphi(\theta) = (\mu_1/\sigma_1^2, -1/2\sigma_1^2, \mu_2/\sigma_2^2, -1/2\sigma_2^2).$$

Table 3.3 gives a comparison of the first- and third-order approximations for a selection of values of n_1, n_2, σ_1^2 and σ_2^2. Except in cases of extreme imbalance, the third-order approximation is remarkably accurate.

Given that the calculations can be relatively easily implemented in a variety of models, it would be useful for theoretical discussions to be able to delineate the inferential basis for approximations based on r^* computed using u in eqn (3.19) or Q in eqn (3.20), and the situation is still somewhat unsatisfactory. The quantity r^* does seem to arise quite naturally from the likelihood ratio, and the marginal distribution of r^* is what gives a pivotal statistic, but beyond this it is I think difficult to see what 'statistical principle' is at work. Perhaps the asymptotic derivation is enough. Some further discussion of this point is given by Pierce and Bellio (2004).

TABLE 3.3. Comparison of the normal approximation to r and to r^*, with r^* using eqn (3.20), for the Behrens–Fisher problem. The table gives the non–coverages of the nominal upper and lower endpoints for a 90% confidence interval, in 100,000 simulations. In all rows $\mu_1 = 2$, $\mu_2 = 0$.

		Exact		0.05	0.95	0.05	0.95
n_1	n_2	σ_1^2	σ_2^2		r		r^*
3	2	2	1	0.015	0.893	0.033	0.965
20	2	2	1	0.127	0.875	0.066	0.934
7	5	2	1	0.069	0.931	0.045	0.950
20	15	2	1	0.057	0.944	0.050	0.950
3	2	4	1	0.104	0.895	0.041	0.959
20	2	4	1	0.109	0.898	0.063	0.938
7	5	4	1	0.058	0.930	0.050	0.949
20	15	4	1	0.057	0.944	0.050	0.950
2	3	2	1	0.124	0.875	0.041	0.959
2	20	2	1	0.155	0.849	0.069	0.933
5	7	2	1	0.074	0.925	0.051	0.949
15	20	2	1	0.057	0.942	0.050	0.949
2	3	4	1	0.137	0.862	0.048	0.951
2	20	4	1	0.162	0.842	0.069	0.933
5	7	4	1	0.077	0.922	0.052	0.948
15	20	4	1	0.058	0.941	0.050	0.949

The similarity of the approximations in the Bayesian and non-Bayesian versions suggests the possibility of choosing a prior so that the resulting posterior distribution has accurate frequentist coverage to some order of approximation. A discussion of the asymptotic normality of the posterior distribution shows that Bayesian and non-Bayesian likelihood inference have the same limiting distributions, so investigating this further requires higher order asymptotics. Welch and Peers (1963) showed for scalar θ that Jeffreys prior $\pi(\theta) \propto \{i(\theta)\}^{1/2}$ ensures that posterior quantiles provide lower confidence bounds with error $O(n^{-3/2})$. Extensions to the nuisance parameter setting have been less successful, as unique priors providing this asymptotic equivalence are not readily available. An overview of these results is given in Reid *et al.* (2003). An interesting foundational point was raised in the discussion and reply of Pierce and Peters (1992), and is illustrated perhaps most clearly in the derivation of the distribution function for r^* from the p^* approximation to the density of $\hat{\theta}$ given in eqn (3.7). A change of variable from $\hat{\theta}$ to r means the Jacobian includes $\partial \ell / \partial \hat{\theta}$, which in turn means that the result involves a very particular sample-space dependence of the log likelihood function.

In the nuisance parameter setting the profile log likelihood plays an important role, and in particular a plot of the profile log likelihood can often be informative (Barndorff-Nielsen and Cox, 1994, §3.5). As argued above, adjustments to the profile log likelihood arise naturally from the asymptotic point of view, although the simple adjustment using $\log|j_{\lambda\lambda}(\psi, \hat{\lambda}_\psi)|$ is unsatisfactory in its dependence on the parameterization. More refined adjustments to the profile log likelihood are based on the same asymptotic arguments as lead to the r^* approximation, and a computationally tractable version based on the arguments that lead to (3.20) is described and illustrated in Fraser (2003).

This discussion has focused on inference for a scalar component of θ, as that leads to the most tractable, and also most effective, approximations. The p-value function approach does not extend easily to inference about a vector parameter, as departures from the 'null' point are multidimensional. Bartlett correction of the log likelihood ratio does provide an improvement of the asymptotic χ^2 approximation, which is another motivation for the primacy of the likelihood ratio over asymptotically equivalent statistics such as the standardized maximum likelihood estimate. Skovgaard (2001) proposes a different form of correction of the log likelihood ratio that is closer in spirit to the r^* approximation. Another possibility is to derive univariate departures in some way, such as conditioning on the direction of the alternative vector and examining its length.

The most important advance in Bayesian inference is the possibility of simulating observations from the posterior density using sophisticated numerical methods. This has driven an explosion in the use of Bayesian methods in practice, particularly for models in which there is a natural hierarchy of effects, some of which can be modelled using priors and hyper-priors. In these settings the inferential basis is straightforward, and the main difficulties are computational, but there is less attention paid to the stability of the inference with respect to the choice of prior than is perhaps desirable. The approximations discussed here can be useful in checking this, and perhaps have a role in assessing convergence of the Markov-chain-based sampling algorithms. From the point of view of the study of theoretical statistics, an approach that incorporates Bayesian inference as a major component should spend, I think, considerable time on the choice of prior and the effect of the prior. The construction of non-informative priors, and the assessment of the sampling properties of inference based on 'popular' priors should be addressed at length.

3.3 Discussion

More traditional approaches to the study of the theory of statistics emphasize, to varying degrees, the topics of point estimation and hypothesis testing. The approach based on likelihood essentially replaces both of these by the p-value function. The use of this function also sidesteps the discussion of one-tailed and two-tailed tests. The conditional approach to vector parameter inference mentioned above extends the discussion in Cox and Hinkley (1974, §3.4iv).

Among the more classical topics, some understanding of some basic concepts is needed at least for several specific applied contexts, and perhaps more generally. For example, much of the work in non-parametric regression and density estimation relies on very generally specified models, and relatively *ad hoc* choices of estimators. In this setting the bias and variance of the estimators, usually of functions, but sometimes of parameters, are the only easily identifiable inferential quantities, and it is of interest to compare competing procedures on this basis. As another example, the notion of the power of a test to detect a meaningful substantive difference plays a prominent role in many medical contexts, including the analysis of data from clinical trials. While the relation of power to sample size is less direct, and less meaningful, than the relation of the length of a confidence interval to sample size, the use of fixed level testing is sufficiently embedded in some fields that students do need to learn the basic definitions of size and power. In other fields the basic definitions of decision theory, including loss functions and utilities, will be important.

I see no role beyond historical for extended discussion of optimality considerations, particularly with regard to testing. The existence and/or construction of most powerful tests under various special circumstances is in my view largely irrelevant to both practice and theory. The development of admissible or otherwise optimal point estimators may be useful in specialized decision theory settings, and perhaps has a role in choosing among various estimators of functions in settings like non-parametric regression, but does not seem to me to be an essential part of the basics of statistical theory.

One concern about the approach based on higher-order asymptotic theory is whether it is excessively specialized, in view of the many developments in modelling for complex settings that have been developed along with the rise in computing power. An important question is whether this theory provides any general concepts that may prove useful in much more complex settings, such as analysis of censored survival data, of longitudinal data, of data with complex spatial dependencies, of graphical models, or hierarchical models, to name just a few. The approach outlined in Section 3.2 emphasizes the primacy of the likelihood function, the isolation of a one-dimensional distribution for inference about the likelihood function, and the isolation of a scalar parameter of interest. This can I think be used much more generally.

In many applications of parametric modelling, the role of the likelihood function is well established, and is often the first step in an approach to constructing inference from very complex models. The results described in the previous section indicate that relying on the first-order normal approximation may be misleading, especially in the presence of large numbers of nuisance parameters. Some adjustment for nuisance parameters seems essential, both for plotting the likelihood and for inference. A Bayesian adjustment is the most easily implemented, but raises the difficulty of the dependence of the results on the prior. This dependence is rather clearly isolated in expressions like eqn (3.15) or eqn (3.16).

The emphasis in Section 3.2 on scalar parameters of interest may be important for discussion of more realistic models, where the identification of this parameter may not be obvious. For example, in the transformed regression model of Box and Cox (1964)

$$y_i^\lambda = x_i^T \beta + \sigma e_i, \quad i = 1, \ldots, n,$$

where $\theta = (\beta, \sigma, \lambda)$, and we assume e_i are independent standard normal variates, it seems likely that components of β are not the real parameters of interest. In work as yet unpublished with Fraser and Wong, we suggest that for a simple linear regression model $x_i = (1, z_i)$, one version of a parameter of interest is the rate of change in the median response, at a fixed value z^0:

$$\psi(\theta) = \left. \frac{d}{dz}(\beta_0 + \beta_1 z)^{1/\lambda} \right|_{z=z^0}.$$

The calculations described in Section 3.2.3 are easily carried out in this setting.

In many settings very complex models are constructed at least in part because there is a vast amount of data available. One example is the use of regression models in environmental epidemiology that adjust for potential confounders using methods based on splines. Another is the use of hierarchical modelling of variance components in models for longitudinal data. The study of asymptotic theory can at least serve as a reminder that the models with very large numbers of nuisance parameters may lead to misleading inference, and that a Bayesian approach is likely to need fairly careful attention to the specification of the prior. The theory can also provide a basis for understanding that the amount of information in the data may be quite different from the apparent sample size. An excellent illustration of this appears in Brazzale (2000, §5.3.1).

Recent work by Claeskens (2004) uses marginal likelihood for variance components in fitting regression splines in an interesting combination of ideas from likelihood asymptotics and non-parametric smoothing methods.

Although the study of theoretical statistics, and particularly foundations, is perhaps not so fashionable today, there are in our journals a great many theoretical, or at least mathematically technical, papers. Yet most graduate departments seem to be struggling with their basic theory courses, trying to make them relevant, yet finding it difficult to escape the more classical structure. This is certainly an evolutionary process, but I think the time is right to accelerate the modernization that was initiated in Cox and Hinkley (1974).

Acknowledgements

I would like to thank Augustine Wong for providing the simulation results for the Behrens–Fisher problem. I am grateful to Don Fraser and Anthony Davison for helpful discussion, and to two referees for very useful comments. It is a privilege and pleasure to acknowledge the enormous influence that David Cox has had on my understanding of statistics.

4
Exchangeability and regression models

Peter McCullagh

4.1 Introduction

Sir David Cox's statistical career and his lifelong interest in the theory and application of stochastic processes began with problems in the wool industry. The problem of drafting a strand of wool yarn to near-uniform width is not an auspicious starting point, but an impressive array of temporal and spectral methods from stationary time series were brought to bear on the problem in Cox (1949). His ability to extract the fundamental from the mundane became evident in his discovery or construction of the eponymous Cox process in the counting of neps in a sample of wool yarn Cox (1955a). Subsequent applications include hydrology and long-range dependence (Davison and Cox 1989, Cox 1991), models for rainfall (Cox and Isham 1988, Rodriguez-Iturbe et $al.$ 1987, 1988), and models for the spread of infectious diseases (Anderson et $al.$ 1989).

At some point in the late 1950s, the emphasis shifted to statistical models for dependence, the way in which a response variable depends on known explanatory variables or factors. Highlights include two books on the planning and analysis of experiments, seminal papers on binary regression, the Box–Cox transformation, and an oddly titled paper on survival analysis. This brief summary is a gross simplification of Sir David's work, but it suits my purpose by way of introduction because the chief goal of this chapter is to explore the relation between exchangeability, a concept from stochastic processes, and regression models in which the observed process is modulated by a covariate.

A stochastic process is a collection of random variables, $Y_1, Y_2, \ldots$, usually an infinite set, though not necessarily an ordered sequence. What this means is that $\mathcal{U}$ is an index set on which Y is defined, and for each finite subset $S = \{u_1, \ldots, u_n\}$ of elements in $\mathcal{U}$, the value $Y(S) = \big(Y(u_1), \ldots, Y(u_n)\big)$ of the process on S has distribution P_S on $\mathbb{R}^S$. This chapter is a little unconventional in that it emphasizes probability distributions rather than random variables. A real-valued process is thus a consistent assignment of probability distributions to observation spaces such that the distribution P_n on $\mathbb{R}^n$ is the marginal distribution of P_{n+1} on $\mathbb{R}^{n+1}$ under deletion of the relevant co-ordinate. A notation such as $\mathbb{R}^n$ that puts undue emphasis on the incidental, the dimension of the observation space, is not entirely satisfactory. Two samples of equal size need not have the same distribution, so we write $\mathbb{R}^S$ rather than $\mathbb{R}^n$ for the set of real-valued functions on the sampled units, and P_S for the distribution.

A process is said to be exchangeable if each finite-dimensional distribution is symmetric, or invariant under co-ordinate permutation. The definition suggests that exchangeability can have no role in statistical models for dependence, in which the distributions are overtly non-exchangeable on account of differences in covariate values. I argue that this narrow view is mistaken for two reasons. First, every regression model is a set of processes in which the distributions are indexed by the finite restrictions of the covariate, and regression exchangeability is defined naturally with that in mind. Second, regression exchangeability has a number of fundamental implications connected with lack of interference (Cox 1958*a*) and absence of unmeasured covariates (Greenland *et al.* 1999). This chapter explores the role of exchangeability in a range of regression models, including generalized linear models, biased-sampling models (Vardi 1985), block factors and random-effects models, models for spatial dependence, and growth-curve models. The fundamental distinction between parameter estimation and sample-space prediction is a recurring theme; see Examples 5 and 6 below.

Apart from its necessity for asymptotic approximations, the main reason for emphasizing processes over distributions is that the unnatural distinction between estimation and prediction is removed. An estimated variety contrast of 50 ± 15 kg/ha is simply a prediction concerning the likely difference of yields under similar circumstances in future seasons. Although the theory of estimation could be subsumed under a theory of prediction for statistical models, there are compelling reasons for maintaining the separation. On a purely theoretical point, estimation in the sense of inference concerning the model parameter may be governed by the likelihood principle, whereas inference in the sense of prediction is not: see Section 4.5.1 below. Second, apart from convenience of presentation and linguistic style, parameter estimation is the first step in naive prediction. The second step, frequently trivial and therefore ignored, is the calculation of conditional distributions or conditional expectations, as in prediction for processes in the standard probabilistic sense. Finally, parameter estimation is equivariant under non-linear transformation: the evidence from the data in favour of $g(\theta) \in S$ is the same as the evidence for $\theta \in g^{-1}S$. Pointwise prediction, in the sense of the conditional mean of the response distribution on a new unit, is not equivariant under non-linear response transformation: the mean of $g(Y)$ is not $g(\mathrm{E}Y)$.

We do not aim to contribute to philosophical matters such as where the model comes from, nor to answer practical questions such as how to select a model within a class of models, how to compute the likelihood function, or how to decide whether a model is adequate for the task at hand. In addition, while the mathematical interpretation is clear, any physical interpretation of the model requires a correspondence between the mathematical objects, such as units, covariates and observations, and the physical objects that they represent. This correspondence is usually implicit in most discussions of models. Despite its importance, the present chapter has little to say on the matter.

4.2 Regression models

4.2.1 Introduction

We begin with the presumption that every statistical model is a set of processes, one process for each parameter value $\theta \in \Theta$, aiming to explore the consequences of that condition for regression models. The reason for emphasizing processes over distributions is that a process permits inferences in the form of predictions for the response on unsampled units, including point predictions, interval predictions and distributional predictions. Without the notion of a process, the concept of further units beyond those in the sample does not exist as an integral part of the mathematical construction, which greatly limits the possibilities for prediction and inference. Much of asymptotic theory, for instance, would be impossible in the absence of a process or set of processes.

To each potential sample, survey or experiment there corresponds an observation space, and it is the function of a process to associate a probability distribution with each of these spaces. For notational simplicity, we restrict our attention to real-valued processes in which the observation space corresponding to a sample of size n is the n-dimensional vector space of functions on the sampled units. The response value is denoted by $Y \in \mathbb{R}^n$. Other processes exist in which an observation is a more complicated object such as a tree or partition (Kingman 1978), but in a regression model where we have one measurement on each unit, the observation space is invariably a product set, such as $\mathbb{R}^n$ or $\{0,1\}^n$, of responses or functions on the sampled units.

The processes with which we are concerned here are defined on the set $\mathcal{U}$ of statistical units and observed on a finite subset called the sample. The entire set or population $\mathcal{U}$ is assumed to be countably infinite, and the sample $S \subset \mathcal{U}$ is a finite subset of size n. The term sample does not imply a random sample: in a time series the sample units are usually consecutive points, and similar remarks apply to agricultural field experiments where the sample units are usually adjacent plots in the same field. In other contexts, the sample may be stratified as a function of the covariate or classification factor. A process P is a function that associates with each finite sample $S \subset \mathcal{U}$ of size n a distribution P_S on the observation space $\mathbb{R}^S$ of dimension n. Let $S \subset S'$ be a subsample, and let $P_{S'}$ be the distribution on $\mathbb{R}^{S'}$ determined by the process. If logical contradictions are to be avoided, P_S must be the marginal distribution of $P_{S'}$ under the operation of co-ordinate deletion, that is, deletion of those units not in S'. A process is thus a collection of mutually compatible distributions of this sort, one distribution on each of the potential observation spaces.

An exchangeable process is one for which each distribution P_S is symmetric, or invariant under co-ordinate permutation. Sometimes the term infinitely exchangeable process is used, but the additional adjective is unnecessary when it is understood that we are dealing with a process defined on a countably infinite set. Exchangeability is a fundamental notion, and much effort has been devoted to the characterization of exchangeable processes and partially exchangeable

processes (De Finetti 1974, Aldous 1981, Kingman 1978). Despite the attractions
of the theory, the conventional definition of exchangeability is too restrictive to
be of much use in applied work, where differences among units are frequently
determined by a function x called a covariate.

Up to this point, we have talked of a process in terms of distributions, not
in terms of a random variable or sequence of random variables. However, the
Kolmogorov extension theorem guarantees the existence of a random variable
Y taking values in $\mathbb{R}^\mathcal{U}$ such that the finite-dimensional distributions are those
determined by P. As a matter of logic, however, the distributions come first and
the existence of the random variable must be demonstrated, not the other way
round. For the most part, the existence of the random variable poses no difficulty,
and all distributional statements may be expressed in terms of random variables.
The process Y is a function on the units, usually real-valued but possibly vector-
valued, so there is one value for each unit.

To avoid misunderstandings at this point, the statistical units are the objects
on which the process is defined. It is left to the reader to interpret this in a suitable operational sense depending on the application at hand. By contrast, the
standard definition in the experimental design literature holds that a unit is 'the
smallest division of the experimental material such that two units may receive
different treatments' (Cox 1958a). The latter definition implies random or delib-
erate assignment of treatment levels to units. At a practical level, the operational
definition is much more useful than the mathematical definition. While the two
definitions coincide in most instances, Example 3 shows that they may differ.

4.2.2 Regression processes

A covariate is a function on the units. It may be helpful for clarity to distinguish
certain types of covariate. A quantitative covariate is a function $x\colon \mathcal{U} \to \mathbb{R}$ or
$x\colon \mathcal{U} \to \mathbb{R}^p$ taking values in a finite-dimensional vector space. This statement
does not exclude instances in which x is a bounded function or a binary function.
A qualitative covariate or factor is a function $x\colon \mathcal{U} \to \Omega$ taking values in a set Ω
called the set of levels or labels. These labels may have no additional structure,
in which case the term nominal scale is used, or they may be linearly ordered or
partially ordered or they may constitute a tree or a product set. The exploitation
of such structure is a key component in successful model construction, but that
is not the thrust of the present work. For the moment at least, a covariate is a
function $x\colon \mathcal{U} \to \Omega$ taking values in an arbitrary set Ω. Ordinarily, of course, the
values of x are available only on the finite sampled subset $S \subset \mathcal{U}$, but we must
not forget that the aim of inference is ultimately to make statements about the
likely values of the response on unsampled units whose x-value is specified. If
statistical models have any value, we must be in a position to make predictions
about the response distribution on such units, possibly even on units whose
covariate value does not occur in the sample.

At this point the reader might want to draw a distinction between estimation
and prediction, but this distinction is more apparent than real. If a variety

contrast is estimated as 50 ± 15 kg/ha, the prediction is that the mean yield for other units under similar conditions will be 35–65 kg/ha higher for one variety than the other, a prediction about the difference of infinite averages. Without the concept of a process to link one statistical unit with another, it is hard to see how inferences or predictions of this sort are possible. Nonetheless, Besag (2002, p. 1271) makes it clear that this point of view is not universally accepted. My impression is that the prediction step is so obvious and natural that its mathematical foundation is taken for granted.

Let $x \colon \mathcal{U} \to \Omega$ be a given function on the units. Recall that a real-valued process is a function P that associates with each finite subset $S \subset \mathcal{U}$ a distribution P_S on $\mathbb{R}^S$, and that these distributions are mutually compatible with respect to subsampling of units. A process having the following property for every integer n is called regression exchangeable or exchangeable modulo x.

(RE) *Two finite samples $S = \{i_1, \ldots, i_n\}$ and $S' = \{j_1, \ldots, j_n\}$ of equal size, ordered such that $x(i_r) = x(j_r)$ for each r, determine the same distribution $P_S = P_{S'}$ on $\mathbb{R}^n$.*

Exchangeability modulo x is the condition that if x takes the same value on two samples, the distributions are also the same. Any distinction between units, such as name or identification number that is not included as a component of x, has no effect on the response distribution. The majority of models that occur in practical work have this property, but Example 3 below shows that there are exceptions. The property is a consequence of the definition of a statistical model as a functor on a certain category, the injective maps, in the sense of McCullagh (2002) or Brøns (2002), provided that the parameter space is a fixed set independent of the design.

Exchangeability in the conventional sense implies that two samples of equal size have the same response distribution regardless of the covariate values, so exchangeability implies regression exchangeability. For any function g defined on Ω, exchangeability modulo $g(x)$ implies exchangeability modulo x, and $g(x) \equiv 0$ reduces to the standard definition of exchangeability. Exchangeability modulo x is not to be confused with partial exchangeability as defined by Aldous (1981) for random rectangular matrices.

The first consequence of exchangeability, that differences between distributions are determined by differences between covariate values, is related, at least loosely, to the assumption of 'no unmeasured confounders' (Greenland *et al.* 1999). This is a key assumption in the literature on causality. At this stage, no structure has been imposed on the set $\mathcal{U}$, and no structure has been ruled out. In most discussions of causality the units have a temporal structure, so the observation on each unit is a time sequence, possibly very brief, and the notion of 'the same unit at a later time' is well defined (Lindley 2002, Singpurwalla 2002, Pearl 2002). The view taken here is that causality is not a property of a statistical model but of its interpretation, which is very much context dependent. For example, Brownian motion as a statistical model has a causal interpretation in terms of thermal molecular collisions, and a modified causal interpretation may be relevant to stock-market applications. In agricultural field work where the plots are in linear order, Brownian motion may be used as a model for one component of the random variation, with no suggestion of a causal interpretation. This is not fundamentally different from the use of time-series models and methods for non-temporal applications (Cox 1949). Thus, where the word causal is used, we talk of a causal interpretation rather than a causal model. We say that any difference between distributions is associated with a difference between covariate values in the ordinary mathematical sense without implying a causal interpretation.

The second consequence of regression exchangeability, that the distribution of Y_i depends only on the value of x on unit i, and not on the values on other units, is a key assumption in experimental design and in clinical trials called lack of interference (Cox 1958a, p. 19). It is known, however, that biological interference can and does occur. Such effects are usually short range, such as root interference or fertilizer diffusion, so typical field trials incorporate discard strips to minimize the effect. This sort of interference can be accommodated within the present framework by including in x the necessary information about nearby plots. Interference connected with carry-over effects in a crossover trial can be accommodated by defining the statistical units as subjects rather than subjects at specific time points; see Example 3.

4.2.3 Interaction

Suppose the model is such that responses on different units are independent, and that $x = (v_0, v)$ with v_0 a binary variable indicating treatment level, and v a baseline covariate or other classification factor. The response distributions at the two treatment levels are $P_{0,v}$ and $P_{1,v}$. It is conventional in such circumstances to define 'the treatment effect' by a function or functional of the two distributions,

$$\text{Treatment effect} = T(P_{1,v}) - T(P_{0,v}) = \tau(v),$$

such that $T(P_{1,v}) = T(P_{0,v})$ implies $P_{1,v} = P_{0,v}$ for all model distributions. For example, T might be the difference between the two means (Cox 1958a), the difference between the two log-transformed means (Box and Cox 1964), the difference between the two medians, the log ratio of the two means, the log odds ratio (Cox 1958c), the log hazard ratio (Cox 1972), or the log variance ratio. In principle, T is chosen for ease of expression in summarizing conclusions and in making predictions concerning differences to be expected in future. Ideally, T is chosen so that, under the model in its simplest form, the treatment effect is constant over the values of v, in which case we say that there is no interaction between treatment and other covariates. In practice, preference is given to scalar functions because these lead to simpler summaries, but the definition does not require this.

If $P_{1,v} = P_{0,v}$ for each v, the model distributions do not depend on the treatment level, and the treatment effect $\tau(v)$ is identically zero. Conversely, a zero treatment effect in the model implies equality of distributions. If $\tau(v)$ is identically zero, we say that there is no treatment effect. The process is then exchangeable modulo the baseline covariates v, i.e. ignoring treatment. This is a stronger condition than regression exchangeability with treatment included as a component of x.

If the treatment effect is constant and independent of v, we say that there is no interaction, and the single numerical value suffices to summarize the difference between distributions. Although the process is not now exchangeable in the sense of the preceding paragraph, the adjustment for treatment is usually of a very simple form, so much of the simplicity of exchangeability remains. If the treatment effect is not constant, we say that there is interaction. By definition, non-zero interaction implies a non-constant treatment effect, so a zero treatment effect in the presence of non-zero interaction is a logical contradiction.

It is possible to define an average treatment effect $\mathrm{ave}\{\tau(v)\}$, averaged with respect to a given distribution on v, and some authors refer to such an average as the 'main effect of treatment'. Such averages may be useful in limited circumstances as a summary of the treatment effect in a specific heterogeneous population. However, if the interaction is appreciable, and in particular if the sign of the effect varies across subgroups, we would usually want to know the value in each of the subgroups. A zero value of the average treatment effect does not imply exchangeability in the sense discussed above, so a zero average rarely corresponds to a hypothesis of mathematical interest. Nelder (1977) and Cox (1984a) argue that statistical models having a zero average main effect in the presence of interaction are seldom of scientific interest. McCullagh (2000) reaches a similar conclusion using an argument based on algebraic representation theory in which selection of factor levels is a permissible operation.

4.3 Examples of exchangeable regression models

The majority of exchangeable regression models that occur in practice have independent components, in which case it is sufficient to specify the marginal distributions for each unit. The first four examples are of that type, but the fifth example shows that the component variables in an exchangeable regression process need not be independent or conditionally independent.

Example 1: Classical regression models In the classical multiple regression model, the covariate x is a function on the units taking values in a finite-dimensional vector space $\mathcal{V}$, which we call the covariate space. Each point $\theta = (\beta, \sigma)$ in the parameter space consists of a linear functional $\beta \in \mathcal{V}'$, where $\mathcal{V}'$ is the space dual to $\mathcal{V}$, plus a real number σ, and the parameter space consists of all such pairs. If the value of the linear functional β at $v \in \mathcal{V}$ is denoted by $v^{\mathrm{T}}\beta$, the value on unit i is $x_i^{\mathrm{T}}\beta$. In the classical linear regression model the response distribution for unit i is normal with mean equal to $x_i^{\mathrm{T}}\beta$ and variance σ^2. The model may be modified in a number of minor ways, for example by restricting σ to be non-negative to ensure identifiability.

From the point of view of regression exchangeability, generalized linear models or heavy-tailed versions of the above model are not different in any fundamental way. For example, the linear logistic model in which $\eta_i = x_i^{\mathrm{T}}\beta$ and Y_i is Bernoulli with parameter $1/\{1 + \exp(-\eta_i)\}$ is an exchangeable regression model in which the parameter space consists of linear functionals on $\mathcal{V}$.

Example 2: Treatment and classification factors   A treatment or classification factor is a function x on the units taking values in a set, usually a finite set, called the set of levels. It is conventional in applied work to draw a strong distinction between a treatment factor and a classification factor (Cox 1984*a*). The practical distinction is an important one, namely that the level of a treatment factor may, in principle at least, be determined by the experimenter, whereas the level of a classification factor is an immutable property of the unit. Age, sex and ethnic origin are examples of classification factors: medication and dose are examples of treatment factors. I am not aware of any mathematical construction corresponding to this distinction, so the single definition covers both. A block factor as defined in Section 4.5 is an entirely different sort of mathematical object from which the concept of a set of levels is missing.

Let Ω be the set of treatment levels, and let $\tau\colon \Omega \to \mathbb{R}$ be a function on the levels. In conventional statistical parlance, τ is called the vector or list of treatment effects, and differences such as

$$\tau(M) - \tau(F) \quad \text{or} \quad \tau(\text{Kerr's pinks}) - \tau(\text{King Edward})$$

are called contrasts. We note in passing that the parameter space $\mathbb{R}^{\Omega}$ has a preferred basis determined by the factor levels, and a preferred basis is essential for the construction of an exchangeable prior process for the effects should this be required. Without a preferred basis, no similar construction exists for a general linear functional β in a regression model.

In the standard linear model with independent components, the distribution of the response on unit i is Gaussian with mean $\tau\{x(i)\}$ and variance σ^2. The parameter space is the set of all pairs (τ, σ) in which τ is a function on the levels and σ is a real number. For each parameter point, condition (RE) is satisfied by the process. Once again, the extension to generalized linear models presents no conceptual difficulty.

Example 3: Crossover design   In a two-period crossover design, one observation is made on each subject under different experimental conditions at two times sufficiently separated that carry-over effects can safely be neglected. If we regard the subjects as the statistical units, which we are at liberty to do, the design determines the observation space $\mathbb{R}^2$ for each unit. The observation space corresponding to a set of n units is $(\mathbb{R}^2)^n$. Let x be the treatment regime, so that (x_{i1}, x_{i2}) is the ordered pair of treatment levels given to subject i. In the conventional statistical model the response distribution for each unit is bivariate Gaussian with covariance matrix $\sigma^2 I_2$. The mean vector is

$$\mathrm{E}\begin{pmatrix} Y_{i1} \\ Y_{i2} \end{pmatrix} = \begin{pmatrix} \alpha_i + \tau_{x_{i1}} \\ \alpha_i + \tau_{x_{i2}} + \delta \end{pmatrix},$$

in which α is a function on the subjects, and δ is a common temporal trend. The parameter space consists of all functions α on the n subjects, all functions τ on the treatment levels, plus the two scalars (δ, σ), so the effective dimension is $n + 3$ for a design with n subjects and two treatment levels.

This model is not regression exchangeable because two units i, j having the same treatment regime $(x_{i1}, x_{i2}) = (x_{j1}, x_{j2})$ do not have the same response distribution for all parameter values: the difference between the two means is $\alpha_i - \alpha_j$. This model is a little unusual in that the parameter space is not a fixed set independent of the design: the dimension depends on the sample size. Nonetheless, it is a statistical model in the sense of McCullagh (2002).

An alternative Gaussian model, with units and observation spaces defined in the same manner, has the form

$$\mathrm{E}\begin{pmatrix} Y_{i1} \\ Y_{i2} \end{pmatrix} = \begin{pmatrix} \tau_{x_{i1}} \\ \tau_{x_{i2}} + \delta \end{pmatrix}, \qquad \mathrm{cov}\begin{pmatrix} Y_{i1} \\ Y_{i2} \end{pmatrix} = \begin{pmatrix} \sigma^2 & \rho\sigma^2 \\ \rho\sigma^2 & \sigma^2 \end{pmatrix},$$

with a fixed parameter space independent of the design. Two samples having the same covariate values also have the same distribution, so condition (RE) is satisfied. The temporal effect δ is indistinguishable from a carry-over effect that is independent of the initial treatment. If there is reason to suspect a non-constant carry-over effect, the model may be extended by writing

$$\mathrm{E}\begin{pmatrix} Y_{i1} \\ Y_{i2} \end{pmatrix} = \begin{pmatrix} \tau_{x_{i1}} \\ \tau_{x_{i2}} + \gamma_{x_{i1}} + \delta \end{pmatrix}, \qquad \mathrm{cov}\begin{pmatrix} Y_{i1} \\ Y_{i2} \end{pmatrix} = \begin{pmatrix} \sigma^2 & \rho\sigma^2 \\ \rho\sigma^2 & \sigma^2 \end{pmatrix}.$$

If there are two treatment levels and all four combinations occur in the design, the difference $\gamma_1 - \gamma_0$ is estimable.

Example 4: Biased sampling We consider a biased-sampling model in which observations on distinct units are independent and, for notational convenience, real-valued. The covariate w associates with the ith unit a bias function w_i such that $w_i(x) \geq 0$ for each real number x, The parameter space is either the set of probability distributions or the set of non-negative measures on the Borel sets in $\mathbb{R}$ such that each integral $\int w_i(x)\,\mathrm{d}F(x)$ is finite (Vardi 1985, Kong *et al.* 2003). For each F in the parameter space, the response distribution on unit i is the weighted distribution such that $\mathrm{d}F_i(x) \propto w_i(x)\,\mathrm{d}F(x)$. Thus, to each point in the parameter space the model associates a process with independent but non-identically distributed components. Two units having the same bias function also have the same distribution, so the process is regression exchangeable.

The simplest example is one in which $w_i(x) = 1$ identically for each unit, in which case the maximum likelihood estimator $\widehat{F}$ is the empirical distribution at the observations. Size-biased sampling corresponds to $w(x) = |x|$ or some power of $|x|$, a phenomenon that arises in a wide range of applications from the wool industry (Cox 1962, §5.4) to stereology and auditing (Cox and Snell 1979). In general, the maximum likelihood estimator $\widehat{F}$ is a distribution supported at the observation points, but with unequal atoms at these points.

Example 5: Prediction and smoothing models Consider the modification of the simple linear regression model in which unit i has covariate value x_i, and, in a conventional but easily misunderstood notation,

$$Y_i = \beta_0 + \beta_1 x_i + \eta(x_i) + \epsilon_i. \tag{4.1}$$

The coefficients (β_0, β_1) are parameters to be estimated, ϵ is a process with independent $N(0, \sigma^2)$ components, and η is a zero-mean stationary process on the real line, independent of ϵ, with covariance function

$$\text{cov}\,\{\eta(x), \eta(x')\} = \sigma_\eta^2 \, K(x, x').$$

If η is a Gaussian process, the response distribution for any finite collection of n units may be expressed in the equivalent distributional form

$$Y \sim N(X\beta, \, \sigma^2 I_n + \sigma_\eta^2 V), \tag{4.2}$$

where $V_{ij} = K(x_i, x_j)$ are the components of a positive semidefinite matrix. For each value of $(\beta, \sigma^2, \sigma_\eta^2)$, two samples having the same covariate values determine the same distribution on the observation space, so this model is regression-exchangeable with non-independent components.

The linear combination $\eta(x_i) + \epsilon_i$ in eqn (4.1) is a convenient way of describing the distribution of the process as a sum of more elementary processes. The treacherous aspect of the notation lies in the possibility that η or ϵ might be mistaken for a parameter to be estimated from the data, which is not the intention. The alternative parametric model with independent components and parameter space consisting of all smooth functions η, is also regression exchangeable, but very different from eqn (4.2).

The simplest way to proceed for estimation and prediction is first to estimate the parameters $(\sigma^2, \sigma_\eta^2, \beta_0, \beta_1)$ by maximum likelihood estimation, or by some closely related procedure such as residual maximum likelihood estimation—REML—for the variance components followed by weighted least squares estimation of the regression parameters. With prediction in mind, Wahba (1985) recommends use of generalized crossvalidation over residual maximum likelihood estimation on the grounds that it is more robust against departures from the stochastic model. Efron (2001) considers a range of estimators and seems to prefer the residual maximum likelihood estimator despite evidence of bias. Suppose that this has been done, and that we aim to predict the response value $Y(i^*)$ for a new unit i^* in the same process whose covariate value is $x^* = x(i^*)$.

Proceeding as if the parameter values were given, the conditional expected value of $Y(i^*)$ given the values on the sampled units is computed by the formula

$$\widehat{Y}_{i^*} = \mathrm{E}\left\{Y(i^*)\,|\,Y\right\} = \beta_0 + \beta_1 x^* + k^* \Sigma^{-1}(Y - \mu), \qquad (4.3)$$

where μ, Σ are the estimated mean and covariance matrix for the sampled units, and $k_i^* = \sigma_\eta^2 K(x^*, x_i)$ is the vector of covariances. The conditional distribution is Gaussian with mean (4.3) and constant variance independent of Y. Interpolation features prominently in the geostatistical literature, where linear prediction is called kriging (Stein 1999).

If η is Brownian motion with generalized covariance function $-|x - x'|$ on contrasts, the prediction function (4.3) is continuous and piecewise linear: if $K(x, x') = |x - x'|^3$, the prediction function is a cubic spline (Wahba 1990, Green and Silverman 1994). Of course, K is not necessarily a simple covariance function of this type: it could be in the Matérn class (Matérn 1986) or it could be a convex combination of simple covariance functions. The cubic and linear splines illustrated in Figure 4.1 are obtained by fitting model (4.2) to simulated data (Wahba 1990, p. 45), in which $\eta(x)$ is the smooth function shown as the dashed line.

In statistical work, the adjective 'Bayesian' usually refers to the operation of converting a probability $\mathrm{pr}(A\,|\,B)$ into a probability of the form $\mathrm{pr}(B\,|\,A)$ by supplying additional information and using Bayes's theorem. The transformation from the joint distribution of (Y^*, Y) as determined by the process (4.2), to the conditional distribution $Y^*\,|\,Y$, does not involve prior information or Bayes's theorem. Nonetheless, it is possible to cast the argument leading to eqn (4.3) in a Bayesian mould, so the majority of authors use the term Bayesian or empirical Bayes in this context (Wahba 1990, Efron 2001). A formal Bayesian analysis begins with a prior distribution π on the parameters $(\beta, \sigma^2, \sigma_\eta^2)$, and uses the likelihood function to obtain the posterior distribution. The process is such that

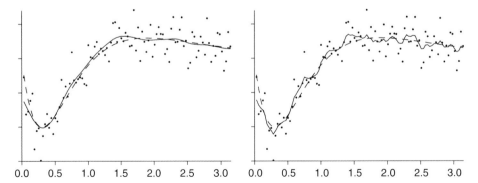

FIG. 4.1. Spline prediction graphs (solid lines) fitted by residual maximum likelihood, cubic on the left, linear on the right. The ideal predictor $\eta(x)$ (dashed
line) is taken from Wahba (1990, p. 45).

the predictive distribution for a new unit has mean (4.3) and constant variance depending on the parameters. The Bayesian predictive distribution is simply the posterior average, or mixture, of these distributions. In this context, the adjective 'Bayesian' refers to the conversion from prior and likelihood to posterior, not to eqn (4.3).

Since η is not a parameter to be estimated, the introduction of extra-likelihood criteria such as penalty functions or kernel density estimators to force smoothness is, in principle at least, unnecessary. In practice, if the family of covariance functions for η includes a smoothness parameter such as the index in the Matérn class, the likelihood function seldom discriminates strongly. Two covariance functions such as $-|x-x'|$ and $|x-x'|^3$ achieving approximately the same likelihood value, usually produce prediction graphs that are pointwise similar. For the data in Figure 4.1, the maximum log residual likelihood values are 42.2 for the model in which $K(x, x') = -|x-x'|$, 42.1 for the 'quadratic' version $|x-x'|^2 \log |x-x'|$, and 41.3 for the cubic. Visually the prediction graphs are very different, so aesthetic considerations may determine the choice for graphical presentation.

To illustrate one crucial difference between an estimator and a predictor, it is sufficient to note that the prediction function (4.3) is not a projection on the observation space in the sense that the least-squares fit is a projection on the observation space. However, it is a projection on a different space, the combined observation-prediction space. Let S_0 be the n sampled units, let S_1 be the m unsampled units for which prediction is required, and let $S = S_0 \cup S_1$ be the combined set. The covariance matrix Σ for the joint distribution in $\mathbb{R}^S$ may be written in partitioned form with components $\Sigma_{00}, \Sigma_{01}, \Sigma_{11}$, and the model matrix may be similarly partitioned, X_0 for the sampled units and X_1 for the unsampled units. The prediction function is linear in the observed value Y_0 and may be written as the sum of two linear transformations,

$$\begin{pmatrix} \widehat{Y}_0 \\ \widehat{Y}_1 \end{pmatrix} = \begin{pmatrix} P_0 & 0 \\ X_1(X_0^\mathrm{T}\Sigma_{00}^{-1}X_0)^{-1}X_0^\mathrm{T}\Sigma_{00}^{-1} & 0 \end{pmatrix} \begin{pmatrix} Y_0 \\ \star \end{pmatrix} + \begin{pmatrix} Q_0 & 0 \\ \Sigma_{10}\Sigma_{00}^{-1}Q_0 & 0 \end{pmatrix} \begin{pmatrix} Y_0 \\ \star \end{pmatrix}, \quad (4.4)$$

where $P_0 = X_0(X_0^\mathrm{T}\Sigma_{00}^{-1}X_0)^{-1}X_0^\mathrm{T}\Sigma_{00}^{-1}$, $Q_0 = I - P_0$, and $\star$ is the unobserved value. Evidently $\widehat{Y}_0 = Y_0$ as it ought. The first transformation is the least-squares projection $P \colon \mathbb{R}^S \to \mathbb{R}^S$ onto $\mathcal{X} \subset \mathbb{R}^S$ of dimension p. The second transformation is a projection $T \colon \mathbb{R}^S \to \mathbb{R}^S$, self-adjoint with respect to the inner product Σ^{-1}, and thus an orthogonal projection. Its kernel consists of all vectors of the form $(x, \star)$, i.e. all vectors in $\mathcal{X} + \mathbb{R}^{S_1}$ of dimension $m + p$, and the image is the orthogonal complement. Direct calculation shows that $PT = TP = 0$ so the sum $P + T$ is also a projection.

Most computational systems take $X_1 = X_0$, so the predictions are for new units having the same covariate values as the sampled units. The first component in eqn (4.4) is ignored, and the prediction graphs in Figure 4.1 show the conditional mean $\widehat{Y}_1$ as a function of x, with Σ estimated in the conventional way by marginal maximum likelihood based on the residuals.

The preceding argument assumes that K is a proper covariance function, so Σ is positive definite, which is not the case for the models illustrated in Figure 4.1. However, the results apply to generalized covariance functions under suitable conditions permitting pointwise evaluation provided that the subspace $\mathcal{X}$ is such that K is positive semidefinite on the contrasts in $\mathcal{X}^0$ (Wahba 1990).

Example 6: Functional response model   Consider a growth-curve model in which the stature or weight of each of n subjects is measured at a number of time points over the relevant period. To keep the model simple, the covariate for subject i is the schedule of measurement times alone. If we denote by $Y_i(t)$ the measured height of subject i at time t, the simplest sort of additive growth model may be written in the form

$$Y_i(t) = \alpha_i + m(t) + \eta_i(t) + \epsilon_i(t),$$

in which α is an exchangeable process on the subjects, m is a smooth random function of time with mean μ, η is a zero-mean process continuous in time and independent for distinct subjects, and ϵ is white noise. All four processes are assumed to be independent and Gaussian. The distributions are such that Y is Gaussian with mean $\mathrm{E}\{Y_i(t)\} = \mu(t)$ and covariance matrix/function

$$\mathrm{cov}\left\{Y_i(t), Y_j(t')\right\} = \sigma_\alpha^2 \delta_{ij} + \sigma_m^2 K_m(t, t') + \sigma_\eta^2 K_\eta(t, t')\delta_{ij} + \sigma_\epsilon^2 \delta_{ij}\delta_{t-t'}.$$

If the functions μ, K_m and K_η are given or determined up to a small set of parameters to be estimated, all parameters can be estimated by maximum likelihood or by marginal maximum likelihood. The fitted growth curve, or the predicted growth curve for a new subject from the same process, can then be obtained by computing the conditional expectation of $Y_{i*}(t)$ given the data. Note that if σ_m^2 is positive, $Y_{i*}(t)$ is not independent of the values on other subjects, so the predicted value is not the fitted mean $\mu(t)$.

The model shown above satisfies condition (RE), so it is regression exchangeable even though the components are not independent. It is intended to illustrate the general technique of additive decomposition into simpler processes followed by prediction for unobserved subjects. It is ultimately an empirical matter to decide whether such decompositions are useful in practice, but real data are likely to exhibit departures of various sorts. For example, the major difference between subjects may be a temporal translation, as in the alignment of time origins connected with the onset of puberty. Further, growth measurements are invariably positive and seldom decreasing over the interesting range. In addition, individual and temporal effects may be multiplicative, so the decomposition may be more suitable for log transformed process. Finally, there may be covariates, and if a covariate is time dependent, i.e. a function of t, the response distribution at time t could depend on the covariate history.

4.4   Causality and counterfactuals

4.4.1   Notation

It is conventional in both the applied and the theoretical literature to write the linear regression model in the form

$$\mathrm{E}(Y_i \mid x) = x_i^{\mathrm{T}}\beta, \qquad \mathrm{var}(Y_i \mid x) = \sigma^2,$$

when it is understood that the components are independent. The notation suggests that $x^{\mathrm{T}}\beta$ is the conditional mean of the random variable Y and σ^2 is the conditional variance, as if x were a random variable defined on the same probability space as Y. Despite the notation and the associated description, that is not what is meant because x is an ordinary function on the units, not a random variable.

The correct statement runs as follows. First, x is a function on the units taking values in $\mathbb{R}^p$, and the values taken by x on a finite set of n units may be listed as a matrix X of order $n \times p$ with rows indexed by sampled units. Second, to each parameter point (β, σ), the model associates a distribution on $\mathbb{R}^n$ by the formula $N(X\beta, \sigma^2 I_n)$, or by a similar formula for generalized linear models. In this way, the model determines a set of real-valued processes, one process for each parameter point. Each process is indexed by the finite restrictions of x, with values in the observation space $\mathbb{R}^n$. No conditional distributions are involved at any point in this construction. At the same time, the possibility that the regression model has been derived from a bivariate process by conditioning on one component is not excluded.

Even though no conditional distributions are implied, the conventional notation and the accompanying description are seldom seriously misleading, so it would be pedantic to demand that they be corrected. However, there are exceptions or potential exceptions.

The linear regression model associates with each parameter point $\theta = (\beta, \sigma)$ a univariate process: it does not associate a bivariate process with a pair of parameter values. As a consequence, it is perfectly sensible to compare the probability $P_{x,\theta}(E)$ with the probability $P_{x,\theta'}(E)$ for any event $E \subset \mathbb{R}^n$, as in a likelihood ratio. But it makes no sense to compare the random variable Y in the process determined by θ with the random variable in the process determined by θ'. A question such as 'How much larger would Y_i have been had β_1 been 4.3 rather than 3.4?' is meaningless within the present construction because the two processes need not be defined on the same probability space. The alternative representation of a linear regression model

$$Y_i = \beta_0 + \beta_1 x_i + \sigma\epsilon_i$$

is potentially misleading on this point because it suggests the answer $(4.3-3.4)x_i$.

4.4.2   Exchangeability and counterfactuals

A counterfactual question is best illustrated by examples such as 'If I had taken aspirin would my headache be gone?' or 'How much longer would Cynthia Crabb have survived had she been given a high dose of chemotherapy rather than a low dose?' The presumption here is that $\mathcal{U}$ consists of subjects or patients, that x is a function representing treatment and other baseline variables, and that Y is the outcome. Formally, $i = Cynthia\ Crabb$ is the patient name, $x(i) = low$ is the treatment component of the covariate, $P_{\{i\}}$ is the survival distribution, and $Y: (\Omega, \mathcal{F}, P) \to \mathbb{R}^{\mathcal{U}}$ is a random variable whose ith component $Y_i(\omega)$ is the outcome for Cynthia Crabb. Since the constructed process is a real-valued function on the units, there is only one survival time for each subject. Within this framework, it is impossible to address a question that requires Cynthia Crabb to have two survival times. For a more philosophical discussion, the reader is referred to Dawid (2000) who reaches a similar conclusion, or Pearl (2000) who reaches a different conclusion.

If the preceding question is not interpreted as a literal counterfactual, it is possible to make progress by using regression exchangeability and interpreting the question in a distributional sense as follows. Among the infinite set of subjects on which the process is defined, a subset exists having the same covariate values as Cynthia Crabb except that they have the high dose of chemotherapy. Conceptually, there is no difficulty in supposing that there is an infinite number of such subjects, identical in all covariate respects to Cynthia Crabb except for the dose of chemotherapy. By regression exchangeability, all such subjects have the same survival distribution. Provided that we are willing to interpret the question as a comparison of the actual survival time of Cynthia Crabb with the distribution of survival times for patients in this subset, the mathematical difficulty of finding Cynthia Crabb in two disjoint sets is avoided. The question may now be answered, and the answer is a distribution that may, in principle, be estimated given sufficient data.

It is clear from the discussion of Dawid (2000) that most statisticians are unwilling to forego counterfactual statements. The reason for this seems to be a deep-seated human need to assign credit or blame, to associate causes with observed effects—this sentence being an instance of the phenomenon it describes. My impression is that most practical workers interpret counterfactual matters such as unit-treatment additivity in this distributional sense, sometimes explicitly so (Cox 1958*a*, 2000). However, Pearl (2000) argues to the contrary, that counterfactual statements are testable and thus not metaphysical.

The directness and immediacy of counterfactual statements are appealing by way of parameter interpretation and model explanation. Another way of trying to make sense of the notion is to invoke a latent variable, a bivariate process with two survival times for each subject, so that all random variables exist in the mathematical sense. The first component is interpreted as the survival time at low dose, the second component is interpreted as the survival time at high dose, and the difference $Y_{i2} - Y_{i1}$ or ratio Y_{i2}/Y_{i1} is the desired counterfactual difference in survival times. The treatment value serves to reduce the bivariate process to a univariate process by indicating which of the two components is observed. The net result is a univariate process whose distributions determine the exchangeable regression model described above. From the present point of view the same process is obtained by an indirect route, so nothing has changed. Unless the observation space for some subset of the units is bivariate, the counterfactual variable is unobservable, so counterfactual prediction cannot arise.

Model constructions involving latent or unobserved random processes are used frequently and successfully in statistical models. The net result in such cases is a univariate marginal process defined on the finite restrictions of the covariate. The introduction of the latent component is technically unnecessary, but it is sometimes helpful as a pedagogical device to establish a mechanistic interpretation (Cox 1992, §5.1). Provided that inferences are restricted to estimation and prediction for this marginal process, the technique is uncontroversial. Counterfactual predictions for the latent bivariate survival process are beyond the observation space on which the marginal process is defined, and thus cannot be derived from the marginal model alone. Nonetheless, with heroic assumptions, such as independence of the two survival components that are unverifiable in the marginal process, counterfactual predictions for the bivariate process may be technically possible. However, the absence of a physical counterpart to the mathematical counterfactual makes it hard to understand what such a statement might mean in practice or how it might be checked.

4.4.3 Exchangeability and causality

Why did Cynthia Crabb receive the low dose when other patients with the same non-treatment covariates received the high dose? A statistical model does not address questions of this sort. However, regression exchangeability is a model assumption implying that, in the absence of a treatment effect, the responses for all patients in Cynthia Crabb's baseline covariate class are exchangeable. As a consequence, all patients whose non-treatment covariate values are the same as those of Cynthia Crabb have the same response distribution. Any departure from exchangeability that is associated with treatment assignment may then be interpreted as evidence against the null model of no treatment effect. In applications where it is a feasible option, objective randomization is perhaps the most effective way to ensure that this model assumption is satisfied.

Like all model assumptions, regression exchangeability may prove unsatisfactory in specific applications. If the treatment assignment is done on doctor's advice on the basis of information not available in x, the exchangeability condition may well be violated. Likewise, if the protocol allows patients to select their own dose level, the choice may be based on factors not included in x, in which case there is little assurance that the patients are exchangeable modulo x.

A central theme in much of causal inference is the attempt to deduce or to predict in a probabilistic sense what would have occurred had the design protocol been different from what it actually was. Since the theory of prediction for processes does not extend beyond the index set on which the process is defined, this sort of prediction requires an explicit broader and perhaps fundamentally different foundation. One can envisage a compound doubly randomized design in which the first arm is a conventional randomized experiment, the response on each individual being five-year survival. In the elective arm, patients are permitted to select the drug or dose, so the response is bivariate. This sort of process, in which the observation space itself depends on the covariate, is certainly unconventional, but it is not fundamentally different from the definition given in Section 4.2. The definition of regression exchangeability is unchanged, but would usually be considered in a modified form in which the conditional distribution of survival times given the chosen treatment is the same as the distribution of survival times for the assigned treatment in the randomized arm. In other words, the survival distribution depends on treatment and other baseline covariates, but not on whether the treatment is randomly assigned or freely selected. With this modified concept of exchangeability in the extended process, it is possible to extrapolate from the randomized experiment, by making predictions for the outcomes under a different protocol.

It is invariably the case in matters of causal assessment that closer inspection reveals additional factors or an intermediate sequence of events that could affect the interpretation of treatment contrasts had they been included in the model, i.e. if a different model had been used. A good example can be found in the paper by Versluis *et al.* (2000) on the sound-causing mechanism used by the snapping shrimp *Alpheus heterochaelis*. Since the shrimp tend to congregate in large numbers, the combined sound is appreciable and can interfere with naval sonar. The loud click is caused by the extremely rapid closure of the large snapper claw, in the sense that a sound is heard every time the claw snaps shut and no sound is heard otherwise. It had been assumed that the sound was caused by mechanical contact between hard claw surfaces, and the preceding statement had been universally interpreted in that way. However, closer inspection reveals a previously unsuspected mechanism, in which the claw is not the source of the sound. During the rapid claw closure a high-velocity water jet is emitted with a speed that exceeds cavitation conditions, and the sound coincides with the collapse of the cavitation bubble not with the closure of the claw.

In light of this information, what are we to say of causal effects? The initial statement that the rapid closure of the claw causes the sound, is obviously correct in most reasonable senses. It satisfies the conditions of reproducibility, consistency with subject-matter knowledge, and predictability by well-established theory, as demanded by various authors such as Bradford Hill (1937), Granger (1988), and Cox (1992). However, its very consistency with subject-matter knowledge invites an interpretation that is now known to be false. Whether or not the statement is legally correct, it is scientifically misleading, and is best avoided in applications where it might lead to false conclusions. For example, the observation that a shrimp can stun its prey without contact, simply by clicking its claw, might lead to the false conclusion that snails are sensitive to sonar. The complementary statement that the closure of the claw does not cause the sound, although equally defensible, is certainly not better.

Rarely, if ever, does there exist a most proximate cause for any observed phenomenon, so the emphasis must ultimately be on processes and mechanisms (Cox 1992). Confusion results when the words 'cause' or 'causal' are used with one mechanism, or no specific mechanism, in mind, and interpreted in the context of a different mechanism. For clinical trials where biochemical pathways are complicated and unlikely to be understood in sufficient detail, the word mechanism is best replaced by protocol. The natural resolution is to avoid the term 'causal' except in the context of a specific mechanism or protocol, which might, but need not, involve manipulation or intervention. Thus, the closure of the claw causes the sound through an indirect mechanism involving cavitation. This statement does not exclude the possibility that cavitation itself is a complex process with several stages.

Unless the protocol is well defined, an unqualified statement concerning the causal effect of a drug or other therapy is best avoided. Thus, following a randomized trial in which a drug is found to increase the five-year survival rate, the recommendation that it be approved for general use is based on a model assumption, the prediction that a similar difference will be observed on average between those who elect to use the drug and those who elect not to use it. Equality here is a model assumption, a consequence of regression exchangeability in the modified sense discussed above. As with all model assumptions, this one may prove to be incorrect in specific applications. Unlike counterfactuals, the assumption can be checked in several ways, by direct comparison in a compound doubly randomized experiment, by comparisons within specific subgroups or by comparing trial results with subsequent performance. In the absence of exchangeability, there is no mathematical reason to suppose that the five-year survival rate among those who elect to use the drug should be similar to the rate observed in the randomized experiment. It is not difficult to envisage genetic mechanisms such that those who elect not to use the drug have the longer five-year survival, but all such mechanisms imply non-exchangeability or the existence of potentially identifiable subgroups.

4.5   Exchangeable block models

4.5.1   Block factor

The distinction between a block factor and a treatment factor, closely related to the distinction between fixed and random effects, is a source of confusion and anxiety for students and experienced statisticians alike. As a practical matter, the distinction is not a rigid one. The key distinguishing feature is the anonymous or ephemeral nature of the levels of a block factor. Cox (1984a) uses the term non-specific, while Tukey (1974) prefers the more colourful phrase 'named and faceless values' to make a similar distinction.

Even if it is more rigid and less nuanced, a similar distinction can be made in the mathematics. A block factor B is defined as an equivalence relation on the units, a symmetric binary function $B \colon \mathcal{U} \times \mathcal{U} \to \{0, 1\}$ that is reflexive and transitive. Equivalently, but more concretely, B is a partition of the units into disjoint non-empty subsets called blocks such that $B(i, j) = 1$ if units i, j are in the same block and zero otherwise. The number of blocks may be finite or infinite. For the observed set of n units, B is a symmetric positive semidefinite binary matrix whose rank is the number of blocks in the sample.

A treatment or classification factor $x\colon \mathcal{U} \to \Omega$ is a list of levels, one for each unit. It may be converted into a block factor by the elementary device of ignoring factor labels, a forgetful transformation defined by

$$B(i,j) = \begin{cases} 1, & \text{if } x(i) = x(j), \\ 0, & \text{otherwise.} \end{cases}$$

If X is the incidence matrix for the treatment factor on the sampled units, each column of X is an indicator function for the units having that factor level, and $B = XX^{\mathrm{T}}$ is the associated block factor. It is not possible to convert a block factor into a treatment factor because the label information, the names of the factor levels, is not contained in the block factor.

Since the information in the block factor B is less than the information in x, exchangeability modulo B is a stronger condition than exchangeability modulo x. A process is called block-exchangeable if the following condition is satisfied for each n. Two samples $\{i_1, \ldots, i_n\}$ and $\{j_1, \ldots, j_n\}$, ordered in such a way that $B(i_r, i_s) = B(j_r, j_s)$ for each r, s, determine the same distribution on $\mathbb{R}^n$. Block exchangeability implies that the label information has no effect on distributions. All one-dimensional marginal distributions are the same, and there are only two distinct two-dimensional marginal distributions depending on whether $B(i,j)$ is true or false (one or zero). More generally, the n-dimensional distribution is invariant under those coordinate permutations that preserve the block structure, i.e. permutations π such that $B(i,j) = B(\pi_i, \pi_j)$ for all i, j,

For a sample of size n, the image or range of X is the same subspace $\mathcal{X} \subset \mathbb{R}^n$ as the range of B in $\mathbb{R}^n$, the set of functions that are constant on each block. In the following linear Gaussian specifications, the distribution is followed by a description of the parameter space:

(i) $Y \sim N(X\beta, \sigma^2 I_n)$, $\quad \beta \in \mathbb{R}^\Omega, \sigma > 0$,
(ii) $Y \sim N(B\gamma, \sigma^2 I_n)$, $\quad \gamma \in \mathcal{X}, \sigma > 0$,
(iii) $Y \sim N(\mu, \sigma^2 I_n + \sigma_b^2 B)$, $\quad \mu \in \mathbf{1}, \sigma > 0, \sigma_b \geq 0$,

in which $\mathbf{1} \subset \mathbb{R}^n$ is the one-dimensional subspace of constant functions. In the sense that they determine precisely the same set of distributions on the observation space, the first two forms are equivalent up to reparameterization. Even so, the models are very different in crucial respects.

By definition in (i), $\beta \in \mathbb{R}^\Omega$ is a function on the treatment levels, so inference for specific levels or specific contrasts is immediate. In (ii), one can transform from γ to $\beta = X^{\mathrm{T}}\gamma$ only if the block labels are available. Block labels are not used in (ii), so the formulation does not imply that this information is available. Nonetheless, in most applications with which I am familiar, the function $x\colon \mathcal{U} \to \Omega$ would be available for use, in which case the two formulations are equivalent. Both are regression exchangeable but neither formulation is block exchangeable.

The third form, the standard random-effects model with independent and identically distributed block effects, is different from the others: it is block exchangeable. The expression may be regarded as defining a process on the finite restrictions of x, or a process on the finite restrictions of the block factor B. In that sense (iii) is ambiguous, as is (ii). In practice, it would usually be assumed that the block names are available for use if necessary, as for example in animal-breeding experiments (Robinson 1991). Given that x is available, it is possible to make inferences or predictions about contrasts among specific factor levels using model (iii). The conditional distribution of the response on a new unit with $x(\cdot) = 1$ is Gaussian with mean and variance

$$\frac{\sigma^2 \mu + n_1 \sigma_b^2 \bar{y}_1}{\sigma^2 + n_1 \sigma_b^2}, \qquad \sigma^2 \left(1 + \frac{\sigma_b^2}{n \sigma_b^2 + \sigma^2} \right), \qquad (4.5)$$

where n_1 is the number of units in the sample for which $x(\cdot) = 1$ and $\bar{y}_1$ is the average response. In practice, the parameter values must first be estimated from the available data. If the function x is unavailable, the blocks are unlabelled so inference for specific factor levels or specific contrasts is impossible. Nonetheless, since each new unit i^* comes with block information in the form of the extended equivalence relation B, it is possible to make predictive statements about new units such that $B(i^*, 4) = 1$, i.e. new units that are in the same block as unit 4 in the sample. The formula for the conditional distribution is much the same as that given above, so the mathematical predictions have a similar form except that the block does not have a name. It is also possible, on the basis of the model, to make predictive statements about new units that are not in the same block as any of the sample units. Whether such predictions are reliable is a matter entirely dependent on specific details of the application.

An alternative version of (iii) may be considered, in which the inverse covariance matrix, or precision matrix, is expressed as a linear combination of the same two matrices, I_n and B. If the blocks are of equal size, the inverse of $\sigma_0^2 I_n + \sigma_1^2 B$ is in fact a linear combination of the same two matrices, in which case the two expressions determine the same set of distributions on $\mathbb{R}^n$, and thus the same likelihood function after reparameterization. However, the second formulation does not determine a process because the marginal $(n-1)$-dimensional distribution after deleting one component is not expressible in a similar form, with a precision matrix that is a linear combination of I_{n-1} and the restriction of B. The absence of a process makes prediction difficult, if not impossible.

It is worth remarking at this point that, for a balanced design, the sufficient statistic for model (iii) is the sample mean plus the between- and within-block mean squares. Even for an unbalanced design, an individual block mean such as $\bar{y}_1$ is not a function of the sufficient statistic. Accordingly, two observation points $y \neq y'$ producing the same value of the sufficient statistic will ordinarily give rise to different predictions in eqn (4.3) or eqn (4.5). In other words, the conclusions are not a function of the sufficient statistic. One of the subtleties of the likelihood principle as stated, for example, by Cox and Hinkley (1974, p. 39) or Berger and Wolpert (1988, p. 19) is the clause 'conclusions about θ', implying that it is concerned solely with parameter estimation. Since eqns (4.3) and (4.5) are statements concerning events in the observation space, not estimates of model parameters or statements about θ, there can be no violation of the likelihood principle. On the other hand, a statement about θ is a statement about an event in the tail σ-field of the process, so it is not clear that there is a clear-cut distinction between prediction and parametric inference.

4.5.2 Example: homologous factors

We consider in this section a further, slightly more complicated, example of an exchangeable block model in which the covariate $x = (x_1, x_2)$ is a pair of homologous factors taking values in the set $\Omega = \{1, \ldots, n\}$ (McCullagh 2000). If there is only a single replicate, the observation Y is a square matrix of order n with rows and columns indexed by the same set of levels. More generally, the design is said to be balanced with r replicates if, for each cell (i, j) there are exactly r units u for which $x(u) = (i, j)$. For notational simplicity, we sometimes assume $r = 1$, but in fact the design need not be balanced and the assumption of balance can lead to ambiguities in notation.

The following models are block exchangeable:

$$Y_{ij} = \mu + \eta_i + \eta_j + \epsilon_{ij},$$
$$Y_{ij} = \mu + \eta_i - \eta_j + \epsilon_{ij},$$
$$Y_{ij} = \eta_i - \eta_j + \epsilon'_{ij}.$$

In these expressions η/σ_η and ϵ/σ_ϵ are independent standard Gaussian processes, so the parameter space for the first two consists of the three components $(\mu, \sigma_\eta^2, \sigma_\epsilon^2)$. In the third model, $\epsilon'_{ij} = -\epsilon'_{ji}$, so the observation matrix Y is skew-symmetric.

These expressions suggest that Y is a process indexed by ordered pairs of integers, and in this respect the notation is misleading. The 'correct' version of the first model is

$$Y(u) = \mu + \eta_{x_1(u)} + \eta_{x_2(u)} + \eta'_{x(u)} + \epsilon(u), \qquad (4.6)$$

making it clear that Y and ϵ are processes indexed by the units, and there may be several units such that $x(u) = (i, j)$. In plant-breeding experiments, the units such that $x(u) = (i, i)$ are called self-crosses; in round-robin tournaments, self-competition is usually meaningless, so there are no units such that $x(u) = (i, i)$. In the absence of replication, the interaction process η' and the residual process ϵ are not separately identifiable: only the sum of the two variances is estimable. However, absence of replication in the design does not imply absence of interaction in the model. To put it another way, two distinct models may give rise to the same set of distributions for a particular design. Aliasing of interactions in a fractional factorial design is a well-known example of the phenomenon.

If there is a single replicate, the three models may be written in the equivalent distributional form as follows:

$$Y \sim N(\mu\mathbf{1},\ \sigma_\eta^2 K + \sigma_\epsilon^2 I_{n^2}),$$
$$Y \sim N(\mu\mathbf{1},\ \sigma_\eta^2 K' + \sigma_\epsilon^2 I_{n^2}),$$
$$Y \sim N(0,\ \sigma_\eta^2 K' + \sigma_\epsilon^2 I'_{n^2}).$$

The matrices K, K', I' are symmetric of order $n^2 \times n^2$ and are given by

$$K_{ij,kl} = \delta_{ik} + \delta_{jl} + \delta_{il} + \delta_{jk},$$
$$K'_{ij,kl} = \delta_{ik} + \delta_{jl} - \delta_{il} - \delta_{jk},$$
$$I'_{ij,kl} = \delta_{ik}\delta_{jl} - \delta_{il}\delta_{jk}.$$

Note that δ_{ik} is the block factor for rows, δ_{jl} is the block factor for columns, and the remaining terms δ_{il}, δ_{jk} are meaningless unless the two factors have the same set of levels. Each of the three model distributions is invariant under permutation, the same permutation being applied to rows as to columns. Accordingly, the models depend on the rows and columns as block factors, not as classification factors.

In the standard Bradley–Terry model for ranking competitors in a tournament, the component observations are independent, and the competitor effect $\{\eta_i\}$ is a parameter vector to be estimated (Agresti 2002, p. 436). Such models are closed under permutation of factor levels and under restriction of levels, but they are not invariant, and thus not block exchangeable. By contrast, all three models shown above are block exchangeable, and competitor effects do not occur in the parameter space. To predict the outcome of a match between competitors i, j, we first estimate the variance components by maximum likelihood. In the second stage, the conditional distribution of $Y(u^*)$ given Y for a new unit such that $x(u^*) = (i, j)$ is computed by the standard formulae for conditional distributions, and this is the basis on which predictions are made. This exercise is straightforward provided that the variance components required for prediction in eqn (4.6) are identifiable at the design. An allowance for errors of estimation along the lines of Barndorff-Nielsen and Cox (1996) is also possible.

4.6 Concluding remarks

Kolmogorov's definition of a process in terms of compatible finite-dimensional distributions is a consequence of requiring probability distributions to be well behaved under subsampling of units. Exchangeability is a different sort of criterion based on egalitarianism, the assumption that distributions are unaffected by permutation of units. Regression exchangeability is also based on egalitarianism, the assumption that two sets of units having the same covariate values also have the same response distribution. The range of examples illustrated in Section 4.3 shows that the assumption is almost, but not quite, universal in parametric statistical models.

Regression exchangeability is a fairly natural assumption in many circumstances, but the possibility of failure is not to be dismissed. Failure means that there exist two sets of subjects having the same covariate values that have different distributions, presumably due to differences not included in the covariate. If the differences are due to an unmeasured variable, and if treatment assignment is determined in part by such a variable, the apparent treatment effect is a combination of two effects, one due to the treatment and the other due to the unrecorded variable. Randomization may be used as a device to guard against potential biases of this sort by ensuring that the exchangeability assumption is satisfied, at least in the unconditional sense.

As a function on the units, a covariate serves to distinguish one unit from another, and the notion in an exchangeable regression model is that differences between distributions must be explained by differences between covariate values. However, a covariate is not the only sort of mathematical object that can introduce inhomogeneities or distributional differences. The genetic relationship between subjects in a clinical trial is a function on pairs of subjects. It is not a covariate, nor is it an equivalence relation, but it may affect the distribution of pairs as described in Section 4.5. Two pairs having the same covariate value may have different joint distributions if their genetic relationships are different. The relevant notion of exchangeability in this context is that two sets of units having the same covariate values and the same relationships also have the same joint distribution.

Exchangeability is a primitive but fundamental concept with implications in a wide range of applications. Even in spatial applications, if we define the relationship between pairs of units to be their spatial separation, the definition in Section 4.5 is satisfied by all stationary isotropic processes. The concept is not especially helpful or useful in the practical sense because it does not help much in model construction or model selection. Nonetheless, there are exceptions, potential areas of application in which notions of exchangeability may provide useful insights. The following are four examples.

In connection with factorial decomposition and analysis of variance, Cox (1984*a*, §5.5) has observed that two factors having large main effects are more likely to exhibit interaction than two factors whose main effects are small. To mimic this phenomenon in a Bayesian model, it is necessary to construct a partially exchangeable prior process in the sense of Aldous (1981) that exhibits the desired property. Does exchangeability allow this? If so, describe such a process and illustrate its use in factorial models.

Given a regression-exchangeable process, one can duplicate an experiment in the mathematical sense by considering a new set of units having the same covariate values as the given set. For a replicate experiment on the same process, the test statistic $T(Y^*)$ may or may not exceed the value $T(Y)$ observed in the original experiment: the two statistics are exchangeable and thus have the same distribution. The excedent probability or p-value is a prediction on the combined sample space $\mathrm{pr}\{T(Y^*) \geq T(Y) \,|\, Y\}$, and as such is not subject to the likelihood principle. The subsequent inference, that a small p-value is evidence against the model or null hypothesis, if interpreted as evidence in favour of specific parameter points in a larger parameter space, is an inference potentially in violation of the likelihood principle. Bearing in mind the distinction between estimation and prediction, clarify the nature of the likelihood-principle violation (Berger and Wolpert 1988).

A mixture of processes on the same observation spaces is a process, and a mixture of exchangeable processes is an exchangeable process. An improper mixture of processes is not a process in the Kolmogorov sense. The fact that improper mixtures are used routinely in Bayesian work raises questions connected with definitions. What sort of process-like object is obtained by this non-probabilistic operation? Is it feasible to extend the definition of a process in such a way that the extended class is closed under improper mixtures? For example, the symmetric density functions

$$f_n(x_1, \ldots, x_n) = n^{-1/2}\Gamma\{(n-\nu)/2\}\pi^{-n/2} \left\{ \sum_{i=1}^{n}(x_i - \bar{x}_n)^2 \right\}^{-(n-\nu)/2}$$

are Kolmogorov-compatible in the sense that the integral of f_{n+1} with respect to x_{n+1} gives f_n. For $n \geq 2$, the ratio f_{n+1}/f_n is a transition density, in fact Student's t on $n - \nu$ degrees of freedom centred at $\bar{x}_n$. In symbols, for $n \geq 2$,

$$X_{n+1} = \bar{x}_n + \left\{ \frac{(n^2-1)s_n^2}{n(n-\nu)} \right\}^{1/2} \epsilon_n,$$

in which s_n^2 is the sample variance of the first n components, and the components of ϵ are independent with distributions $\epsilon_n \sim t_{n-\nu}$. However, f_n is not integrable on $\mathbb{R}^n$, so these functions do not determine a process in the Kolmogorov sense, and certainly not an exchangeable process. Nonetheless, the transition densities permit prediction, either for one value or averages such as $\overline{X}_\infty = \bar{x}_n + s_n\epsilon_n/[(n-1)/\{n(n-\nu)\}]^{1/2}$.

In the preceding example, the transition density $p_n(x;t) = f_{n+1}(x,t)/f_n(x)$ is a function that associates with each point $x = (x_1, \ldots, x_n)$ a probability density on $\mathbb{R}$. In other words, a transition density is a density estimator in the conventional sense. By necessity, the joint two-step transition density $p_n^2(x;t_1,t_2) = f_{n+2}(x,t_1,t_2)/f_n(x)$ is a product of one-step transitions

$$p_n^2(x;t_1,t_2) = p_n(x;t_1)\,p_{n+1}\{(x,t_1),t_2\}.$$

For an exchangeable process, this density is symmetric under the interchange $t_1 \leftrightarrow t_2$. Both marginal distributions of p_n^2 are equal to p_n, so two-step-ahead prediction is the same as one-step prediction, as is to be expected in an exchangeable process. Ignoring matters of computation, a sequence of one-step predictors determines a two-step predictor, so a one-step density estimator determines a two-step density estimator. For commercial-grade kernel-type density estimators, it appears that these estimators are not the same, which prompts a number of questions. Is the difference between the two estimators an indication that density estimation and prediction are not equivalent activities? If so, what is the statistical interpretation of the difference? Is the difference appreciable or a matter for concern? If it were feasible to compute both estimators, which one would be preferred, and for what purpose?

Acknowledgement

This research was supported in part by the US National Science Foundation.

5
On semiparametric inference

Andrea Rotnitzky

5.1 Introduction

Modern datasets are often rich, high-dimensional structures but scientific questions of interest concern a low-dimensional functional $\beta = \beta(F)$ of the supposed distribution F of the data. Specification of realistic parametric models for the mechanism generating high-dimensional data is often very challenging, if not impossible. Non- and semiparametric models in which the data-generating process is characterized by parameters ranging over a large, non-Euclidean, space and perhaps also a few meaningful real-valued parameters, meet the challenge posed by these high-dimensional data because they make no assumptions about aspects of F of little scientific interest. Inference about β under semiparametric models is therefore protected from the possibility of misspecification of uninteresting components of F.

Loosely speaking, a semiparametric estimator of β is one that converges in probability to $\beta(F)$ under all distributions F allowed by the semiparametric model. Modern semiparametric theory is fundamentally an asymptotic theory. It addresses the following questions: do semiparametric estimators of β exist? When they exist, what are general methods for constructing them? What is a suitable notion of lower asymptotic bound and information in semiparametric models? How well does a semiparametric estimator perform—does it succeed in extracting all the information available in the data for β, at least asymptotically, and if not, how much information does it lose? How much efficiency is lost through relaxing assumptions and assuming a semiparametric rather than a parametric model? Under which conditions is there no information loss? When can we find asymptotically normal estimators of β that do not require estimation of large irregular nuisance parameters? Are there general techniques for constructing asymptotically optimal estimators, when such estimators exist?

In the last two decades there has been an explosive growth in the number of semiparametric models that have been proposed, studied and applied, primarily in biostatistics and econometrics. This growth has in great part been due to the ever-increasing availability of computing resources that have made the fitting of these models tractable, the development of an optimality theory leading to an understanding of the structure of inference and therefore providing systematic answers to several of the above questions, and to the parallel development of asymptotic theory, specifically empirical process theory, useful for obtaining the limiting distribution of semiparametric estimators that depend on estimators of high-dimensional nuisance parameters.

A key notion in the theory of efficient semiparametric estimation is that of the semiparametric variance bound of a parameter. Loosely speaking, this is the variance of the most concentrated limiting distribution one could hope a $\sqrt{n}$-consistent estimator to have. Informally, $\widehat{\beta}$ is a $\sqrt{n}$-consistent estimator of β if $\sqrt{n}\left\{\widehat{\beta} - \beta\left(F\right)\right\}$ converges to a non-degenerate distribution under any F allowed by the semiparametric model, where n is the sample size. Unfortunately, the bound is sometimes meaningless. In many important models, the parameter β of interest has a finite variance bound but no estimator is consistent for β uniformly over all the laws allowed by the model. This pessimistic asymptotic result has severe negative consequences in finite samples. It implies that no 'valid' interval estimator for the components of β exists whose length shrinks to zero in probability as the sample size increases. By 'valid' we mean here that for each sample size and under all laws allowed by the model the coverage is no smaller than the prescribed level. The essence of this difficulty lies in the fact that estimation of β requires additional estimation of some infinite-dimensional, 'irregular' nuisance parameter that cannot be well estimated over the model due to the curse of dimensionality.

Despite, or perhaps because of, the aforementioned difficulties, semiparametric theory is of much value, for it provides the tools for determining in a given problem if $\sqrt{n}$-consistent estimators of β that do not require estimation of large, irregular, nuisance parameters exist. When they exist, the theory also provides guidance as to how to compute estimators that asymptotically extract much of the information in the data. When they do not exist, the theory can be used to:

(i) identify the parts of the data-generating process that inevitably need to be smoothed;

(ii) determine how much smoothness is needed to obtain $\sqrt{n}$-consistent estimators of β; and

(iii) point the way to the construction of such estimators when the smoothness conditions hold.

These theoretical developments can, indeed, be useful in practice. A statistician can use (i) and (ii) in conjunction with a subjective assessment in the problem at hand both of the validity of the smoothness requirements and of the possibility of effectively borrowing the information necessary for smoothing, to determine if interval estimators based on the estimators in (iii) can be expected to be approximately valid. It remains largely unresolved as to how to proceed when the smoothness requirements are not met and/or information cannot be borrowed for smoothing. One possibility that has been recently examined by Robins (2004) is mentioned in Section 5.5.

In this chapter we review some key elements of semiparametric theory. Section 5.2 gives a number of examples that illustrate the usefulness of semiparametric modelling. Section 5.3 gives a non-technical account of the formulation of the semiparametric variance bound and ways for calculating it. In Section 5.4 we discuss the curse of dimensionality. We show how some key steps in the calculation of the bound can help to determine if in any given problem one can find estimators that do not require estimation of irregular nuisance parameters and hence are not afflicted by the curse of dimensionality. In Section 5.5 we discuss a possibility for approaching inference when estimation of irregular parameters is inevitable, and raise some unresolved questions.

There exist only a handful of surveys of semiparametric theory. Newey (1990) gives a non-technical introduction to semiparametric bounds with illustrations from econometric applications. Bickel *et al.* (1993) produced the first book to give a rigorous treatment of the theory of semiparametric inference and it remains a major reference for the study of the topic. It provides the calculation of the bound in numerous examples, including missing and censored data problems, truncation/bias sampling problems and measurement error and mixture models. Newer developments are surveyed by van der Vaart (2000) and van der Vaart (1998, Chapter 25), placing special emphasis on the technical tools needed for the derivation of the asymptotic distribution of semiparametric estimators. van der Laan and Robins (2003) focus on inference in semiparametric models useful for the analysis of realistic longitudinal studies with high-dimensional data structures that are either censored or missing. The present chapter borrows material from all these references.

A survey of the many journal articles on specific semiparametric models is beyond the scope of this chapter. In addition, many important topics are not covered here, in particular a discussion of the various proposals for efficient estimation under smoothness conditions, such as maximum semiparametric profile likelihood, maximum penalized likelihood, maximum sieve likelihood, and one-step estimation. van der Vaart (2000), van der Geer (2000), and Owen (2001) are three references where surveys on these methods can be found.

Semiparametric testing theory is much less developed and is not treated here. We refer the reader to van der Vaart (1998, §25.6), who shows that tests about regular parameters based on efficient estimators are locally efficient in an appropriate sense, and Murphy and van der Vaart (2000) who discuss semiparametric likelihood ratio tests and related intervals. Bickel *et al.* (2004) give a promising new general testing framework based on semiparametric efficient score statistics, applicable, for example, in constructing goodness-of-fit tests of semiparametric models with substantial power for specified directions of departure.

We limit our discussion to models for independent identically distributed observations. By and large, the efficiency theory so far available is for this case. Optimality results with dependent observations have been developed using problem-specific arguments in some settings, such as for certain time series models (Drost *et al.* 1994) and Markov chains (Greenwood and Wefelmeyer 1995). However, recent work of Bickel and Kwon (2001), following the ideas of Bickel (1993) and Levit (1978), has opened up new horizons for defining and calculating efficiency bounds in a broad class of semiparametric models for dependent data where the analogue of a largest nonparametric model with independent, identically distributed data can be defined. Further developments in this direction seem promising.

Finally, we limit our discussion to inference for regular parameters, that is, parameters for which we can at least hope to find $\sqrt{n}$-consistent estimators. There are many problems where the parameter of interest is 'irregular', such as a conditional mean or a density function. The theory of minimax rates of convergence for irregular parameters over large models and the construction of adaptive estimators that estimate the parameter better when it is easier to estimate, are areas of active research. See Hoffmann and Lepski (2002) and its discussion for references on adaptive estimation, and van der Geer (2000, Chapter 7, 9) and van der Vaart (2000, Chapter 8) for general strategies for determining minimax rates of convergence.

5.2 Semiparametric models

Informally, a semiparametric model is one in which the collection $\mathcal{F}$ of possible distributions that could have generated the observations $X_1, \ldots, X_n$ is indexed by a parameter ranging over a big set, that is, a set larger than the parameter set of any parametric model. If, as we shall assume throughout, the observations $X_1, \ldots, X_n$ are a random sample—that is, independent copies—of some random structure X, then a semiparametric model assumes that the distribution P^* of X belongs to a collection $\mathcal{F} = \{P_\theta : \theta \in \Theta\}$ of measures on a measurable space $(\mathcal{X}, \mathcal{A})$ where Θ is some large non-Euclidean set. By a random structure X we mean a measurable map on some underlying probability space that takes values in an arbitrary sample space $\mathcal{X}$. The special model in which $\mathcal{F}$ does not restrict the distribution of X, except perhaps for some regularity conditions, is usually referred to as a non-parametric model. Semiparametric theory studies the problem of making inference about the value at P^* of a map $\beta : \mathcal{F} \to \mathcal{B}$ where $\mathcal{B}$ is a normed linear space. A special case important for applications, which this chapter will focus on, is that in which $\mathcal{B} = \mathbb{R}^k$ and $\mathcal{X}$ is a Euclidean sample space. Then P is in one-to-one correspondence with a cumulative distribution function F, the semiparametric model $\mathcal{F}$ can be regarded as a collection of distribution functions $\{F_\theta : \theta \in \Theta\}$ and $\beta^* \equiv \beta(P^*) \equiv \beta(F^*) \equiv \beta(\theta^*)$ is a $k \times 1$ parameter vector.

Throughout the discussion below we use $*$ to indicate the true data-generating mechanism and the associated true parameter values, we write $\mathcal{F}(\Theta)$ instead of $\mathcal{F}$ whenever we need to emphasize the parameter set, and we use the subscript F in E_F and var_F to indicate that the expectation and variance are calculated under F.

Example 1: Cumulative distribution function Suppose that we wish to estimate $F^*(t_0) = \mathrm{pr}(X \le t_0)$ based on a random sample $X_1, \ldots, X_n$, under the non-parametric model $\mathcal{F}_1$ that does not restrict the distribution of X. Here,

$\beta : \mathcal{F}_1 \to \mathbb{R}$ is defined as $\beta(F) = F(t_0)$. Because the distribution of X is unrestricted, one might expect that the empirical cumulative distribution function, $\widehat{F}(t_0) = n^{-1} \sum_{i=1}^n I(X_i \leq t_0)$ is a 'most efficient' estimator of $F^*(t_0)$ under the non-parametric model. Levit (1975) was the first to formalize and prove this assertion.

Suppose that due to the nature of the problem in hand we know the value μ^* of the mean of X and we still want to estimate $F^*(t_0)$. Then model $\mathcal{F}_2 = \{F : \mathbb{E}_F(X) = \mu^*\}$ encodes our knowledge about the data-generating process. The empirical cumulative distribution function $\widehat{F}(t_0)$ is a semiparametric estimator in this model but it fails to exploit the information about $F^*(t_0)$ entailed by knowledge of the mean of F^*. Semiparametric theory indicates how to construct estimators that exploit such information.

In the following examples, the semiparametric models are indexed with a Euclidean parameter α of interest and an infinite-dimensional nuisance parameter η ranging in some non-Euclidean set η, so $\beta(F_{\alpha,\eta}) = \alpha$.

Example 2: Symmetric location Suppose that we know that the distribution of a continuous random variable X is symmetric and we wish to estimate its centre of symmetry α^*. Then we consider the model $\mathcal{F}(\Theta)$ where $\Theta = \alpha \times \eta$ in which α is the real line, η is the collection of all densities η symmetric around 0, that is, with $\eta(u) = \eta(-u)$ for all u and

$$\mathcal{F}(\Theta) = \{F_{\alpha,\eta} : f_{\alpha,\eta}(x) = \eta(x - \alpha), \alpha \in \alpha, \eta \in \eta\}.$$

Stein (1956) showed that the information about α^* is the same whether or not the distributional shape η^* is known, a remarkable finding that was the catalyser of earlier investigations in semiparametric theory. Later, we will give a precise definition of information and formalize this statement.

Example 3: Partial linear regression Consider a model for n independent copies of $X = (Z, V, Y)$ that assumes that

$$E(Y \mid Z, V) = \omega^*(V) + \alpha^* Z,$$

where ω^* is an arbitrary unknown function and Z is a dichotomous treatment indicator. In this model, $\mathcal{F}(\alpha \times \kappa \times \omega \times \pi)$ is the collection of distributions

$$f_X(x; \alpha, \kappa_1, \kappa_2, \omega, \pi) = \kappa_1 \{y - \alpha z - \omega(v) \mid v, z\} \pi(v)^z \{1 - \pi(v)\}^{1-z} \kappa_2(v),$$

where $\alpha \in \alpha$, the real line, and $\kappa = (\kappa_1, \kappa_2) \in \kappa = \kappa_1 \times \kappa_2$, κ_1 is the index set for the collection of arbitrary distributions of $\varepsilon(\alpha, \omega) = Y - \alpha Z - \omega(V)$ given Z and V, κ_2 is the set of all distributions of V, π is the set of all conditional distributions of Z given V, and ω is the set of arbitrary functions of V. Estimation of α^* in this model under smoothness conditions on ω^* or variations of it imposing additional restrictions on the error distribution has been extensively studied. See, for example, Engle *et al.* (1984), Heckman (1986), Rice (1986), Bickel and Ritov (1988), Robinson (1988), Chen (1988), Andrews (1991), Newey (1990),

Bickel *et al.* (1993), Robins *et al.* (1992), Donald and Newey (1994), van der Geer (2000), and Robins and Rotnitzky (2001).

In biostatistical and econometric applications, the model $\mathcal{F}\left(\boldsymbol{\alpha} \times \boldsymbol{\kappa} \times \boldsymbol{\omega} \times \boldsymbol{\pi}\right)$ is often postulated in order to analyse follow-up observational studies in which n independent copies of $X = (Y, Z, V)$ are recorded, V being a high-dimensional vector of potential confounding factors, many of which are continuous. In the absence of additional confounders and measurement error, the difference

$$\mathrm{E}\left(Y \mid Z = 1, V\right) - \mathrm{E}\left(Y \mid Z = 0, V\right) = \alpha$$

is equal to the average causal treatment effect obtained when comparing treatment $Z = 1$ with treatment $Z = 0$ among subjects with covariates V. The model assumes that there is no treatment–confounder interaction, but makes no assumption about the form of the functional dependence of $\mathrm{E}\left(Y \mid Z = 0, V\right)$ on the confounders V, which is not of scientific interest.

If the data had been obtained from a randomized study with known randomization probabilities $\pi^{*}\left(v\right)$, then the resulting model would be $\mathcal{F}(\boldsymbol{\alpha} \times \boldsymbol{\kappa} \times \boldsymbol{\omega} \times \{\pi^{*}\})$. A major advantage of conducting inferences under this model as opposed to under those that also make parametric assumptions on $\omega^{*}\left(V\right)$ and on the error distribution, is that model $\mathcal{F}\left(\boldsymbol{\alpha} \times \boldsymbol{\kappa} \times \boldsymbol{\omega} \times \{\pi^{*}\}\right)$ is guaranteed to be correctly specified with $\alpha^{*} = 0$ under the null hypothesis H_0 that treatment has no causal effect on any subject in the population. Thus, using semiparametric estimators of α^{*} and their estimated standard error we can construct a test of H_0 whose actual level is guaranteed to be nearly equal to its nominal level, at least in large samples. In Section 5.4 we show how semiparametric theory can be used to show that estimators of α^{*} that do not suffer from the curse of dimensionality exist in model $\mathcal{F}\left(\boldsymbol{\alpha} \times \boldsymbol{\kappa} \times \boldsymbol{\omega} \times \{\pi^{*}\}\right)$ but not in model $\mathcal{F}\left(\boldsymbol{\alpha} \times \boldsymbol{\kappa} \times \boldsymbol{\omega} \times \boldsymbol{\pi}\right)$.

Many semiparametric models have been proposed in response to the need to formulate realistic models for the generating mechanisms of complex high-dimensional data configurations arising in biomedical and econometric studies. Such complexities arise either by design or because the intended, ideal, dataset is not fully observed. Examples include coarsened—that is, censored or missing— data, measurement error/mixture data, and truncated/biased sampling data. The first two are examples of the important class of models usually referred to as information-loss models, in which $X = q\left(Y\right)$ where Y is some incompletely observed structure, generally of high dimension, and q is a many-to-one function. A model $\mathcal{F}^{\mathrm{latent}}$ for the law F_{Y} of Y is assumed that induces a model $\mathcal{F}^{\mathrm{obs}}$ for the distribution $P_{F_{\mathrm{Y}}}$ of the observed data X. The goal is to make inference about some $\beta\left(F_{\mathrm{Y}}\right)$. Inference is especially sensitive to model misspecification because the information encoded in $\mathcal{F}^{\mathrm{latent}}$ is used to impute the unobserved part of Y. Consequently, semiparametric modelling becomes especially attractive in this setting.

Coarsened data models have $X = q\left(C, U\right)$, where C is a missingness or censoring variable, U is the intended data and $q\left(c, u\right)$ indicates the part of u that is observed when $C = c$. The model $\mathcal{F}^{\mathrm{latent}}$ is defined by a model $\mathcal{F}^{\mathrm{full}}$ for

F_{U} and a model $\mathcal{G}$ for the coarsening mechanism, $G\left(C \mid U\right)$. The estimand of interest is usually a parameter of F_{U}. For example, consider the full-data structure $U = \left(T, \overline{V}\left(T\right)\right)$, where $\overline{V}\left(t\right) \equiv \{V\left(u\right) : 0 \leq u \leq t\}$ is a high-dimensional multivariate time-dependent data structure measured until a failure time T. In clinical and epidemiological studies U is often right-censored in a subset of the study participants, owing to the end of the study or to dropout. The observed data then consist of independent realizations of

$$X = \Phi\left(C, U\right) = \left(\tilde{T} = \min\left(T, C\right), \Delta = I\left(T \leq C\right), \overline{V}\left(\tilde{T}\right)\right),$$

where C is a censoring variable. Consider estimation of the marginal survival function $S\left(t_0\right) = \mathrm{pr}\left(T > t_0\right)$ and of the regression parameters α^* in the time-independent proportional hazards model $\mathrm{pr}\{T > t \mid V\left(0\right)\} = S_0\left(t\right)^{\exp\{\alpha^* V\left(0\right)\}}$, where S_0 is unrestricted. Semiparametric theory teaches us that the Kaplan–Meier estimator of $S\left(t_0\right)$ and the maximum partial likelihood estimator of α^* (Cox 1975) are semiparametric efficient when the intended data are comprised of just T, for estimation of $S\left(t_0\right)$, and of $(T, V\left(0\right))$, for estimation of α^*. These estimators fail to exploit the information in the time-dependent process V, however, and are consistent only under the strong and often unrealistic assumption that censoring is non-informative, that is, $G\left(C \mid U\right) = G\left(C\right)$ (Begun *et al.* 1983). Semiparametric theory indicates how to construct estimators of $S\left(t_0\right)$ and of α^* that allow the analyst to adjust appropriately for informative censoring due to measured prognostic factors V, while simultaneously quantifying the sensitivity of inference to non-identifying assumptions concerning residual dependence between the failure time and censoring due to unmeasured factors (Scharfstein and Robins 2002). It also points the way to the construction of estimators that recover information by non-parametrically exploiting the correlation between the process V and the failure times of censored subjects. An important advantage of such procedures in clinical trials is that they lead to tests of the hypothesis $\alpha^* = 0$ that exploit the information on the post-treatment data V without compromising the validity of the usual intention-to-treat null hypothesis of no effect of assigned treatment on survival (Robins and Rotnitzky 1992).

5.3   Elements of semiparametric theory

5.3.1   Variance bounds

The theory of semiparametric inference is based on a geometric analysis of likelihoods and scores. As we shall see, results may be obtained without detailed probabilistic calculation. The theory starts with the formulation of the semiparametric bound. This notion was first introduced by Stein (1956) who, motivated by the symmetric location problem, gave the following intuitive definition.

A parametric submodel of a semiparametric model $\mathcal{F}$ for independent and identically distributed data assumes that F^* belongs to a subset $\mathcal{F}_{\mathrm{par}}$ of $\mathcal{F}$

indexed by a Euclidean parameter. A semiparametric consistent and asymptotically normal estimator of a Euclidean parameter $\beta \equiv \beta(F^*)$ cannot have asymptotic variance smaller than the Cramér–Rao bound in any parametric submodel of $\mathcal{F}$. Consequently, such an estimator cannot have variance smaller than the supremum of the Cramér–Rao bounds for all parametric submodels. This supremum is referred to as the semiparametric variance bound and is denoted here by $C(F)$. To be precise about the definition of the bound, however, it is necessary to compute the supremum only over regular parametric submodels in which the likelihood is a smooth function and the parameter space is open, so that the Cramér–Rao bound is well defined.

The bound $C(F)$ is the variance of the 'best limiting distribution' one can hope for $\sqrt{n}$-consistent estimators that are also regular. Regular estimators are $\sqrt{n}$-consistent estimators whose convergence to their limiting distribution is uniform in shrinking parametric submodels with parameters differing from a fixed parameter value by $O(n^{-1/2})$. The class excludes superefficient estimators and more generally, estimators with excellent behaviour under one law at the expense of excessive bias in neighbourhoods of that law. The following extension due to Begun *et al.* (1983) of Hajek's representation theorem (Hajek 1970) to semiparametric models establishes that the normal distribution $N(0, C(F))$ with mean zero and variance $C(F)$ is the best limiting distribution one could hope for a regular estimator of β.

Theorem 5.1. (Representation Theorem) *Let $\widehat{\beta}$ be a regular estimator of $\beta(F)$ in model $\mathcal{F}$. Then the limiting distribution of $\sqrt{n}\left\{\widehat{\beta} - \beta(F)\right\}$ under F is the same as the distribution of $W + U$, where W and U are independent and $W \sim N\{0, C(F)\}$.*

It is important to note that the theorem assumes the existence of a regular estimator. In some problems the bound $C(F)$ is finite but meaningless because no $\sqrt{n}$-consistent estimator of $\beta(F)$ exists. This situation is further examined in Section 5.4.

A regular estimator of $\beta(F)$ is said to be globally semiparametric efficient in $\mathcal{F}$ if

$$\sqrt{n}\left\{\widehat{\beta} - \beta(F)\right\} \overset{\mathcal{L}(F)}{\to} N\{0, C(F)\}, \tag{5.1}$$

for all F in the model, and where $\overset{\mathcal{L}(F)}{\to}$ denotes convergence in distribution under F. The estimator is said to be *locally semiparametric efficient* at the submodel $\mathcal{F}_{\mathrm{sub}}$ if it is regular in model $\mathcal{F}$ and satisfies eqn (5.1) for all F in the submodel $\mathcal{F}_{\mathrm{sub}}$.

5.3.2   Tangent spaces

Estimators that are $\sqrt{n}$-consistent for parameters β with finite variance bound may or may not exist but never exist for parameters with an infinite bound

(Chamberlain 1986). An important issue is therefore the characterization of parameters β for which the bound is finite, as these are the only ones for which one can hope to construct a $\sqrt{n}$-consistent estimator. Koshevnik and Levit (1976) and Pfanzagl and Wefelmeyer (1982) defined a class of estimands, the pathwise-differentiable parameters, for which the bound is finite. The class naturally arises from the following argument.

Suppose that $\beta(F) = \mathrm{E}_F\{\psi(X)\}$ for some known $k \times 1$ vector function $\psi(X)$ with $\mathrm{var}_F\{\psi(X)\} < \infty$ for all F in the model. Then, in any regular parametric submodel, say, with a set of laws $\mathcal{F}_{\mathrm{par}} = \{F_t : t \in \mathbb{R}^p\}$ and $F_{t=0} = F^*$, for which $\beta(F_t)$ is a differentiable function of t, one has

$$\left. \frac{\partial}{\partial t^{\mathrm{T}}} \beta(F_t) \right|_{t=0} = \mathrm{E}_F\{\psi(X) S_t(F^*)^{\mathrm{T}}\}, \qquad (5.2)$$

where $S_t(F^*)$ is the score for t in the submodel $\mathcal{F}_{\mathrm{par}}$ evaluated at $t = 0$ and ^{T} denotes matrix transposition. When interchange of differentiation and integration is possible and $f(x; t)$ is a differentiable function of t, then eqn (5.2) can be derived by differentiating $\beta(F_t) = \int \psi(x) f(x; t) \, \mathrm{d}x$ under the integral sign. More generally, Ibragimov and Has'minskii (1981, Lemma 7.2) showed that eqn (5.2) remains valid in any regular parametric submodel. Expression (5.2) implies that the Cramér–Rao bound $C_{\mathrm{par}}(F^*)$ for $\beta(F^*)$ in the parametric submodel is

$$\mathrm{E}_{F^*}\{\psi(X) S_t(F^*)^{\mathrm{T}}\} \mathrm{E}_{F^*}\{S_t(F^*) S_t(F^*)^{\mathrm{T}}\}^{-1} \mathrm{E}_{F^*}\{S_t(F^*)\psi(X)^{\mathrm{T}}\},$$

the predicted value from the population least squares fit of $\psi(X)$ on $S_t(F^*)$ under F^*. Equivalently,

$$C_{\mathrm{par}}(F^*) = \mathrm{var}_{F^*}[\Pi_{F^*}\{\psi(X) \mid \Lambda_{\mathrm{par}}(F^*)\}],$$

where $\Lambda_{\mathrm{par}}(F)$ denotes the set $\{c^{\mathrm{T}} S_t(F) : c$ any conformable vector of constants$\}$ and where, for any closed linear space Ω in the Hilbert space $\mathcal{L}_2(F)$ of random variables $b(X)$ with finite variance and with covariance inner product $\langle b_1(X), b_2(X) \rangle = \mathrm{E}_F\{b_1(X) b_2(X)\}$, the quantity $\Pi_F\{\psi(X) \mid \Omega\}$ denotes a $k \times 1$ vector with jth entry equal to the projection of the jth entry of $\psi(X)$ into Ω; here $j = 1, \ldots, k$. A calculation shows that if $\cup_{\mathcal{F}_{\mathrm{par}}} \Lambda_{\mathrm{par}}(F)$ is a linear space, then

$$C(F) = \sup_{\mathcal{F}_{\mathrm{par}}} \mathrm{var}_F[\Pi_F\{\psi(X) \mid \Lambda_{\mathrm{par}}(F)\}] = \mathrm{var}_F[\Pi_F\{\psi(X) \mid \Lambda(F)\}],$$

where

$$\Lambda(F) \text{ is the closure of the linear span } \cup_{\mathcal{F}_{\mathrm{par}}} \Lambda_{\mathrm{par}}(F). \qquad (5.3)$$

The space $\Lambda(F)$ is called the *tangent space* for model $\mathcal{F}$ at F. The projection theorem for Hilbert spaces implies that the projection of $\psi(X)$ into $\Lambda(F)$ exists. Consequently, since the elements of $\Lambda(F)$ have finite variance, the bound $C(F)$ for $\beta(F) = \mathrm{E}_F\{\psi(X)\}$ is finite.

Example 1 (continued) The parameter $\beta(F) = F(t_0)$ can be written as $\beta(F) = \mathrm{E}_F\{\psi(X)\}$ with $\psi(X) = I(X \leq t_0)$. In model $\mathcal{F}_1$ any mean zero function $a(X)$ with finite variance is a score in a parametric submodel. For example, $a(X)$ is the score at $t = 0$ in the submodel $f(x;t) = c(t) f^*(x) G\{ta(x)\}$ where $G(u) = 2(1 + \mathrm{e}^{-2u})^{-1}$ and $c(t)$ is a normalizing constant. Thus the tangent space $\Lambda(F^*)$ in the nonparametric model $\mathcal{F}_1$ is equal to $\mathcal{L}_2^0(F^*)$, the space of all mean zero finite variance functions of X under F^*. The projection of $\psi(X) = I(X \leq t_0)$ into $\Lambda(F^*)$ is $I(X \leq t_0) - \beta(F^*)$. Consequently, the variance bound for $\beta(F^*)$ is $C_1(F^*) = \beta(F^*)\{1 - \beta(F^*)\}$ and hence the empirical cumulative distribution function, having asymptotic variance equal to $C_1(F^*)$, is semiparametric efficient.

Consider now estimation in model $\mathcal{F}_2$. Since $\mathrm{E}_{F_t}(X - \mu^*) = 0$ under any parametric submodel $f(x;t)$, with $t \in \mathbb{R}^p$, say passing through F^* at $t = 0$, then differentiating with respect to t, we obtain $\mathrm{E}_{F_t}\{(X - \mu^*) S_t(F^*)^{\mathrm{T}}\} = 0$, provided $\mathrm{var}_{F_t}(X) < \infty$. It is then natural to conjecture that $\Lambda(F^*)$ in model $\mathcal{F}_2$ is equal to the subset of $\mathcal{L}_2^0(F^*)$ comprised of functions of X that are uncorrelated with $X - \mu^*$. This conjecture is indeed true. The projection of $\psi(X) = I(X \leq t_0)$ into $\Lambda(F^*)$ is now equal to

$$I(X \leq t_0) - \mathrm{E}_{F^*}\{I(X \leq t_0)(X - \mu^*)\} \mathrm{var}_{F^*}(X)^{-1}(X - \mu^*)$$

and the variance bound equals

$$C_2(F^*) = \beta(F^*)\{1 - \beta(F^*)\} - \mathrm{E}_{F^*}\{I(X \leq t_0)(X - \mu^*)\}^2 \mathrm{var}_{F^*}(X)^{-1}.$$

If $X > t_0$ with positive probability, then $C_2(F^*) < C_1(F^*)$, so knowledge of $\mathrm{E}_{F^*}(X)$ improves the efficiency with which we can estimate $\beta(F^*)$. The estimator given by

$$\widetilde{F}(t_0) = n^{-1}\sum_{i=1}^{n}\{I(X_i \leq t_0) - \widehat{\tau}(X_i - \mu^*)\}, \quad \widehat{\tau} = \frac{\sum_{i=1}^{n} I(X_i \leq t_0)(X_i - \mu^*)}{\sum_{i=1}^{n}(X_i - \mu^*)^2},$$

is semiparametric efficient. The construction in this example mimics the construction in parametric models of efficient estimators of a parameter of interest when a nuisance parameter is known.

5.3.3 Pathwise differentiable parameters, gradients and influence functions

In the preceding subsection, $C(F)$ was finite because it could be represented as the variance of the projection of the random variable $\psi(X)$ into $\Lambda(F)$. Indeed, identity (5.2) was the sufficient condition for $C(F)$ to be finite. Consequently, any parameter $\beta(F)$ for which there exists a $k \times 1$ random vector $\psi_F(X)$ with finite variance under F such that eqn (5.2) holds for all regular parametric submodels has a finite variance bound. Such a parameter $\beta(F)$ is called a *regular* or *pathwise differentiable parameter* at F. The class of regular parameters essentially includes all estimands that admit a regular estimator. This follows from van der Vaart (1988), who showed that differentiability of $\beta(F_t)$ and the existence of at least one regular estimator of $\beta(F)$ are sufficient for regularity of $\beta(F)$.

The function $\psi_F(X)$ is called a *gradient* of $\beta(F)$. Gradients are not uniquely defined. For example, if $\psi_F(X)$ is a gradient, then so is $\psi_F(X) + M$ for any constant vector M. However, all gradients of $\beta(F)$ must have identical projection into the tangent space $\Lambda(F)$. This projection, which in turn is also a gradient, is called the *canonical gradient* or *efficient influence function*, and is denoted by $\psi_{F,\text{eff}}(X)$. It follows from eqn (5.3) that the variance bound for a regular parameter $\beta(F)$ equals the variance of its canonical gradient.

Gradients are intimately connected with a special, very broad, class of regular estimators—the regular asymptotically linear estimators. An estimator $\widehat{\beta}$ is asymptotically linear at F if and only if there exists a mean zero, finite variance function $\varphi_F(X)$ such that when X has distribution F, then

$$\sqrt{n}\left\{\widehat{\beta} - \beta(F)\right\} = \frac{1}{\sqrt{n}}\sum_{i=1}^{n}\varphi_F(X_i) + o_p(1).$$

The function $\varphi_F(X)$ is called the influence function of $\widehat{\beta}$, the name being a reminder that, up to first order, $\varphi_F(X_i)$ measures the influence of X_i on $\widehat{\beta}$ (Hampel 1974).

An important class of asymptotically linear estimators comprises those that are solutions to unbiased estimating functions. Specifically, suppose that in a model indexed by (α, η) we can find a function $m(x; \alpha)$ depending on α but not on the nuisance parameter η such that

$$\mathrm{E}_{\alpha^*,\eta^*}\{m(X;\alpha^*)\} = 0 \tag{5.4}$$

at the true parameters α^* and η^*, regardless of the value of η^*. Then, under suitable regularity conditions, estimators $\widehat{\alpha}$ of α^* solving

$$\sum_{i=1}^{n} m(X_i;\alpha) = 0 \tag{5.5}$$

are asymptotically linear at F_{α^*,η^*} with influence function given by

$$\varphi_{F_{\alpha^*,\eta^*}}(X) = \left[-\partial\mathrm{E}_{F_{\alpha^*,\eta^*}}\{m(X;\alpha)\}/\partial\alpha\big|_{\alpha=\alpha^*}\right]^{-1} m(X;\alpha^*). \tag{5.6}$$

Newey (1990, Theorem 2.2) showed that an asymptotically linear estimator $\widehat{\beta}$ with influence function $\varphi_F(X)$ satisfying certain regularity conditions is regular if and only if $\varphi_F(X)$ satisfies eqn (5.2) with $\varphi_F(X)$ replacing $\psi(X)$. Consequently, influence functions of regular asymptotically linear estimators of $\beta(F)$ are gradients quite generally. We shall see in Section 5.4 that this result and identity (5.6) yield a method for assessing whether estimators that do not suffer from the curse of dimensionality exist under a given model.

5.3.4 The semiparametric efficient score

The tangent space of semiparametric models indexed by a parameter α of interest and an infinite-dimensional nuisance parameter η, has more structure that can be exploited in calculating the set of gradients and the bound for $\beta(F_{\alpha,\eta}) = \alpha$. Specifically, suppose that $\theta = (\alpha, \eta)$ and $\Theta = \alpha \times \eta$, $\alpha \subseteq \mathbb{R}^k$ and η is a non-Euclidean space. The submodel $\mathcal{F}(\{\alpha^*\} \times \eta)$ in which α is fixed at its true value and $\eta \subseteq \eta$ is unknown is called the nuisance submodel. Its tangent space at F^*, denoted $\Lambda_{\mathrm{nuis}}(F^*)$, is called the nuisance tangent space. The scores $S_{\eta_t}(F^*)$ in regular parametric submodels of $\mathcal{F}(\{\alpha^*\} \times \eta)$ are called the nuisance scores. In addition, the submodel $\mathcal{F}(\alpha \times \{\eta^*\})$ in which the nuisance parameter η is fixed at its true value but α is unknown is a parametric model with score denoted by $S_\alpha(F^*)$. It follows that if $\psi_{F^*}(X)$ is any gradient of $\beta(F^*)$ in model $\mathcal{F}(\alpha \times \eta)$, then

$$0 = \left.\frac{\partial}{\partial t}\beta(\alpha^*, \eta_t)\right|_{t=0} = \mathrm{E}_{F^*}\left\{\psi_{F^*}(X)\, S_{\eta_t}(F^*)^{\mathrm{T}}\right\}, \qquad (5.7)$$

$$I_k = \left.\frac{\partial}{\partial \alpha}\beta(\alpha, \eta^*)\right|_{\alpha=\alpha^*} = \mathrm{E}_{F^*}\left\{\psi_{F^*}(X)\, S_\alpha(F^*)^{\mathrm{T}}\right\}, \qquad (5.8)$$

where I_k is the $k \times k$ identity matrix. Thus gradients of α are orthogonal to the nuisance tangent space and have covariance with the score for α equal to the identity. Indeed, quite generally, the reverse is also true: any function $\psi_{F^*}(X)$ that satisfies eqns (5.7) and (5.8) is a gradient (Bickel *et al.* 1993, §3.4, Proposition 1).

  The previous derivation yields the following useful characterization of the efficient influence function. Define

$$S_{\alpha,\mathrm{eff}}(F^*) = S_\alpha(F^*) - \Pi_{F^*}\left\{S_\alpha(F^*) \mid \Lambda_{\mathrm{nuis}}(F^*)\right\}$$

and

$$\psi_{F^*}^*(X) = \mathrm{var}_{F^*}\left\{S_{\alpha,\mathrm{eff}}(F^*)\right\}^{-1} S_{\alpha,\mathrm{eff}}(F^*),$$

where $\Pi_{F^*}(\cdot \mid \cdot)$ was defined in Section 5.3.2. By construction, $\psi_{F^*}^*(X)$ satisfies eqns (5.7) and (5.8) and it is an element of $\Lambda(F^*)$. Therefore, $\psi_{F^*}^*(X)$ is equal to the efficient influence function $\psi_{F^*,\mathrm{eff}}(X)$. The vector $S_{\alpha,\mathrm{eff}}(F^*)$ is called the efficient score, and its variance is called the semiparametric information bound. The semiparametric efficient score is the generalization to semiparametric models of the effective score in a parametric model. Its interpretation is identical to that in parametric models. When η is known, the variation in the score $S_\alpha(F^*)$ corresponds to the information for α. When η is unknown, a part of the information about α is lost, and this corresponds to the lost part of the variation in the score $S_\alpha(F^*)$ resulting from adjustment for its regression on the infinite-dimensional score for η.

Example 2 (continued) A nuisance submodel indexed by t with truth at $t = 0$ is of the form $f(x;t) = \eta(x - \alpha^*;t)$. Thus, a nuisance score is of the form

$$g(X - \alpha^*) = \left.\frac{\partial \log \eta(X - \alpha^*;t)}{\partial t}\right|_{t=0},$$

where $g(u) = g(-u)$. This suggests that $\Lambda_{\text{nuis}}(F^*)$ is the set of mean zero, finite variance, functions $g(X - \alpha^*)$ with $g(u) = g(-u)$. This conjecture is indeed true (see Bickel *et al.* 1993, §3.2, Example 4). Consequently, any anti-symmetric function $q(u)$, that is, with $q(-u) = -q(u)$, such that $q(X - \alpha^*)$ has finite variance has $E_{F^*}\{q(X - \alpha^*)g(X - \alpha^*)\} = 0$. The score for α is $S_\alpha(F^*) = \eta_u^*(X - \alpha^*)/\eta^*(X - \alpha^*)$, where $\eta_u^*(u) = d\eta^*(u)/du$ provided the derivative exists. Since a symmetric function has an antisymmetric derivative, we conclude that $\eta_u^*(u)/\eta^*(u)$ is antisymmetric. Consequently,

$$\Pi_{F^*}\{S_\alpha(F^*)|\Lambda_{\text{nuis}}(F^*)\} = 0, \quad S_\alpha(F^*) = S_{\alpha,\text{eff}}(F^*).$$

Thus, the information for estimating the center of symmetry is the same whether or not the shape η^* is known and it is equal to $I(\eta^*) = E_{F^*}\{S_\alpha(F^*)\} = \int\{\eta_u^*(u)\}^2/\eta^*(u)\,du$. Beran (1974) and Stone (1975) have shown that it is possible to construct globally semiparametric efficient estimators $\hat\alpha$ of α^* under the minimal condition that $I(\eta^*)$ is finite.

Estimators like $\hat\alpha$ in Example 2, which are efficient in a model $\mathcal{F}(\alpha \times \{\eta^*\})$ with η^* known even though their construction does not require knowledge of η^*, are called *adaptive*. A necessary condition for adaptive estimation is orthogonality of $S_\alpha(F^*)$ with the nuisance tangent space. Bickel (1982) discussed sufficient conditions for adaptation.

5.4 The curse of dimensionality

5.4.1 Introduction

We noted earlier that the variance bound of a parameter of interest is sometimes finite but meaningless. Ritov and Bickel (1990) and Robins and Ritov (1997) gave several examples of this situation, which occurs when estimation of the parameter of interest also entails additional estimation of a high-dimensional irregular nuisance parameter such as a conditional mean function. The usual approach to this problem is to impose minimal smoothness conditions on the irregular parameters under which $\sqrt{n}$-consistent estimators of the parameter of interest exist, and to treat non-existence results in the absence of smoothness conditions as mathematical curiosities. In a sequence of papers, Robins and colleagues (Robins and Ritov 1997, Robins *et al.* 2000, Robins and Wasserman 2000, van der Laan and Robins 2003) have taken the opposite point of view; see also Bickel and Ritov (2000). They have argued that even when the smoothed model is known to be true, an asymptotic theory based on the larger model that does not assume smoothness is a more appropriate guide to small-sample performance. This is due to the curse of dimensionality: for example, if a conditional mean on a highly multivariate covariate needs to be estimated, smoothness conditions are not relevant even when satisfied because no two units have covariate values close enough to allow the borrowing of information necessary for smoothing. Robins (2004) takes an intermediate position, noting the possibility of developing an asymptotic theory based on a model that assumes sufficient smoothness on the nuisance parameters to ensure the existence of estimators that converge in probability to the parameter of interest at some rate but not sufficient to ensure the existence of $\sqrt{n}$-consistent estimators. Section 5.5 expands on this.

5.4.2   Estimation avoiding the curse of dimensionality

In some models indexed by (α, η), where α is a Euclidean parameter and η is infinite dimensional, it is possible to find unbiased estimating functions, that is, functions of α and X satisfying eqn (5.4) for all η^*. Under regularity conditions, when such function exists, whatever the value of η^* there exist a regular estimator $\hat{\alpha}$ of α^* solving eqn (5.5). The estimator $\hat{\alpha}$ has the enormous advantage of not suffering from the curse of dimensionality because its construction does not depend on estimation of nuisance parameters. Under some moment conditions, there exists also a sequence $\sigma_{\mathrm{n}}(\alpha, \eta)$, $n = 1, 2, \ldots$, such that

$$\sup_{(\alpha,\eta)} \left| \mathrm{pr}_{(\alpha,\eta)} \left\{ \frac{\sqrt{n}\,(\hat{\alpha} - \alpha)}{\sigma_{\mathrm{n}}(\alpha, \eta)} < t \right\} - \Phi(t) \right| \underset{n\to\infty}{\to} 0, \tag{5.9}$$

where Φ is the standard normal cumulative distribution function, $\mathrm{pr}_{(\alpha,\eta)}$ denotes the probability calculated under $F_{\alpha,\eta}$ and to avoid distracting complications we take the parameter α to be scalar. If $\hat{\alpha}$ satisfies eqn (5.9) and we can find an estimator $\hat{\sigma}_n$ such that

$$\sup_{(\alpha,\eta)} \mathrm{pr}_{(\alpha,\eta)} \left\{ \left| 1 - \frac{\hat{\sigma}_n}{\sigma_n(\alpha, \eta)} \right| > \varepsilon \right\} \underset{n\to\infty}{\to} 0, \tag{5.10}$$

then $I_n = \hat{\alpha} \pm \hat{\sigma}_n \Phi^{-1}(1 - \tau/2)/\sqrt{n}$ is an asymptotic $(1 - \tau)$ confidence interval; that is, it satisfies

$$\sup_{F_{\alpha,\eta} \in \mathcal{F}} \left| \mathrm{pr}_{(\alpha,\eta)}(\alpha \in I_n) - (1 - \tau) \right| \underset{n\to\infty}{\to} 0. \tag{5.11}$$

Uniform convergence rather than pointwise convergence,

$$\mathrm{pr}_{(\alpha,\eta)}(\alpha \in I_n) \underset{n\to\infty}{\to} 1 - \tau \text{ at each } (\alpha, \eta),$$

is relevant when the concern is, as it is of course in practice, to ensure that the actual coverage of the interval at a fixed and, possibly large, n, is close to the nominal level. If pointwise but not uniform convergence holds for I_n, then there does not exist an n at which the coverage probability of I_n is guaranteed to be close to its nominal level under all laws $F_{\alpha,\eta}$ allowed by the model.

Under mild conditions, when $\widehat{\alpha}$ satisfies eqn (5.9) the bootstrap estimator $\widehat{\sigma}_n$ of the standard error of $\sqrt{n}\,(\widehat{\alpha} - \alpha)$ will satisfy eqn (5.10). Thus, such an estimator can be used to center confidence intervals with length shrinking to zero at rate $O_p\left(n^{-1/2}\right)$.

A central question of practical interest is therefore the investigation of conditions under which we can hope to find unbiased estimating functions that depend only on α. Identity (5.6) and the comment following it give a clue to the answer. Specifically, if such $m\,(X;\alpha)$ exists, then apart from a constant factor, $m\,(X;\alpha^*)$ must be equal to a gradient $\psi_{F_{\alpha^*,\eta^*}}\,(X)$ of α at F_{α^*,η^*}. Since $\widehat{\alpha}$ is consistent regardless of the value of η, then $\psi_{F_{\alpha^*,\eta^*}}\,(X)$ must satisfy

$$\mathrm{E}_{F_{\alpha^*,\eta}}\left\{\psi_{F_{\alpha^*,\eta^*}}\,(X)\right\} = 0 \text{ for all } \eta. \tag{5.12}$$

This condition is therefore necessary for the existence of an unbiased estimating function. Indeed, taking $m\,(X;\alpha) = \psi_{F_{\alpha,\eta^*}}\,(X)$, we see that the condition is also sufficient for eqn (5.4) to hold. As we saw earlier, gradients are orthogonal to the nuisance tangent space, so in the search for unbiased estimating functions one can restrict attention to $\Lambda_{\mathrm{nuis}}\,(F_{\alpha^*,\eta^*})^{\perp}$, the collection of mean zero, finite variance functions of X which are orthogonal to $\Lambda_{\mathrm{nuis}}\,(F_{\alpha^*,\eta^*})$.

Example 3 (continued) It follows from Bickel *et al.* (1993, §4.3) that in model $\mathcal{F}\,(\boldsymbol{\alpha} \times \boldsymbol{\kappa} \times \boldsymbol{\omega} \times \{\pi^*\})$ with π^* known,

$$\Lambda_{\mathrm{nuis}}\,(F_{\alpha^*,\kappa^*,\omega^*,\pi^*})^{\perp} = \{m_h\,(X;\alpha^*) = \{Z - \pi^*\,(V)\}\,\{(Y - \alpha^* Z)\,h_1\,(V) + h_2\,(V)\}:$$
$$h\,(\cdot) = (h_1\,(\cdot),h_2\,(\cdot)) \text{ unrestricted}\} \cap \mathcal{L}_2\,(F_{\alpha^*,\kappa^*,\omega^*,\pi^*}).$$

All elements of this set have mean zero under $F_{\alpha^*,\kappa,\omega,\pi^*}$, regardless of the values of $\kappa = (\kappa_1, \kappa_2)$ and ω. Under moment conditions, $\widehat{\alpha}$ solving $\sum_i m_h\,(X_i;\alpha) = 0$ for arbitrary h satisfies eqn (5.9) where η stands for all nuisance parameters in the model.

In models like $\mathcal{F}\,(\boldsymbol{\alpha} \times \boldsymbol{\kappa} \times \boldsymbol{\omega} \times \{\pi^*\})$ in Example 3, in which all elements of $\Lambda_{\mathrm{nuis}}\,(F_{\alpha^*,\eta^*})^{\perp}$ have mean zero under $F_{\alpha^*,\eta}$ for any η, one has the additional advantage of being able to compute locally efficient estimators of α that do not suffer from the curse of dimensionality. Specifically, suppose that, as in the previous example, the set $\Lambda_{\mathrm{nuis}}\,(F_{\alpha^*,\eta^*})^{\perp}$ is indexed by some parameter h, possibly infinite-dimensional. Then $S_{\alpha,\mathrm{eff}}\,(F_{\alpha^*,\eta^*})$ corresponds to some optimal index h_{eff} that often depends on α^* and η^*. Suppose that one uses an estimator $\widehat{h}_{\mathrm{eff}}$ that converges in some norm to h_{eff} under some guessed dimension-reducing model for h_{eff}, which we call the working model. Under regularity conditions, the quantity $\widehat{\alpha}$ that solves $\sum_i m_{\widehat{h}_{\mathrm{eff}}}\,(X_i;\alpha) = 0$ has the same limiting distribution as the solution of $\sum_i m_{h_{\mathrm{lim}}}\,(X_i;\alpha) = 0$, where h_{lim} is the probability limit of $\widehat{h}_{\mathrm{eff}}$. The key to the validity of this result is the unbiasedness of the 'estimated' estimating function, i.e. the fact that $\overline{m}^*_{h_{\mathrm{eff}}}\,(\alpha^*) = 0$, where for any h and any α, $\overline{m}^*_h\,(\alpha) \equiv \mathrm{E}_{F_{\alpha^*,\eta^*}}\left\{m_h\,(X;\alpha)\right\}$. The estimator $\widehat{\alpha}$ has the same asymptotic distribution as the solution to the efficient score equation under the working model and remains asymptotically normal otherwise. Thus, $\widehat{\alpha}$ is locally efficient at the working model.

Example 3 (continued) The efficient score for α in model $\mathcal{F}(\alpha \times \kappa \times \omega \times \{\pi^*\})$ is

$$S_{\alpha,\text{eff}}\left(F_{\alpha^*,\kappa^*,\omega^*,\pi^*}\right) = \{Z - \pi^*(V)\}\{Y - \alpha^* Z - \omega^*(V)\} c^*(V)^{-1},$$

where

$$c^*(V) = \pi^*(V)\sigma^*(0,V) + \{1 - \pi^*(V)\}\sigma^*(1,V), \quad \sigma^*(Z,V) = \text{var}_{F^*}(Y \mid Z, V).$$

Note that $S_{\alpha,\text{eff}}\left(F_{\alpha^*,\kappa^*,\omega^*,\pi^*}\right) = m_{h_{\text{eff}}}(X;\alpha^*)$, where $h_{\text{eff},1}(V) = c^*(V)^{-1}$ and $h_{\text{eff},2}(V) = -\omega^*(V) c^*(V)^{-1}$ (Bickel *et al.* 1993, §4.3). Let $\widetilde{\alpha}$ be a preliminary estimator of α, say solving eqn (5.5) using an arbitrary $m_h(X;\alpha)$ in $\Lambda_{\text{nuis}}(F_{\alpha,\kappa^*,\omega^*,\pi^*})^\perp$. Let $\widehat{\omega}(v)$ be the fitted value at $V = v$ in the regression of $Y_i - \widetilde{\alpha} Z_i$ on V_i, $i = 1, \ldots, n$ under some working model for $\omega^*(V)$, and let

$$\widehat{c}(v) = \pi^*(v)\widehat{\sigma}(0,v) + \{1 - \pi^*(v)\}\widehat{\sigma}(1,v),$$

where the quantity $\widehat{\sigma}(z,v)$ is the fitted value at $(Z,V) = (z,v)$ in the regression of $\{Y_i - \widetilde{\alpha} Z_i - \widehat{\omega}(V_i)\}^2$ on $(Z_i, V_i,)$, $i = 1, \ldots, n$, under some working model for $\text{var}_{F^*}(Y \mid Z, V)$. Under regularity conditions, the estimator $\widehat{\alpha}$ solving $\sum_i m_{\widehat{h}_{\text{eff}}}(X_i;\alpha) = 0$ where $\widehat{h}_{\text{eff}}$ is defined like h_{eff} but with $\widehat{\omega}$ and $\widehat{\sigma}$ instead of ω^* and σ^*, is a locally semiparametric efficient at the working models.

Robins and Ritov (1997) warned that local efficiency must be interpreted with caution if the working model is large. They noted that the variance bound may be an overly optimistic measure of the actual sampling variability of the estimator under some laws allowed by the working model: formally, $\sigma_n^2(\alpha, \eta)/C(F_{\alpha,\eta}) \to 1$ as $n \to \infty$, but not uniformly in (α, η). For instance, suppose that in the previous example the working model only specifies that ω^* and $\sigma^*(z, \cdot)$ are continuous functions. Suppose that the sample is randomly split into two halves, and separate multivariate kernel smoothers $\widehat{\omega}_j$ and $\widehat{\sigma}_j(z, \cdot)$ of ω^* and $\sigma^*(z, \cdot)$ are computed on each half-sample, for $j = 1, 2$. Suppose that $\widehat{\alpha}$ is a solution to $\sum_i m_{\widehat{h}_{\text{eff}}(i)}(X_i;\alpha) = 0$, where $\widehat{h}_{\text{eff}}(i)$ is the estimator of h_{eff} that uses $\widehat{\omega}_j$ and $\widehat{\sigma}_j(z, \cdot)$ based on the data in the half-sample that does not contain unit i. Under some moment conditions, $\widehat{\alpha}$ is locally efficient at the working model. However, the variance of $\sqrt{n}(\widehat{\alpha} - \alpha^*)$ is larger than the variance bound under some laws allowed by the working model. This reflects the fact that at any given sample size n, there exist very wiggly continuous functions ω^* and c^* that cannot be well estimated by $\widehat{\omega}$ and $\widehat{c}$.

Bickel (1982) has characterized a class of models where eqn (5.12) holds for all gradients, and for which, therefore, locally efficient estimation is feasible. These are models such that the collection $\mathcal{F}_{\text{nuis}} = \{F_{\alpha^*,\eta} : \eta \in \eta\}$ is convex in η, that is, any mixture of two distributions in $\mathcal{F}_{\text{nuis}}$ is also in $\mathcal{F}_{\text{nuis}}$. Identity (5.12) holds in such models because

$$\overline{f}(x;t) \equiv tf(x;\alpha^*,\eta) + (1-t)f(x;\alpha^*,\eta^*), \quad 0 \le t \le 1,$$

belongs to $\mathcal{F}_{\text{nuis}}$ for every pair (η^*,η). Consequently

$$\left. \frac{\partial \log \overline{f}(x;t)}{\partial t} \right|_{t=0} = \frac{f(x;\alpha^*,\eta^*)}{f(x;\alpha^*,\eta)} - 1$$

is a nuisance score. As any gradient $\psi_{F_{\alpha^*,\eta^*}}(X)$ is orthogonal to nuisance scores, we see that

$$0 = \mathrm{E}_{\alpha^*,\eta^*}\left[\psi_{F_{\alpha^*,\eta^*}}(X)\left\{\frac{f(x;\alpha^*,\eta^*)}{f(x;\alpha^*,\eta)} - 1\right\}\right] = \mathrm{E}_{\alpha^*,\eta}\left\{\psi_{F_{\alpha^*,\eta^*}}(X)\right\}.$$

Convexity of the nuisance model is a sufficient but not necessary condition for eqn (5.12) to hold for all gradients. For example, the nuisance model in model $\mathcal{F}(\boldsymbol{\alpha} \times \boldsymbol{\kappa} \times \boldsymbol{\omega} \times \{\pi^*\})$ of Example 3 is not convex but, as noted earlier, all gradients of α satisfy eqn (5.12).

Example 3 (continued) Consider model $\mathcal{F}(\boldsymbol{\alpha} \times \boldsymbol{\eta} \times \{\omega^*\} \times \{\pi^*\})$ in which ω^* is known. This is a special case of the conditional mean model $\mathcal{F}_{\text{cond}}(\boldsymbol{\alpha} \times \boldsymbol{\eta})$ defined by the sole restriction that $\mathrm{E}_{F^*}(Y \mid W) = d(W;\alpha^*)$ for some known function $d(W; \cdot)$. An easy calculation shows that the nuisance model is convex. Chamberlain (1987) showed that $\Lambda_{\text{nuis}}(F_{\alpha^*,\eta^*})^{\perp}$ equals

$$\{m_h(X;\alpha^*) = h(W)\{Y - d(W;\alpha^*)\} : h(\cdot) \text{ unrestricted}\} \cap \mathcal{L}_2(F_{\alpha^*,\eta^*}).$$

As predicted by the theory, the elements of this set have mean zero under $F_{\alpha^*,\eta}$ for any η.

5.4.3 Unfeasible estimation due to the curse of dimensionality

Unfortunately, unbiased estimating functions that depend on α only do not always exist. In many models $\Lambda_{\text{nuis}}(F_{\alpha^*,\eta^*})^{\perp}$ is of the form

$$\{m_h(X;\alpha^*,\rho(\eta^*)) : h \text{ in some index set}\} \cap \mathcal{L}_2(F_{\alpha^*,\eta^*}), \tag{5.13}$$

where $\rho(\eta)$ is some non-Euclidean function of η and

$$\overline{m}_h^*\{\alpha,\rho(\eta)\} \equiv \mathrm{E}_{F_{\alpha^*,\eta^*}}[m_h\{X;\alpha,\rho(\eta)\}]$$

satisfies $\overline{m}_h^*\{\alpha^*,\rho(\eta^*)\} = 0$ but $\overline{m}_h^*\{\alpha^*,\rho(\eta)\} \ne 0$ for $\eta \ne \eta^*$.

Example 3 (continued) It follows from results in Bickel *et al.* (1993, §4.3) that in model $\mathcal{F}(\boldsymbol{\alpha} \times \boldsymbol{\kappa} \times \boldsymbol{\omega} \times \boldsymbol{\pi})$ with π^* unknown,

$$\Lambda_{\text{nuis}}(F_{\alpha^*,\kappa^*,\omega^*,\pi^*})^{\perp} = \{m_h(X;\alpha^*,\omega^*,\pi^*) = [Z - \pi^*(V)][Y - \alpha^* Z - \omega^*(V)]h(V) :$$
$$h(\cdot) \text{ unrestricted}\} \cap \mathcal{L}_2(F_{\alpha^*,\kappa^*,\omega^*,\pi^*}).$$

Here $\eta^* = (\kappa^*,\omega^*,\pi^*)$, $\rho(\eta^*) = (\omega^*,\pi^*)$ and $\overline{m}_h^*(X;\alpha^*,\omega^\dagger,\pi^\dagger) \ne 0$ if $\omega^\dagger \ne \omega^*$ and $\pi^\dagger \ne \pi^*$.

Estimators satisfying eqn (5.9) may not exist when $\Lambda_{\text{nuis}} \left(F_{\alpha^*, \eta^*} \right)^{\perp}$ is of the form of eqn (5.13). Indeed, the situation may be much worse. For example, Ritov and Bickel (1990) showed that in model $\mathcal{F} \left(\boldsymbol{\alpha} \times \boldsymbol{\kappa} \times \boldsymbol{\omega} \times \boldsymbol{\pi} \right)$ of Example 3 there exists no uniformly consistent estimator of α^*, that is, one satisfying the condition

$$\sup_{(\alpha, \eta)} \text{pr}_{(\alpha, \eta)} \left(|\widehat{\alpha} - \alpha| > \varepsilon \right) \xrightarrow[n \to \infty]{} 0 \text{ for all } \varepsilon > 0. \tag{5.14}$$

The essence of this is seen in the following heuristic argument where, to avoid distracting technicalities, we assume that V is scalar and uniformly distributed on the interval $(0, 1)$. For each fixed n, one can find a partition of the interval $(0, 1)$, with partition bins $B_m, m = 1, \ldots, M$, such that with very high probability, no two units have values of V in the same bin. Inside each bin, the observed data capture neither the association between the confounder and the outcome within each treatment level $\text{E} \left(Y \mid Z = z, V = \cdot \right)$, $z = 0, 1$, nor the association between the confounder and treatment $\text{pr} \left(Z = z \mid V = \cdot \right)$. Thus, inside each bin, there is essentially confounding by the variable V, as the following example illustrates.

Example 4   Divide B_m into two subintervals B_{1m} and B_{2m} of equal length. Consider a set of 2^M probability distributions, $f_j (x)$, $j = 1, \ldots, 2^M$, satisfying in each bin B_m one of the two possibilities shown in Table 5.1. Because at most one V is observed in each bin, with a sample of size n a data analyst will not be able to distinguish the possibility that one of the 2^M laws generated the data from the possibility that the distribution satisfying $\text{E} \left(Y \mid Z = 0, V \right) = 2.8$, $\text{E} \left(Y \mid Z = 1, V \right) = 8.2$, and $\text{pr}(Z = 1 \mid V) = 0.5$ generated the data. However, $\alpha^* = 0$ under any of the 2^M former laws while $\alpha^* = 5.4$ under the latter. Thus, $\alpha^* = 0$ cannot be discriminated from $\alpha^* = 5.4$.

A negative consequence of the lack of uniform consistency is that there exists no asymptotic $1 - \tau$ confidence interval I_n satisfying eqn (5.11) with length shrinking to zero, not even at a rate slower that $O_p \left(n^{-1/2} \right)$. For if such I_n existed, its midpoint $\widehat{\alpha}$ would satisfy eqn (5.14).

5.4.4   Semiparametric inference and the likelihood principle

Example 3 illustrates the important general point raised by Robins and Ritov (1997) that in very high-dimensional models there may exist no procedure with

TABLE 5.1. Possible values of $\text{E} \left(Y \mid Z = z, V = v \right)$ and $\text{pr} \left(Z = z \mid V = v \right)$ in bin B_m; see Example 4.

	Possibility 1				Possibility 2			
	$v \in B_{1m}$		$v \in B_{2m}$		$v \in B_{1m}$		$v \in B_{2m}$	
z	0	1	0	1	0	1	0	1
$\text{E} \left(Y \mid Z = z, V = v \right)$	1	1	10	10	10	10	1	1
$\text{pr} \left(Z = z \mid V = v \right)$	0.8	0.2	0.2	0.8	0.2	0.8	0.8	0.2

good frequentist properties that obeys the likelihood principle. Specifically, in Example 3, let

$$\mathbf{W} = \{(Z_i, V_i) : i = 1, \ldots, n\}, \quad \mathbf{Y} = \{Y_i : i = 1, \ldots, n\}.$$

In model $\mathcal{F}(\boldsymbol{\alpha} \times \boldsymbol{\kappa} \times \boldsymbol{\omega} \times \boldsymbol{\pi})$ the likelihood equals $f(\mathbf{Y} \mid \mathbf{W}; \alpha, \omega, \kappa_1) f(\mathbf{W}; \pi, \kappa_2)$. The preceding discussion implies that whether or not the distribution of the S-ancillary statistic $\mathbf{W}$ (Cox and Hinkley 1974) is known has severe consequences for inference. If (π^*, κ_2^*) are unknown, we have just seen that no uniform consistent estimator and hence, no asymptotic confidence interval exists. In contrast, when (π^*, κ_2^*) is known, there exist both estimators satisfying eqn (5.9) and asymptotic confidence intervals with length shrinking to zero at rate $O_p(n^{-1/2})$. This is so because, as noted in Section 5.4.2, such procedures exist when the randomization probabilities π^* are known even if κ_2^* is unknown. Now, in the model with (π^*, κ_2^*) known, any procedure that obeys the likelihood principle must give the same inference for α^* for any known value of (π^*, κ_2^*). But this implies that such procedure can also be used for inference about α^* in model $\mathcal{F}(\boldsymbol{\alpha} \times \boldsymbol{\kappa} \times \boldsymbol{\omega} \times \boldsymbol{\pi})$ with (π^*, κ_2^*) unknown. Since no well-behaved procedure exists in model $\mathcal{F}(\boldsymbol{\alpha} \times \boldsymbol{\kappa} \times \boldsymbol{\omega} \times \boldsymbol{\pi})$ we conclude that any inferential procedure that works well in the model with (π^*, κ_2^*) known must violate the likelihood principle and depend on the randomization probabilities π^* or on the law κ_2^* of V, or both.

5.4.5 Double robustness

The discussion above implies that in Example 3, both in theory and practice, the size of model $\mathcal{F}(\boldsymbol{\alpha} \times \boldsymbol{\kappa} \times \boldsymbol{\omega} \times \boldsymbol{\pi})$ must be reduced in order to obtain well-behaved procedures. One possibility is to consider inference under the submodel $\mathcal{F}(\boldsymbol{\alpha} \times \boldsymbol{\kappa} \times \boldsymbol{\omega}_{\mathrm{sub}} \times \boldsymbol{\pi})$, where $\boldsymbol{\omega}_{\mathrm{sub}}$ is a small subset of $\boldsymbol{\omega}$. If $\boldsymbol{\omega}_{\mathrm{sub}}$ is sufficiently small, then there exist estimators of α that satisfy eqn (5.9) where η stands for all nuisance parameters in the model. For example, if $\boldsymbol{\omega}_{\mathrm{sub}}$ is a parametric subset $\{\omega(\cdot; \psi) : \psi \in \mathbb{R}^q\}$ of $\boldsymbol{\omega}$ then the first component $\widehat{\alpha}_{\mathrm{ls}}$ of the possibly non-linear least squares estimator $\left(\widehat{\alpha}_{\mathrm{ls}}, \widehat{\psi}_{\mathrm{ls}}\right)$ of (α, ψ) satisfies eqn (5.9) under mild regularity conditions. A second possibility is to assume that π^* lies in a small, perhaps parametric, subset $\boldsymbol{\pi}_{\mathrm{sub}}$ of $\boldsymbol{\pi}$. Estimators of α^* satisfying eqn (5.9) exist in model $\mathcal{F}(\boldsymbol{\alpha} \times \boldsymbol{\kappa} \times \boldsymbol{\omega} \times \boldsymbol{\pi}_{\mathrm{sub}})$. For example, the estimator $\widehat{\alpha}_{\mathrm{prop}}$ solving

$$\sum_i (Y_i - \alpha Z_i) \left\{ Z_i - \pi\left(V_i; \widehat{\xi}\right) \right\} = 0,$$

where $\widehat{\xi}$ is the maximum likelihood estimator of ξ under the submodel $\boldsymbol{\pi}_{\mathrm{sub}} = \{\pi(\cdot; \xi) : \xi \in \mathbb{R}^q\}$, satisfies eqn (5.9) under mild regularity conditions.

Neither $\widehat{\alpha}_{\mathrm{ls}}$ nor $\widehat{\alpha}_{\mathrm{prop}}$ is entirely satisfactory: the first is inconsistent if $\omega^* \notin \boldsymbol{\omega}_{\mathrm{sub}}$, and the second is inconsistent if $\pi^* \notin \boldsymbol{\pi}_{\mathrm{sub}}$. Fortunately, estimators exist that are uniformly asymptotically normal and unbiased in the larger model $\mathcal{F}(\boldsymbol{\alpha} \times \boldsymbol{\eta} \times ((\boldsymbol{\omega} \times \boldsymbol{\pi}_{\mathrm{sub}}) \cup (\boldsymbol{\omega}_{\mathrm{sub}} \times \boldsymbol{\pi})))$ that assumes that one of the models

π_{sub} or ω_{sub}, but not necessarily both, is correctly specified. Such estimators have been called doubly robust (Robins *et al.* 2000). For example, the estimator $\widehat{\alpha}_{\mathrm{dr}}$ obtained by solving the equation

$$\sum_i \left\{ Y_i - \alpha Z_i - \omega \left(V_i; \widehat{\psi}_{\mathrm{ls}} \right) \right\} \left\{ Z_i - \pi \left(V_i; \widehat{\xi} \right) \right\} = 0$$

is doubly robust.

Doubly robust estimators have been derived independently by various authors in specific problems (Brillinger 1983, Ruud 1983, 1986, Duan and Li 1987, 1991, Lipsitz *et al.* 1999, Robins *et al.* 2000) but do not always exist. An open problem in semiparametric theory is to characterize the problems in which doubly robust inference is feasible. Some initial progress is reported in Robins and Rotnitzky (2001).

5.5 Discussion

Doubly robust inference, even if feasible, does not resolve the difficulties posed by the curse of dimensionality. It is not honest to severely reduce the size of the model if we believe that the initial semiparametric model encodes our entire knowledge about the data-generating process. Dimension-reducing strategies such as double robustness based on parametric submodels lead to estimation procedures, in particular point and interval estimators, that perform well if we happen to be lucky enough to be working with data that arose from a law in the reduced model. But if not, such procedures may lead to quite misleading inferences.

What then should we do? As Example 4 illustrates, there is very little one can do when the model is so gigantic that data-generating processes with extreme anomalies can occur under the model. However, in most realistic problems one can reasonably exclude such anomalies and assume that the data-generating process does have some regularity. For instance, in Example 3, one may be willing to assume that the functions ω^* and π^* are Holder ν for some $\nu < 1$; recall that a map $u \mapsto g(u)$ between two normed spaces is said to be Holder ν, for $\nu < 1$, if and only if there exists a constant K such that for all u_1, u_2, $\|g(u_1) - g(u_2)\| \le K \|u_1 - u_2\|^{\nu}$. This raises the following important unresolved issues:

(a) Consider a semiparametric model indexed by a Euclidean parameter α and a non-Euclidean parameter η. To protect against model misspecification, one may take a conservative position and assume some but not much smoothness on η or some of its components, thereby losing the ability to construct $\sqrt{n}$-consistent estimators of α. Almost nothing is known about how to estimate α once the smoothness requirements on the nuisance parameter η needed for $\sqrt{n}$-consistent estimation of α are relaxed. A first challenging question is: how do we construct estimators that achieve a minimax optimal rate of convergence under a large smooth model $\mathcal{F}_{\text{smooth}}$? By this we mean estimators whose mean squared error converges at the fastest possible rate to zero in the worse possible scenario. Formally, we want to find estimators $\widehat{\alpha}$ of α such that for some sequence τ_n converging to zero as $n \to \infty$,

$$\limsup_{n \to \infty} \sup_{F_{\alpha,\eta} \in \mathcal{F}_{\text{smooth}}} \mathrm{E}_{F_{\alpha,\eta}} \left(\tau_n^{-2} \|\widehat{\alpha} - \alpha\|^2 \right) < \infty \qquad (5.15)$$

and the normalizing sequence τ_n cannot be improved in order, in the sense that if $\widetilde{\alpha}$ is another estimator for which there exists a sequence ς_n such that eqn (5.15) holds with $\widetilde{\alpha}$ and ς_n instead of $\widehat{\alpha}$ and τ_n, then τ_n converges to zero at least as fast as ς_n.

(b) More importantly, how do we construct honest confidence regions I_n for α^*, that is, ones that satisfy eqn (5.11) with $\mathcal{F}_{\text{smooth}}$ instead of $\mathcal{F}$, whose size converges to zero as $n \to \infty$? We may even be more ambitious and ask that our honest regions converge to zero at the fastest possible rate, satisfying

$$\sup_{F_{\alpha,\eta} \in \mathcal{F}_{\text{smooth}}} \mathrm{pr}_{(\alpha,\eta)} \left[\varphi_n^{-2} \text{ volume} (I_n) > \delta \right] \to 0 \text{ as } n \to \infty$$

for each $\delta > 0$, where the constant φ_n converges to zero at the fastest possible rate.

We may even hope to be able to improve on the estimators sought in (a). Over the last decade the nonparametrics community has spent considerable effort investigating adaptive estimators whose convergence rate is as fast as possible given the unknown smoothness of the true underlying irregular object being estimated, a density or conditional expectation for instance. It is therefore not too ambitious to hope to find rate-adaptive estimators of α^* as well. However, we doubt the relevance to practice of adaptive estimators. Since adaptive variance estimators for adaptive estimators of irregular parameters do not exist, we suspect that the same would be true for estimators of the variance of adaptive estimators of α^*. But without an assessment of their uncertainty, the usefulness of adaptive estimators will be limited. In biomedical applications, for instance, it is common practice to combine evidence from various studies without access to the raw data. For meta-analysis it is far more important to have a valid confidence interval, even if large, than a good point estimator of α^* without assessment of its uncertainty.

The reader may feel that the questions raised in (a) and (b) are of mathematical interest but of little relevance to practice. We disagree: indeed, we view the answers as one possible way to reconcile asymptotic theory with finite sample performance. One possibility is to calibrate the assumed degree of smoothness with the sample size and the dimension of the model. For instance, it may be known that the data at hand have been generated by η^* with smoothness obeying the demands for $\sqrt{n}$-consistent estimation of α^*. However, with the available sample size the ability to estimate η^* may be severely limited even if the smoothness conditions are imposed. Consequently, the real accuracy with which α can be estimated may be much lower than that predicted by the theory, and may indeed resemble more the accuracy that the theory would have predicted if one had assumed much less smoothness of η. One could then calibrate inference on α^* by assuming a less restrictive model. Precise calibration would be practically impossible, but a sensitivity analysis to various degrees of assumed smoothness may be a reasonable strategy to give an indication of the real magnitude of uncertainty.

We conclude by noting that Robins (2004) and Robins and van der Vaart (2004), following Small and McLeish (1994) and Waterman and Lindsay (1996), have outlined a promising new theory of higher-order scores and Bhattacharyya bases that may yield fruitful ways to resolve the questions posed above.

Acknowledgements

This work was partially funded by grants from the United States National Institutes of Health. It was partially completed at the Department of Economics at Di Tella University, to which the author is grateful for providing a stimulating environment. She also thanks James Robins for illuminating discussions on the consequences of the curse of dimensionality and, in particular, for suggesting Example 4.

6

On non-parametric statistical methods

Peter Hall

6.1 Parametric and nonparametric statistics

David Cox and I have collaborated on just one paper (Cox and Hall 2002), and I feel sure that we view it in quite different ways. The work is on inference for simple random effects models in non-normal distributions, and David would, I feel sure, see it as an account of statistical methodology in cases that are close to normal, but do not actually enjoy that distribution. In particular, 'local' or 'approximate' normality can be used to justify the moment-based approach taken in our work. For my part, I view the research as an account of straight non-parametric inference, albeit more easy to justify in the case of distributions with light tails.

The appropriateness of this dual view of contemporary statistical inference is far more common than one might be led to believe, given the tension that sometimes is said to exist between finite- and infinite-parameter perspectives of the statistical universe (see, for example, Breiman 2001). It is trite to say that the dual view has been with us for a very long while. Thus, the central limit theorems of Laplace, known early in the nineteenth century, can be viewed as providing a local view of inference about a mean, as the population distribution strays from normality.

Additionally, most non-parametric methods are closely linked, in either spirit or concept, to much older parametric approaches. For example, the distinctly non-parametric passion for bump hunting, which is often a search for inhomogeneity in the form of subpopulations, is close in spirit to the much older art of fitting mixture models and identifying significant components. Consider, for example, David Cox's work on exploring homogeneity by fitting normal mixture models (Cox 1966). He viewed the marked presence of an additional normal population to be a 'descriptive feature likely to indicate mixing of components'. Non-parametric approaches to bump hunting date from Silverman's (1981) work on multimodality.

To take another example, many non-parametric methodologies for dimension reduction (for example Friedman and Stuetzle 1981, Friedman *et al.* 1984, Huber 1985) have at their heart relatively conventional techniques for principal component analysis.

These remarks do not in any way detract from the novelty, often very profound, of new developments in non-parametric statistics, for example in bump hunting or dimension reduction. Rather, they point to a very healthy willingness on the part of modern statistical scientists to borrow, where necessary, the ingenious and simple ideas of their predecessors. However, the evidence does indicate that the suggestion that two opposing cultures exist in modern statistical science may need to be evaluated cautiously. Arguably, remarks about opposing cultures are made more to provoke discussion and debate than to reflect real division or to indicate genuinely divergent methodological development, in the past, present, or future. Discussions of the potential future of statistics, addressing both parametric and non-parametric ideas, include those of Hall (2001) and Raftery *et al.* (2001). Surveys of the past include Titterington and Cox (2001).

6.2 Computation and non-parametric statistics

Non-parametric statistics has benefited far more than its parametric counterpart from the exponential increase in computing power that Moore's law (Moore 1965) has visited upon us for the last 40 years. This is at least partly because much of non-parametric statistics was not really feasible until the desktop-computer age, which dawned in the 1970s and early 1980s. However, the mathematical details of permutation methods, of techniques for bandwidth choice in curve estimation (see, for example, Pitman 1937, Woodroofe 1970), and of related topics in non-parametric statistical inference, were worked through in the pre-dawn of the modern computer age. Of necessity, in this era they were treated much more as intellectual and technical exercises, lacking authoritative numerical demonstration, than as ready-to-implement practical methods.

However, these exercises placed us firmly on the right path, and the evolution of statistics benefited greatly from discussion of methodologies that, in some cases, we were not in a position to apply for more than thirty years. Even when Simon (1969) was strenuously advocating permutation methods, he sought acceptance not by showing how to implement the technique, which in 1969 was still not quite computationally feasible for many of us, but by discussing it in the context of gambling experiments. The current 'reality' view of academic statistics, which focuses firmly on practical implementation using available computing technology, would have been quite out of place in that era. In important respects, the present climate of research has moved too far in the direction of demanding immediate practical justification for every methodological development. In particular we have been seduced by a view, certainly false but nevertheless firmly held in some quarters, that we already have all the computing resources that we should seriously contemplate for today's research.

Almost unquestionably, the part of parametric statistics that has benefited most from advances in computing is Bayesian statistics. Non-parametric Bayesian statistics has undergone its own contemporary revolution, contributing significantly to areas as widely separated as signal denoising and survival analysis. Recent examples include work of Damien and Walker (2002) on Bayesian non-parametric methods for the analysis of survival data, Denison *et al.* (2002) on Bayesian smoothing, Tardella (2002) on Bayesian non-parametric methods for analysing capture-recapture data, Wood *et al.* (2002) and Yau *et al.* (2003) on model and variable selection in non-parametric regression, Ghosh and Ramamoorthi (2003) on systematic approaches to Bayesian non-parametric methods, Guo *et al.* (2003) on non-parametric Bayesian methods for information theory, Hasegawa and Kozumi (2003) on the estimation of Lorenz curves in econometrics, Maras (2003) on signal denoising, and Scaccia and Green (2003) on non-parametric Bayesian methods for estimating growth curves.

These articles have been selected because they mainly address Bayesian methods in non-parametric settings. There are many more papers treating both parametric and non-parametric techniques of Bayesian type, or both frequentist and Bayesian ideas applied to the development of non-parametric methods. One of the attractions of the *Biometrika* centenary volume (Titterington and Cox 2001) is that it does not cross-categorize statistics into Bayesian and non-Bayesian approaches; classification along broad methodological lines proved adequate for the volume, as it should.

Some contemporary Bayesian methods share the local frequency-based outlook that is typical of many non-parametric techniques. Such methods, arguably more than others, are fielded in the contexts of very large, or very complex, or very high-dimensional, datasets. These features combine to produce the relatively computationally intensive nature of non-parametric methods.

The features also confer a relatively high degree of adaptivity, which makes non-parametric methods particularly flexible. For example, some non-parametric approaches accommodate non-stationarity very easily; and, especially in spatial problems, where global parametric models can be both cumbersome and restrictive, the reliance of non-parametric approaches on little more than local relationships can be a real boon. Thus, for instance, non-parametric methods are increasingly popular in studies of spatial seismic data (e.g. Choi and Hall 1999, Ogata 2001, Estevez-Perez *et al.* 2002, Zhuang *et al.* 2002), where in the past unreliable results have sometimes been obtained through fitting inappropriate parametric models. However, non-parametric techniques applied to spatial data can demand very high levels of computation, especially when bootstrap methods are employed to develop confidence statements.

Several of the applications we have discussed, and a multitude of others, for example in genomics, involve machine-recorded data in quantities that have seldom been seen in the past. Moreover, the inexpensive way in which many parameters, rather than just a few, can be recorded by a machine often massively increases the number of dimensions. All these aspects add significantly to the computational demands placed on contemporary non-parametric technologies, as well as to the many new circumstances that those techniques have to address.

6.3 Mathematics and non-parametric statistics

6.3.1 Introduction

Twenty-five years ago, when I did not have a permanent job and was starting to become interested in theoretical problems associated with non-parametric statistics, a senior colleague warned me of the dangers of following this path. The future of statistics, he assured me, lay in its computational side, and theory was becoming a sideline. Statistical methodologies would be explored numerically, largely by simulation. Not only were theoretical statisticians becoming increasingly irrelevant, they were actually impeding the progress of statistics.

He was not alone, among colleagues in Australia, in making remarks such as this. Similar views were sometimes expressed on the other side of the world, too. Julian Simon was to counsel that non-parametric statistics, and in particular the bootstrap, 'devalues the knowledge of conventional mathematical statisticians, and especially the less competent ones'. This 'priesthood,' he argued, 'with its secret formulaic methods is rendered unnecessary' (draft material for a later edition of Simon 1993).

The sentiments I heard 25 years ago are still expressed here, in Australia, from time to time. There is little doubt that during this period, in my country, the academic pursuit of statistics has become, overall, less mathematical and more exploratory in nature. This is less true in other some parts of the world. For example, the theoretical appendices of papers in the *Journal of the American Statistical Association*, and perhaps even *Biometrika*, are more likely, today, to include a discussion of minimax optimality, or to use a relatively contemporary result from functional analysis, than they were a quartercentury ago.

This technical sophistication is widely acknowledged by the statistics community in the context of, say, wavelet methods, which are themselves founded on relatively new and advanced mathematical technology. However, although I appreciate that there is some disagreement with my arguments, I suggest that the usefulness of relatively advanced mathematical theory for elucidating the properties of far less sophisticated non-parametric methodologies, is also high.

In fact, this usefulness is increasing, as escalating computer power allows us to do far more complex things to datasets, and to address much more complex problems. There is, or at least there should be, more demand than ever for mathematical statisticians who can render a complex statistical problem sufficiently simple, or abstract, to make it mathematically solvable; and at the same time retain the essential features that enable us to capture critical intuition and to link the mathematical solution directly to the results of practical implementation. David Cox has lamented the disinclination of contemporary statistics postgraduate students, in the United Kingdom, to study theoretical problems in statistics: 'It would be good if there were a few more doctoral students saying that they'd like to continue for a while to do some theory.' (Cox 2004, interviewed by Helen Joyce).

We shall see several instances where the elucidation of non-parametric methods relies heavily on mathematics, the latter drawn from areas of number theory, algebraic geometry and operator theory. I appreciate that it is often the case, particularly in applied problems, that relatively simple mathematical arguments can convey a great deal of information. But I'd like to make the point, not well enough appreciated in my view, that there is often a necessity for relatively sophisticated, even complex, mathematical arguments if we are to properly understand a wide range of practical issues arising in statistics.

6.3.2 Number theory

There are a number of examples of applications of number-theoretic methods to statistics. See, for example, Patil (1963), De la Cal (1989), Hirth (1997), and Athreya and Fidkowski (2002). However, the majority are, perhaps, somewhat removed from either applications or statistical inference. Let us address three other examples, the connections of which to practicality or to statistics are arguably stronger. A fourth example will emerge in the Appendix.

Example 1: Digitizing an image   Consider laying an infinite straight line across a regular $n \times n$ grid, or lattice, of points or vertices, in the plane. It is unimportant whether grid points are arranged in squares or triangles or hexagons, but the square case is more familiar to most of us, so let us settle on that one. Colour each grid vertex above the line white, and each vertex below the line black. Now remove the line; allow the value of n to increase, keeping fixed the distances between adjacent vertices; and attempt to recover the position of the line.

If the line has a rational slope, θ say, relative to the grid axes, and if it does not pass through one of the vertices of the grid, then neither the slope nor the intercept of the line can be consistently recovered, even if $n = \infty$. On the other hand, if the slope is irrational then we can recover both the slope and the intercept, with an error that decreases to zero as n increases. The rate at which we can recover the line depends on the 'type' of irrational number that θ is; see the last paragraph of this example for discussion.

On the other hand, if we place the line across an $n \times n$ square in which there are $O(n^2)$ *randomly* distributed 'vertices', for example the points of a homogeneous Poisson process within an $n \times n$ region, then we can always recover the line as n increases, regardless of its slope.

Thus, the cases of information on a random grid of points, and information on a regular grid with the same number of points, are quite different. The intuition here is that, in the rational-gradient case with vertices on a grid, the vertices are always at least a certain fixed distance from the line, unless the line happens to actually pass through a vertex. The vertices get arbitrarily close, sufficiently far out along the line, if θ is irrational, but this happens even for rational θ if the vertices are arranged in a random way. Hence, in both the latter cases we can readily approximate the line, but in the rational-gradient case with no vertices on a grid, we cannot.

These issues underpin some of the problems facing digital technologies as they attempt to compete with their analogue predecessors. For a given amount of 'information' about boundaries in an image, obtained through colours on randomly distributed film grains, in the analogue case, and pixel values, for a digital image, the analogue method always has a significant edge. This advantage can be important, for example in subminiature imaging devices.

In practice, of course, the boundary represented by the line is not straight. It is a curve, the spatial variation of which cannot realistically be modelled parametrically. However, 'estimators' that are based, like conventional non-parametric statistical smoothers, on local properties, can be used to approximate the boundary. Even in deterministic cases, familiar arguments from non-parametric statistics offer a very good foundation on which to develop properties of such boundary approximations. See, for example, Hall and Raimondo (1998).

To conclude this example, let us address the issue of different 'types' of irrational numbers, drawing connections between the quality of approximation and results in number theory. Go back to the case of a straight line placed across the grid, where we colour vertices above the line black, and those below, white. There are many ways of constructing good approximations to the line within the $n \times n$ square, using only the pattern of black and white colours. Let D_n denote the Hausdorff distance of the true line from any one of these good approximations. If the slope of the line is an 'algebraic irrational,' meaning that it can be expressed as a real root of an equation involving rational coefficients and exponents, then the convergence rate of D_n to zero can be no better than $O(n^{-1-\epsilon})$ as n increases, for any $\epsilon > 0$. This property is equivalent to Roth's (1955) Theorem, for which Klaus Roth received a Fields Medal, and it has application to statistical studies of the performance of imaging methods. More generally, the type of an irrational number θ is characterized by the rate of convergence, to θ, of that number's continued-fraction expansion. We shall discuss this further in the last three paragraphs of our treatment of Example 2.

Example 2: Estimating multiperiodic functions Some stars in the heavens emit radiation that may plausibly be modelled as a superposition of periodic functions. In particular, if Y_i denotes the intensity of the noisy signal observed at time X_i, then

$$Y_i = g(X_i) + \epsilon_i \,,$$

where the errors ϵ_i have zero means, g represents the true signal, and

$$g(x) = \mu + \sum_{j=1}^{p} g_j(x) \,, \quad -\infty < x < \infty \,, \tag{6.1}$$

with μ denoting a constant, g_j a smooth, non-vanishing, real-valued periodic function with minimal period θ_j, and $0 < \theta_1 < \cdots < \theta_p < \infty$. If $p = 1$ then we drop μ from the right-hand side of eqn (6.1); if $p \geq 2$ then we centre g_j by asking that $\int_0^{\theta_j} g_j = 0$.

It is believed that, in at least some instances, the multiperiodic character of some stellar radiation may arise because the radiation is emitted by one or more single-period sources. If there is more than one such source then they may be so close together that they cannot be resolved. Some 30 years ago it was discovered that so-called cataclysmic variable stars are actually pairs of nearby binary stars, the radiation of each varying in a periodic fashion with periods measured in hours. In other instances, multiperiodicity probably arises for quite different reasons.

There is a variety of ways of estimating $\theta_1, \ldots, \theta_p$, and the associated functions $g_1, \ldots, g_p$, but none of them will give consistent results unless each ratio θ_i/θ_j, for $1 \leq i < j \leq p$, is an irrational number. Indeed, this condition is necessary and sufficient for identifiability, in particular for the value of p to be as small as possible subject to g being uniquely representable, in terms of periodic functions, by eqn (6.1). However, the nemesis of rational versus irrational numbers, and its implications for estimator performance, arise even in the uniperiodic case, as we shall shortly relate.

One class of estimator of period is based on the periodogram, of which the square is

$$A(\omega)^2 \equiv A_{\cos}(\omega)^2 + A_{\sin}(\omega)^2,$$

where ω denotes a potential frequency and

$$A_{\mathrm{cs}}(\omega) = \frac{1}{n} \sum_{j=1}^{n} Y_j \, \mathrm{cs}(\omega X_j) \quad \text{or} \quad A_{\mathrm{cs}}(\omega) = \frac{1}{n} \sum_{j=1}^{n} (Y_j - \bar{Y}) \, \mathrm{cs}(\omega X_j),$$

with cs denoting either cos or sin. In general, in the uniperiodic case, where $p = 1$ in eqn (6.1) and the period is simply θ, a graph of $A(\omega)$ has spikes in the vicinities of points $2k\pi/\theta$, for integers k. The heights of these spikes decrease quite quickly with increasing $|k|$.

Of course, these properties are analogous to their better-known parametric counterparts, where g is expressed as a finite superposition of trigonometric series. See, for example, Quinn and Thompson (1991), Quinn (1999), and Quinn and Hannan (2001). The difference in the present setting is that we make only non-parametric assumptions about g; it need satisfy only smoothness conditions, rather than the more explicit shape constraint required for a low-dimensional Fourier representation.

However, the properties described two paragraphs above rely on an approximately uniform distribution of the observation times, X_i. If the distribution of the X_is is itself periodic, then the period of the density, $f = f_n$, of the X_is is confounded, in the context of the periodogram estimator, with the periods of the functions g_j.

Unfortunately, in the setting of stars data, f is almost certain to be periodic. Stars cannot be observed during daylight hours, and so the observation times can hardly be uniformly distributed; they are more likely to have a periodic distribution, with period equal to one day.

Thus the period of the distribution of observation times will usually be known. There are still complications, however. First, even in the uniperiodic setting, if the period of g is to be estimable solely from data on the locations of peaks in the periodogram then it should be an irrational number on a scale where units are days. Second, the spikes in the periodogram now occur with greater multiplicity than before. When $p = 1$ they arise at points of the form $2k\pi + (2\ell\pi/\theta)$, where k, ℓ are arbitrary integers and θ denotes the period of g. Some information about θ is available in the heights of periodogram peaks, although we shall not address that topic here.

The condition that the period of g, assumed to be a uniperiodic function, be irrational, is relatively strong. Using other estimation methods, for example those based on smoothing and least-squares, rather then the periodogram, the assumption is unnecessary. However, the periodogram-based approach is particularly popular in astronomy, dating from Deeming (1975) in that context. Recent work employing the method includes that of Alcock *et al.* (2003), Kilkenny *et al.* (2003) and Oh *et al.* (2004).

For all estimator types, if the observation times X_i are not randomly distributed then they should be 'jittered' in some way, to ensure that, modulo θ, they are reasonably well distributed across the support of g.

In many problems of this type, including estimation of the θ_js in the multiperiodic case, rates of convergence depend on the 'type' of irrational number θ, where the latter could denote a period ratio or, in the uniperiodic case, the period itself. The type is governed by the continued-fraction expansion of θ,

$$\theta = \nu_0 + \cfrac{1}{\nu_1 + \cfrac{1}{\nu_2 + \cfrac{1}{\nu_3 + \cfrac{1}{\nu_4 + \cdots}}}},$$

where $\nu_0, \nu_1, \ldots$, are positive integers. The mth 'convergent' of θ is defined to be the ratio

$$\theta_m = \frac{p_m}{q_m} = \nu_0 + \cfrac{1}{\nu_1 + \cfrac{1}{\nu_2 + \cdots \frac{1}{\nu_m}}},$$

where $p_m, q_m > 0$ are relatively prime. One might fairly say that 'θ_m is the most accurate rational approximation to θ whose denominator does not exceed q_m.'

Roughly speaking, periods for which θ_m converges to θ relatively slowly give poor rates of convergence. The rate can be arbitrarily slow, at least along a subsequence, depending on the value of θ. So, while knowing that θ is irrational does guarantee consistency, it does not tell us much about rates.

Nevertheless, we can say something helpful in an 'average' sense. To adopt a teleological stance for a moment, not altogether out of line with the problem of estimating the periods of heavenly radiation, let us suppose that when God constructed the universe He or She chose the periods of multiperiodic periodic-variable stars in a stochastic way, by selecting them according to a continuous distribution on the real line. Suppose too that astronomers happen upon periodic-variable stars in a similarly random manner. Then, writing $m = m(r)$

for the largest integer such that $q_m \leq r$, it may be shown that with probability 1, $q_{m(r)}$ equals $O(r)$, multiplied by a logarithmic factor. From this and related results it can be proved that, for the assumed randomly constructed universe, period estimators converge at rate $n^{-3/2}$ as sample size, n, increases. The necessary number-theoretic properties are given by, for example, Khintchine (1964).

Example 3: Estimation in errors-in-variables problems   Many non-parametric function estimators are based on orthogonal series. The best-known recent examples are wavelet estimators, but especially in engineering settings, methods founded on trigonometric series estimators are common; see, for example, Hart (1997, §3.3). Their accuracy is largely determined by the rate of convergence, to zero, of Fourier coefficients.

When Fourier-series methods are used in deconvolution problems, in particular to estimate response curves when there are experimental errors in the explanatory variables (see, for example, Efromovich 1994, 1999), the inverses of Fourier coefficients appear in formulae for variances. For example, if a Fourier-series expansion to p terms is used, in the context of symmetric experimental errors and a sample of size n, then a variance formula may contain a series such as

$$s(p) = n^{-1} \sum_{j=1}^{p} a_j^{-2}, \tag{6.2}$$

where $a_j = \int_{\mathcal{I}} \cos(jx) \, f(x) \, dx$ denotes the jth cosine term in a Fourier expansion of the density, f, of explanatory-variable error, and $\mathcal{I} = [-\pi, \pi]$.

In the context of Berkson's (1950) errors-in-variables model it seems quite realistic to suppose that f is symmetric and compactly supported. Without loss of generality, the scale is chosen so that f is supported within $\mathcal{I}$. Usually the support is not known exactly, although plausible outer bounds for the support can be deduced either empirically or from physical knowledge of the process that gave rise to the errors. A simple model for an f with support inside $\mathcal{I}$ might be

$$f(x) = C \left(\theta\pi - |x| \right)^\rho, \quad |x| \leq \theta\pi,$$

where $0 < \theta < 1$, $\rho \geq 0$ and $C = C(\theta, \rho) > 0$ is chosen to ensure that f integrates to 1. Then f is 'rougher' for smaller ρ.

For simplicity, let us consider just the cases $\rho = 0$ or 1. Here, simple calculus shows that

$$|a_j| \asymp \begin{cases} j^{-1} \, |j\theta - \langle j\theta \rangle|, & \text{if } \rho = 0, \\ j^{-2} \, |\tfrac{1}{2} \, j\theta - \langle \tfrac{1}{2} \, j\theta \rangle|, & \text{if } \rho = 1, \end{cases}$$

where $\langle x \rangle$ denotes the integer nearest to x and the notation $a_j \asymp b_j$ means that a_j / b_j is bounded away from zero and infinity as $j \to \infty$. Clearly, a_j vanishes infinitely often if θ is rational. Then, $s(p)$, defined in eqn (6.2), is not even finite; we must confine attention to the case of irrational θ. So, the 'irrational nemesis' is with us again.

Provided θ is an algebraic irrational (see Section 6.3.1) it may be proved that a_j^{-2} is dominated by constant multiples of j^4 or j^6, in the respective cases $\rho = 0$ or 1. This leads to bounds for $s(p)$. Delving at greater depth into the mysteries of number theory, we find that if the continued-fraction expansion of θ (see Section 6.3.2) uses only a finite number of distinct integers, and in particular if θ equals the square root of a rational number, then $s(p) \leq \mathrm{const.}\, p^{2\rho+2}$, for all $p \geq 1$. See Siegel (1942).

Thus, once again, results in number theory play a central role in determining properties of estimators.

6.3.3 Algebraic geometry

Suppose a population consists of p different subpopulations, and that the sampled data from each subpopulation are vectors of length k. For example, the 'full' or 'mixed' population might represent people who visit a doctor, concerned that they could be suffering a certain illness; and the two subpopulations might represent people who suffer the illness, and those who do not have the illness, respectively. In this setting, $p = 2$. For each presenting patient the doctor records a vector of k continuous measurements. However, the doctor can determine for certain whether a patient is ill or not only by invasive surgery, which is both expensive and potentially dangerous.

Even though we can never be certain whether a given patient actually has the illness, it is of practical interest to know the distributions of symptom vectors for the subpopulation of ill patients, and for the subpopulation of well patients; and it is also desirable to know the percentage of presenting patients in the population who are actually ill, even though these patients cannot be identified individually. How much structure must we assume in order to be able to accurately estimate these unknowns from data?

It is conventional to solve such problems in a parametric setting, by fitting a model to the k-variate distributions of each of the p subpopulations. See Titterington *et al.* (1985), and MacLachlan and Peel (2002), for discussion of the parametric setting. However, even there the number of parameters can be large. For example, if the models are Gaussian then in each subpopulation there are k parameters for the mean and $\frac{1}{2}k(k+1)$ parameters for variance and covariance, so there are $\frac{1}{2}pk(k+3)$ parameters in all. Even if p is as small as 2, i.e. in the case of the doctor–patient example above, there are still $k(k+3)$ unknowns to be estimated. It is convenient, and often necessary, to make simplifying assumptions to reduce this level of complexity.

For example, if we suppose that the marginal distributions of each subpopulation are mutually independent then the number of parameters needed to describe the variance, in a Gaussian model, is only k, and the total number of parameters reduces, when $p = 2$, to only $4k$. However, when the marginals are independent then, provided $p = 2$, the model is fully identified in a non-parametric sense as long as $k \geq 3$ (Hall and Zhou 2003). Moreover, all the distribution functions, and the mixture proportions, can be estimated root-n consistently.

The methods of proof used by Hall and Zhou (2003) shed no light on the case of general $p \geq 3$, however. It is tempting to conjecture that for each p there is a least value of k, k_p, say, such that, whenever $k \geq k_p$, the problem is non-parametrically identified; in particular, it can be consistently estimated under only smoothness conditions on the distributions. That is, in a curious counterpoint to the more conventional 'curse of dimensionality', the problem becomes simpler as the number of dimensions increases.

One approach to solving this problem is to adopt a constructive viewpoint, and consider a method that might potentially be employed for inference. Write the k-variate, p-population mixture model, with independent marginals, as

$$\pi_1 \prod_{i=1}^{k} F_{1i} + \cdots + \pi_p \prod_{i=1}^{k} F_{pi} = G. \qquad (6.3)$$

Here, the π_js are mixing probabilities and the F_{ji}s are marginal distribution functions. We observe data having the k-variate distribution G on the right-hand side, and so we can estimate G. Likewise, we can estimate any ℓ-variate subdistribution, $G_{i_1 \ldots i_\ell}$, obtained by integrating out the other $k - \ell$ marginals of G:

$$\pi_1 \prod_{m=1}^{\ell} F_{1 i_m} + \cdots + \pi_p \prod_{m=1}^{\ell} F_{p i_m} = G_{i_1 \ldots i_\ell}, \quad 1 \leq \ell \leq k, 1 \leq i_1 < \cdots < i_\ell \leq k.$$
$$(6.4)$$

Therefore, if k can be chosen so large that all the marginals F_{ji}, and all the mixture proportions π_j, can be expressed as smooth functionals of the ℓ-variate distributions $G_{i_1 \ldots i_\ell}$, for $1 \leq \ell \leq k$, then we shall have shown that for each p, if k is sufficiently large, all marginals F_{ji}, and all joint distributions, can be identified, and estimated, from data on G.

In the sense that the order of the p populations can always be permuted, there are always at least $p!$ solutions to this problem. Supposing that no two of the π_is are identical, we can remove this redundancy by insisting that $\pi_1 < \cdots < \pi_p$; we shall make this assumption below.

A lower bound to k_p is not difficult to find, although the argument involves a little more number theory. It can be shown that

$$k_p \geq K_p \equiv \inf \left\{ k : 2^k - 1 \geq k\,p + 1 \right\}. \qquad (6.5)$$

Particular values of K_p are $3, 4, 5, 5, 5, 6, \ldots, 6$ for $p = 2, 3, \ldots, 10$, respectively; and $K_p \sim (\log p)/(\log 2)$ as p increases. See the Appendix for details.

Obtaining an upper bound to k_p, and thus proving that the independent-marginal mixture model in eqn (6.3) is always identifiable, from a non-parametric viewpoint, if k is sufficiently large, is more difficult. It is here that the theory of birational maps, applied to algebraic varieties, makes an entrance (Elmore *et al.* 2003). This approach enables it to be shown that k_p is finite, and in fact is no larger than a quantity that equals $\{1 + o(1)\}\, 6p \log p$ as p increases.

More generally, applications of algebraic geometry to statistics include those to problems involving contingency tables, graphical models, and multivariate analysis. The Institute of Mathematics and its Applications, based at the University of Minnesota, will host a Special Thematic Year, from September 2006 to June 2007, on these and other applications of algebraic geometry.

6.3.4 Operator theory

Perhaps the best-known area of contemporary non-parametric statistics where operator theory has found application is the study of functional principal component analysis. There, the spectral expansion of a symmetric linear operator is the functional-data analogue of a decomposition into principal components.

Indeed, if X is a random function on an interval $\mathcal{I}$, and if

$$K(u, v) = \operatorname{cov}\{X(u), X(v)\}$$

denotes the covariance function, then we may expand K in the form

$$K(u, v) = \sum_{j=1}^{\infty} \theta_j \, \psi_j(u) \, \psi_j(v), \tag{6.6}$$

where $\theta_1 \geq \theta_2 \geq \cdots \geq 0$ is an enumeration of the eigenvalues of the symmetric, linear operator with kernel K; that is, of the transformation, κ, that takes a function ψ to the function $\kappa\psi$, defined by

$$(\kappa\psi)(u) = \int K(u, v)\, \psi(v)\, dv.$$

The corresponding orthonormal eigenvectors are $\psi_1, \psi_2, \ldots$, and in particular ψ_j is a solution of the equation $\kappa\psi_j = \theta_j\, \psi_j$. The functions ψ_j are the principal components of the distribution of X. See Rice and Silverman (1991), Ramsay and Silverman (1997, Chapter 6), and Ramsay and Silverman (2002) for accounts of functional principal components, including the empirical setting discussed immediately below, and Indritz (1963, Chapter 4) for an accessible development of theory for linear operators, in particular of properties of expansions such as eqn (6.6).

Of course, in practice we do not know either the functions ψ_j or the numbers θ_j; we need to estimate them from data. The empirical analogue of K, based on a dataset $X_1, \ldots, X_n$, is, of course, $\widehat{K}$ defined by

$$\widehat{K}(u, v) = \frac{1}{n} \sum_{i=1}^{n} \{X_i(u) - \bar{X}(u)\}\{X_i(v) - \bar{X}(v)\}, \tag{6.7}$$

where $\bar{X} = n^{-1}\sum_i X_i$. Analogously to eqn (6.6) we may write

$$\widehat{K}(u, v) = \sum_{j=1}^{\infty} \hat{\theta}_j \, \widehat{\psi}_j(u) \, \widehat{\psi}_j(v),$$

where $\hat{\theta}_1 \geq \hat{\theta}_2 \geq \cdots \geq 0$ are eigenvalues of the operator with kernel $\widehat{K}$, and the respective eigenvectors are $\widehat{\psi}_1, \widehat{\psi}_2, \ldots$.

The explicitness of eqn (6.7) ensures that we may easily discover properties of $\widehat{K}$, viewed as an approximation to K. However, no such obvious tools are available when we attempt to elucidate the far more important problem of how accurate $\hat{\theta}_j$ and $\widehat{\psi}_j$ are as approximations to θ_j and ψ_j, respectively. For this, we need to know how the closeness of $\widehat{K}$ to K translates into properties of the differences $\hat{\theta}_j - \theta_j$ and $\widehat{\psi}_j - \psi_j$.

Results of this type may be developed, but they require tools that are not widely known to the statistics community. In particular, while 'Taylor expansions' of $\hat{\theta}_j - \theta_j$ and $\widehat{\psi}_j - \psi_j$ may be obtained, they require relatively elaborate manipulation of properties of linear operators. Thus, operator theory lies at the heart of the elucidation of properties of functional principal component analysis.

However, it is not just in the setting of functional principal component analysis that operator theory has important contributions to make to statistics. It makes critical appearances in connection with a range of modern deconvolution problems, in as well as in spline-smoothing and related problems.

For example, in the context of deconvolution, consider the task of estimating a function α by inverting a functional transformation, such as

$$\beta(u) = \int \gamma(u, v)\, \alpha(v)\, dv\,, \tag{6.8}$$

where β and γ may be estimated from data. Many practical problems may be phrased in this setting. They range from instrumental variables problems in econometrics (e.g. Blundell and Powell 2003) to deconvolution problems in biology (e.g. Troynikov 1999). See Donoho (1995) for a discussion of wavelet methods in settings such as these.

We may rewrite eqn (6.8), in many cases equivalently, as

$$b(u) = \int g(u, v)\, \alpha(v)\, dv\,, \tag{6.9}$$

where $b(u) = \int \gamma(w, u)\, \beta(w)\, dw$ and $g(u, v) = \int \gamma(w, u)\, \gamma(w, v)\, dw$. If β and γ are estimable then so too are b and g, and so we seek to invert eqn (6.9) rather than eqn (6.8). The advantage of eqn (6.9) is that the function g is guaranteed to be symmetric; γ, in eqn (6.8), typically is not.

Statistical methods based on eqn (6.9), where data are used to estimate g, are versions of Tikhonov, or quadratic, regularization; see, for example, Tikhonov (1963). By passing from eqn (6.8) to eqn (6.9) we have set up the deconvolution problem as one involving operator-theoretic methods. In order to study properties of those methods one should press into service the sorts of techniques, discussed earlier in this subsection, for exploring the properties of empirical approximations to deterministic operators.

6.4   Conclusion

We have discussed the evolution of non-parametric statistics, noting that its extraordinary growth over the last quartercentury has been intimately linked to a corresponding expansion in computing power. Particularly through developments in non-parametrics, statistics makes demands on computing, and on numerical skills, that are much greater ever before. But it is no contradiction to say that, at the same time, the mathematical methods needed to gain insight into the properties and performance of non-parametric statistics are advanced and challenging, and that the demands that statistics makes of mathematics remain especially high.

Appendix: Derivation of eqn (6.5)

Let us view eqn (6.4) as representing kp unknown functions F_{ji}, expressed in terms of the estimable functions $G_{i_1...i_\ell}$; and endeavour to solve the equations for the unknowns. If $\pi_1,\ldots,\pi_p$ are given, we require at least kp such equations in all. The number is $kp + 1$, rather than $kp + p - 1$, since the π_js are scalars rather than functions. Now, the total number of equations of the type in eqn (6.4) is $2^k - 1$. We may uniquely write $2^k - 1 = kp_k + q_k$, where p_k and q_k are integers and $1 \leq q_k \leq k - 1$. To appreciate why q_k cannot vanish, see Gleason *et al.* (1980, problem 6 of 1972)—number theory makes an appearance, again.

In particular, $2^k - 1 \geq kp_k + 1$. Therefore, if k_r denotes the infimum of values of k such that $p_k \geq r$, then K_p is the least value of k for which we can hope to estimate the univariate distributions F_{ji}, and mixing proportions π_j, for a given number p of populations, by solving the simultaneous equations that arise from eqn (6.4). Particular values of (p_k, q_k) are $(2,1)$, $(3,3)$, $(6,1)$, $(10,3)$ for $k = 3,\ldots,6$, respectively. Hence, $K_p = 3, 4, 5, 5, 5, 6, \ldots, 6$ for $p = 2, 3, \ldots, 10$, respectively. The result in eqn (6.5) also follows.

7
Some topics in social statistics

David Firth

7.1 Introduction

The term 'social statistics' has various meanings in current usage, including:

(a) statistics reported by national and other official statistics agencies, based on censuses and surveys;

(b) the collation and social-scientific interpretation of data from official and other sources, a good recent example being Halsey and Webb (2000);

(c) the development of methods for drawing samples from a population and for making inferences from such samples about well-defined population quantities;

(d) statistical methods of particular value in substantive research fields such as criminology, demography, economics, education, geography, politics, psychology, public health, social policy and sociology.

This chapter will be concerned mainly with statistical methods that fall within meaning (d), a specialization that largely reflects the author's own interests and limitations; meanings (a) and (c) will also be touched upon, briefly. The broad aim will be to summarize some recent themes and research topics and to identify particular areas in which further work seems likely to be fruitful. It should be obvious that (d) does not describe a specific subset of statistical methods: research problems in social-scientific disciplines are very varied in character, and benefit from the application of general statistical principles for their solution. Likewise, methods developed for specific social-science contexts often have much wider applications (e.g. Clogg 1992).

On the day when this introduction was written, the BBC's UK national news programmes gave prominent coverage to some interesting, just-published research that exemplifies some of the issues that commonly arise in connection with social-scientific statistical work and its interpretation. Head *et al.* (2004) report on a large, prospective cohort study of non-industrial London-based civil servants, in which the relationship between alcohol dependence and characteristics of the work environment is explored. Of particular interest was to establish the role played by two distinct psycho-social notions of stress at work, these being 'job strain' (characterized by high employer demands and low employee control) and effort–reward imbalance. The analysis involved measurement of the two types of stress and of alcohol dependence, and the careful use of a sequence of multiple logistic regressions to establish and measure association between stress and alcohol dependence and to control for other factors such as age and job grade. The main substantive finding was that, for men, effort–reward imbalance at work is positively related to alcohol dependence. This finding is, from the public-health viewpoint, suggestive rather than conclusive: a *causal* link between effort–reward imbalance and alcoholism is not established; and the generalizability of the results to groups other than London-based, white-collar, civil-servant study participants is unclear. At a more technical level, though ultimately no less important, are questions about the robustness of the measuring instruments involved—summaries of questionnaire responses, and formal logistic regressions—to possible failure of the detailed modelling assumptions upon which they are based. Finally, an interesting aspect of the BBC's reporting of this research is that it neglected entirely the main result just mentioned, which was found only for men, and instead focused exclusively on the incidental finding that among *women* working in Whitehall the prevalence of alcohol dependence is markedly greater for higher than for lower job grades; the same was not found for men. Aside from considerations of newsworthiness, which are hard to judge, it seems clear that a large part of the appeal of this incidental finding is the simplicity of the statistics involved: alcohol dependence affected only 4% of women employed in the lowest grade, rising to 14% in the highest, and this gradient persists even after adjusting for age. In contrast, the 'main' finding of the study required a combination of rather complex statistical analyses whose results demand a relatively large table for their proper presentation. For the effective communication of statistical results to the broader public, and perhaps also to policymakers, transparency of method is highly desirable; this creates a difficult tension with the tendency for statistical methods to become ever more elaborate in the pursuit of realism and scientific validity.

The coverage of the present article is far from an exhaustive survey of the field, and many topics will be mentioned only rather superficially with the aim of providing an entry to the relevant literature. The collection of vignettes published recently in the *Journal of the American Statistical Association* (e.g., Fienberg 2000, Sobel 2000, Beck 2000, Raftery 2000, Browne 2000) provides useful commentary and references on many aspects that will be neglected here.

7.2   Aspects of statistical models in social science

In social science as in other research fields, statistical models are central to the interpretation of data and can play a key role also in study design. For general discussion of types of statistical model and the various roles played by models in statistical analysis, see, for example, Cox and Snell (1981), Cox (1990) or Cox (1995). In this section some aspects will be discussed that are of particular relevance to current social research.

7.2.1 Multilevel models

The use of 'multilevel'—or 'hierarchical' or 'mixed effects'—models of dependence, in which residual variation has more than one component in order to
reflect clustering or other structure in the study design, has become commonplace in recent years with the advent of suitable software packages. For a recent
introductory treatment see Snijders and Bosker (1999). A fairly general formulation is the *generalized linear mixed model* in which the mean of response vector
y is related through a link function to linear predictor

$$\eta = X\beta + Zu,$$

in which β is a vector of regression parameters of interest, and u is a vector of
random effects having some assumed distribution, for example $u \sim N(0, \Sigma_u)$.
A commonly encountered special case is the hierarchical model with random
intercepts, for example if cases are students (j) within schools (i) a two-level
model is

$$\eta_{ij} = x_{ij}^{\mathrm{T}}\beta_1 + x_i^{\mathrm{T}}\beta_2 + u_i,$$

in which β_1 captures the effect of student-level covariates in vector x_{ij}, β_2 the
effect of school-level covariates in vector x_i, and u_i represents a residual effect
for school i. More complex versions might allow that schools vary also in their
dependence on x_{ij},

$$\eta_{ij} = x_{ij}^{\mathrm{T}}(\beta_1 + v_i) + x_i^{\mathrm{T}}\beta_2 + u_i,$$

in which case it is usually necessary on grounds of invariance to allow u_i and v_i
to be correlated.

Models of this kind have been found very useful, and can rightly be regarded
as essential, in many situations where the data are collected in identifiable clusters (schools in the example above). The prevailing culture in much of empirical
social science is such that a regression analysis that does *not* use multilevel models might be seen as flawed, or at best be viewed with suspicion along the lines
of 'do they not know about multilevel?' This has had some unfortunate consequences, ranging from the use of multilevel models in inappropriate contexts to
the often no-less-serious reliance on 'black box' approaches to analysis.

An example of an inappropriate context would be a comparative study of,
say, political attitudes in 15 or so European nations. This typically would involve extensive survey data from each nation. To regard the nations as 'clusters',
drawn randomly from a larger population of such nations, and then proceed to
a multilevel model in which nation effects are random, seems misguided in two
respects. First, nations are distinct entities, often with very different political
structures, and it is unclear what a notional 'superpopulation' of nations from
which the 15 or so are drawn would represent. Secondly, this is a context in which
the information per nation is large but the number of nations—and hence the
information available to estimate the variance of any random effect u_i, even accepting that there is a meaningful notion of 'superpopulation'—is small. It makes

more sense in this context to estimate *fixed* nation effects. The random-effects formulation is most valuable at the other extreme, where clusters are a large sample from a clear superpopulation and the information per cluster is modest.

The 'black box' aspect arises from the need for specialized computer programs, and is especially severe in the case of non-linear models, for example generalized linear mixed models with non-identity link function. Whereas *linear* mixed models with normally distributed random components yield an analytically explicit likelihood, in non-linear cases the likelihood

$$L(\beta, \Sigma) = \mathrm{E}_u \{ p(y \mid \beta, u) \}$$

in general requires computation of an integral in as many dimensions as the dimensionality of u. In the simplest cases—for example a 2-level model with random intercepts—the problem reduces to a set of one-dimensional integrals that can be approximated accurately by, for example, adaptive quadrature. More generally though, the integral is irreducible and resort must be made either to simulation methods (e.g. Gilks *et al.* 1996, Evans and Swartz 2000) or to 'penalized quasi-likelihood' based on a large system of estimating equations motivated by Laplace-approximation arguments (Breslow and Clayton 1993); in either case the quality of the resultant approximation is unclear without a great deal of careful checking.

There is a need for more transparent approaches that perhaps sacrifice some statistical efficiency. One such that merits further study is explored in Cox and Solomon (2003, §§4.2–4.4). For example, in the case of a 2-level logistic regression with random intercept and random slope,

$$\mathrm{logit}\, \pi_{ij} = \eta_{ij} = \beta_0 + u_i + (\beta_1 + v_i)x_{ij},$$

the suggested procedure is to fit separate logistic regressions

$$\mathrm{logit}\, \pi_{ij} = \beta_{0i} + \beta_{1i}x_{ij}$$

to obtain estimates $(\widehat{\beta}_{0i}, \widehat{\beta}_{1i})$ for each cluster i, then examine the empirical covariance matrix of these estimates to assess whether variation between clusters is predominantly internal (i.e. binomial variation within a cluster) or attributable to real cluster differences; in the latter case (β_0, β_1) is estimated by the unweighted mean of the separate estimates, while in the former case a weighted mean or the fixed-effects-only maximum likelihood estimate is used. This procedure generalizes straightforwardly to more complex models, e.g. multiple logistic regression, and, as is argued by Cox and Solomon (2003), typically involves only a small loss of efficiency relative to 'optimal' methods that can involve substantially more computation. In practice, care is needed in the method of calculation of the separate $(\widehat{\beta}_{0i}, \widehat{\beta}_{1i})$, since the information in each cluster i may be modest; in logistic regression, for example, separate maximum likelihood for each i would at least occasionally yield infinite-valued estimates, and some form of shrinkage is needed. But a relatively intuitive method such as this provides a useful check, at least, on the results obtained from more efficient but less-transparent approaches. Other non-likelihood approaches may have similar merits, for example the use of a pseudo-likelihood as described in Cox and Reid (2004) based on one- or two-dimensional marginal distributions of the response variable, or the somewhat related estimating-equation approach of Zeger *et al.* (1988).

7.2.2 Model as prediction engine: small-area estimation

In some applications a statistical model is assumed only in order to generate predictions, either of future values of a variable or of values hidden on account of cost of observation or other reasons. One such context is the production of estimates for small areas, based on regional or national survey data in which perhaps not all areas are represented; the importance attached to such estimates by policymakers, as a guide in the distribution of public funds for example, appears to be growing. The state-of-the-art is authoritatively surveyed in Rao (2003). In the United Kingdom, for example, the Office for National Statistics (ONS) uses regression models, either linear or generalized-linear, to estimate, from large-scale survey data involving a variable y of interest, the mean value of y for every small area (where 'small area' might mean a local-government ward, of which there are around 8000 in England). Variables appearing in the linear predictor are external to the survey and are available for every local area; these 'predictor' variables would usually include census summaries and administrative data. In typical ONS applications the predictors used are a small subset, perhaps five to ten in number, selected from a larger set of 'candidate' predictors compiled on grounds of supposed relevance; current ONS procedure is to use stepwise subset selection based on significance tests. The use of statistical models in this way is quite different from their use in the more usual social-science context where primary interest is in establishing regularities and explaining them. Here the model is merely a device for pooling information from 'similar' small areas, with similarity being defined in effect by the selected predictors. Various methodological questions arise, including:

(i) Should pooling of information take place also across areas that are geographically close? In view of the rather arbitrary nature of area boundaries this seems desirable. Some work in this direction appears in Ghosh *et al.* (1998), where a generalized linear model is combined with a spatially correlated error process; the Bayesian analysis suggested requires substantial computation and the careful specification of prior distributions.

(ii) How much can be gained by using parameter-shrinkage methods—such as ridge regression or variants—in place of subset selection? Better still, perhaps, would be to consider the use of hybrid shrinkage/selection methods such as the lasso (Tibshirani 1996). It seems clear on general grounds that such methods should yield more stable predictions than approaches based on selection alone.

(iii) How is stability over time best achieved? This is important in terms of the reporting and use of small-area estimates, where changes from one year to the next should ideally be attributable to real change 'on the ground' rather than to changes in the statistical model used for prediction. The avoidance of subset selection should help, but in addition some explicit smoothing of the model's evolution through time seems desirable.

(iv) How should interaction structure among the predictors be explored and, where appropriate, exploited? In some applications it might be better to use a recursive-partitioning—or 'regression tree'—approach, the piecewise-constant nature of which makes it relatively safe for predictive extrapolation, rather than to construct the more usual kind of linear predictor.

(v) How should the uncertainty of model-derived small-area estimates be assessed, and reported? The present ONS approach, which in essence assumes the selected predictive model to be the data-generating process, seems likely to be rather optimistic in its assessment of precision. A more realistic assessment might involve crossvalidatory estimation of predictive accuracy. Conceivably the outcome could be a set of very wide intervals, at least for some areas, making the estimates effectively useless; it is surely important to report such imprecision where it is known to exist.

(vi) How is consistency across two or more levels of aggregation best achieved? For some purposes it is important that aggregates of model-based estimates agree with 'raw' survey estimates for well-defined regions that contain many small areas. Possibilities include *ad hoc* scaling of the model-based estimates, and the forced inclusion of region effects in the model.

There is substantial scope on at least some of these points for interplay with recent epidemiological work on geographical mapping of disease incidence (e.g. Green and Richardson 2002).

7.2.3   Model as measuring instrument

Quite frequently in social-scientific work a statistical model is used in a rather formal way as a device that, in effect, defines a scale of measurement. Three examples that illustrate various aspects will be discussed briefly here. A key consideration in such settings is usually the extent to which measurements made using the model remain useful when the model fails in detail to match the population or process being measured.

Example: Comparative social fluidity   In intergenerational studies of class mobility, the notion of *fluidity* captures the extent to which the social class of a son, say, can be predicted from his father's class; if father's class and son's class are statistically independent, there is perfect fluidity. Empirically, evidence of non-fluidity corresponds to evidence of non-unit odds ratios in a square table that cross-classifies a sample of father–son pairs by their respective social classes. For the comparison of social fluidity across nations or through time, much use has been made in recent years of the *log-multiplicative* or *uniform difference* model (Erikson and Goldthorpe 1992, Xie 1992), in which the cell probability p_{ijt} for father's class i and son's class j in table t, corresponding to nation t or time-point t, is given by

$$\log p_{ijt} = \alpha_{it} + \beta_{jt} + \gamma_t \delta_{ij}. \tag{7.1}$$

In this model the three-way association is structured as a common *pattern* of log odds ratios $\{\delta_{ij}\}$ whose amplitude is modulated by non-negative table-specific parameters $\{\gamma_t\}$, with, say, γ_1 set to 1 for identifiability. The relative values of γ_t then become primary objects of interest: for example, if $\gamma_2 > \gamma_1$ then under this model all of the odds ratios in table 2 are further from unity than the corresponding odds ratios in table 1, indicating unambiguously that table 1 represents the more fluid society. In many applications of model (7.1) there is significant lack of fit to the data, meaning that the assumption of a common association pattern $\{\delta_{ij}\}$ does not hold. Of interest then is whether the estimated values of $\{\gamma_t\}$ continue to represent relative fluidity. The general theory for the large-sample regular behaviour of maximum likelihood estimates in misspecified models is given in Cox (1961): in particular, if $u(\theta)$ is the score function for model parameters θ, the maximum likelihood estimator of θ has as its limit in probability the value θ^* such that $E\{u(\theta^*)\} = 0$. After some reparameterization the expected score equations for δ_{ij} and γ_t in model (7.1) can be written as

$$\sum \frac{\pi_{i+t}\pi_{+jt}}{\pi_{++t}} \left\{ \exp(\delta_{ijt}) - \exp(\gamma_t \delta_{ij}) \right\} \gamma_t = 0 \quad \text{for all } i, j,$$

and

$$\sum \frac{\pi_{i+t}\pi_{+jt}}{\pi_{++t}} \left\{ \exp(\delta_{ijt}) - \exp(\gamma_t \delta_{ij}) \right\} \delta_{ij} = 0 \quad \text{for all } t.$$

Both sets of equations imply weighted averaging of the true log odds ratios δ_{ijt} to determine the 'approximating' log odds ratios $\gamma_t \delta_{ij}$. The weights depend on the marginal distributions via $\pi_{i+t}\pi_{+jt}/\pi_{++t}$. The weights in the first set of equations are also proportional to the multipliers γ_t: a table with γ_t close to zero gets relatively little weight in the combination of δ_{ijt} across tables to determine the 'common' log odds-ratio structure $\{\delta_{ij}\}$. Even under modest lack of fit, these properties can completely invalidate the standard interpretation of the estimated $\{\gamma_t\}$. The use of odds ratios is often motivated by non-dependence upon marginal distributions, which here is violated. The weights given to odds-ratio patterns from different tables are essentially arbitrary. When model (7.1) fails to fit, then, relative fluidity should be assessed by other methods. A promising approach would seem to be to re-express the *saturated* model—for which lack of fit is not an issue—using parameters with direct interpretation in terms of a notional stochastic process; there are parallels here with a similar problem in molecular biology, namely the measurement of evolutionary distance between species using DNA or protein sequence data (e.g. Li 1997, Chapters 3 and 4).

Example: Factor scores Factor analysis has a long history as a model for measurement: see, for example, Knott and Bartholomew (1999). The standard theory in which errors are multivariate normal is well understood, and in the case of the *one-factor* model

$$y_j = \lambda_j f + u_j,$$

with the $\{u_j\}$ independent given f, interpretation is straightforward: the y_j are independent measurements of the assumed common factor f.

In some application contexts the values of f, the *factor scores*, are of primary interest. A recent example appears in the work of Noble *et al.* (2000), where the one-factor model was used in order to combine several indicators of local-level deprivation into a single-number summary for each area; this was done separately for each of several 'domains' of deprivation, such as income, health, education, etc. The one-factor model was used in a rather formal way, essentially as a device for obtaining the factor score for each area. In addition to its rotation-free interpretation, a further advantage of the one-factor model in this regard is that the two most standard methods for obtaining factor scores give results that are identical up to a constant scale factor. Noble *et al.* (2000) report, for at least one of the domains studied, that the one-factor model exhibits lack of fit, which is significant at conventional levels, but that no interpretable second factor is apparent. The fitted one-factor model must then be viewed as the best-fitting single-factor approximation, in the Kullback–Leibler sense, to the true, underlying multivariate distribution. Questions that arise in this context are:

(i) Are the standard methods for obtaining factor scores in any sense optimal or even valid when the one-factor model fails to fit? If not, what is a better method?

(ii) Regardless of lack of fit, how should outliers be handled? The factor score for an area is simply a fixed weighted combination of that area's values of the manifest indicators $\{y_j\}$. The one-factor model implies a particular joint distribution for the $\{y_j\}$, and 'outlier' here means any area whose y-values clearly depart from that joint distribution. For example, it may be that y_1 for some area is not predicted as it should be by the remaining y_j, in which case it would seem sensible to downweight y_1 in the factor-score calculation for that area. Recent work on robust approaches to factor analysis (e.g. Pison *et al.* 2003) focuses more on model estimation than on factor scores, but seems likely still to be relevant here.

It might reasonably be argued, of course, that factor analysis should really be avoided in situations like this, and that it would be preferable either to use a set of indicator weights on which there is wide agreement, or to use an agreed set of indicators for which equal weighting is broadly acceptable (cf. Cox *et al.* 1992). In general, this is surely right. In the case of the deprivation indices reported by Noble *et al.* (2000), however, with eligibility for substantial regeneration funds at stake, no such consensus was available and the weights were determined by factor analysis as a method of last resort.

Example: Relative index of inequality The notion of a relative index of inequality has been used recently in the study of social inequalities in health, especially following the influential work of Kunst and Mackenbach (1995). The broad purpose of such an index is to compare rates of incidence, for example of death or disease, between those having lowest and highest socio-economic status. The resultant measure of social inequality is typically used for comparative purposes, or to study time trends. For a recent example see Davey Smith *et al.* (2002), where it is shown that in Britain socio-economic inequality in mortality rates continued to rise during the 1990s.

In the simplest setting, every individual has a notional socio-economic rank x, scaled to take values between 0 (lowest) and 1 (highest). The rate of incidence of the outcome of interest, such as a specific type of ill health or death, is $f(x)$ per unit of exposure, and the relative index of inequality is defined as $f(0)/f(1)$. In practice, $f(x)$ is unknown and must be estimated from available data, which may be data for the whole population of interest or for a sample. Moreover, x itself typically is not fully observed; rather, individuals are categorized into ordered social classes, and so measurement of x is interval-censored. The standard procedure in common use (e.g. Kunst and Mackenbach 1995, Hayes and Berry 2002) is then as follows:

1. For each of the k ordered social classes ($i = 1, \ldots, k$), let c_i be the fraction of the population in class i or lower (with $c_0 = 0$ and $c_k = 1$).

2. For each class i let $x_i = (c_i + c_{i-1})/2$ be the median social rank for that class, and r_i the rate of incidence of the outcome of interest.

3. Estimate $f(x)$ by linear regression, possibly weighted for different class sizes, of r_i on x_i. This yields a straight-line estimate $a + bx$, say.

4. Compute the estimated relative index of inequality as $a/(a + b)$.

This approach assumes that $f(x)$ is linear, and seems a reasonable method of estimation under that assumption. If, however, $f(x)$ in reality is non-linear, the standard procedure will, in general, induce a bias in the estimated relative index of inequality. The magnitude and direction of such bias depend on the nature of the non-linearity. If $\alpha + \beta x$ denotes the notional straight-line fit to all pairs $\{x, f(x)\}$ in the population of interest, to first order the bias is $\alpha/(\alpha + \beta) - f(0)/f(1)$.

There is no strong reason in principle to expect $f(x)$ to be linear, except perhaps as an approximation in situations where inequality is known to be both slight and monotonic in x. In practice, even when a graph of r_i versus x_i appears roughly straight, there is very often clear evidence of departure from linearity; typical applications involve the use of census, registry or large-scale survey data, in which even modest departures from linearity are readily detected by standard tests.

In general, i.e. when $f(x)$ is not assumed linear, estimation of the relative index of inequality based on a straight-line fit is inconsistent. Some care is needed here over the notion of consistency. The standard notion, based on behaviour as data size goes to infinity with all else fixed, is not useful in the present context, on account of the interval-censored observation of x. For, if the class widths $c_i - c_{i-1}$ are fixed, information on $f(0)$ and $f(1)$ increases with the amount of data only under strong parametric assumptions about $f(x)$, such as linearity. Consistency should, rather, be framed in terms of behaviour under ideal data conditions, which for relative index of inequality estimation would require increasing amounts of data in vanishingly narrow intervals close to both $x = 0$ and $x = 1$. Under such conditions, and with the weakest possible assumption that $f(x)$ is continuous at $x = 0$ and $x = 1$, increasingly precise estimation of $f(0)$ and $f(1)$ becomes possible. The simplest consistent estimator of the relative index of inequality in this asymptotic framework is r_1/r_k, which discards data from all classes other than the lowest and highest. The main appeal of the 'standard' relative index of inequality calculation described above is its simplicity. It is of interest to study the extent, if any, to which a more elaborate method would be worthwhile. Possible elaborations include:

(i) development of an approach that is consistent in the above sense and that also exploits assumed smoothness of $f(x)$ to make use of data from all k classes;

(ii) the use of maximum likelihood based on a Poisson approximation—justified by general arguments as in Brillinger (1986), for example—which properly takes into account the interval-censored observation of x, in place of least-squares fitting to class midpoints;

(iii) an integrated treatment of 'standardizing' variables such as age, which are usually present and important in studies of this kind; and

(iv) improved assessment of precision, which properly distinguishes between sampling variability and lack of fit of the straight-line model.

These are the subject of recent work by J. Sergeant of Nuffield College, Oxford.

7.2.4 Role of sampling weights

Survey data from probability samples have associated 'sampling weights', which are equal to or proportional to $1/\pi_i$ for each sampled case i, with π_i the first-order inclusion probability. The use of such weights in 'enumerative' applications of survey sampling, which aim to estimate one or more well-defined functions of all population values, is essential and well understood. For example, the population mean of variable y from a sample S is estimated approximately unbiasedly by

$$\sum_{i \in S}(y_i/\pi_i)/\sum_{i \in S}(1/\pi_i),$$

which can be viewed as weighted least squares based on the simple model $E(y_i) = \mu$. The weights $1/\pi_i$ correct for systematic under- or over-representation of population members in the sample, in notional repeated sampling. The use of such weights in conjunction with enumerative estimators derived from linear and generalized-linear models is surveyed in Firth and Bennett (1998). Care is needed if the range of proportionate variation of π_i is very large, with one or more of the π_i close to zero, for then a small number of sampled cases may dominate the estimate. There are clear parallels here with notions of importance sampling and auxiliary variables in Monte Carlo simulation for approximating integrals (e.g. Hesterberg 1996). For *analytical* work using survey data the use of sampling weights seems less clear-cut. Pfeffermann (1993) provides a useful review. For concreteness, suppose that interest is in the dependence of y on x, to be summarized through the conditional mean function $m(x) = \mathrm{E}(y \mid x)$. A central line of argument in the survey-sampling literature—see Pfeffermann (1993) or Chambers and Skinner (2003)—is that if, for example, B is the *population* least-squares value of β in the linear model $y = \alpha + \beta x + \text{error}$, then weighted least squares estimation of β in the same model from the sample, with weights $1/\pi_i$, yields an asymptotically unbiased estimator of B; and this result holds regardless of whether $m(x)$ is actually linear in x. This is sometimes given as a robustness rationale for the use of sampling weights with analytical models. In effect, though, this argument converts the analytical objective into an enumerative one, and its relevance to inference on $m(x)$ is unclear. If $m(x)$ is linear, then standard arguments point to the use of $1/\operatorname{var}(y \mid x)$ for weights that maximize efficiency, rather than the sampling weights. If $m(x)$ is *not* linear, then B is merely one of many possible incomplete summaries of $m(x)$. For example, if the true $m(x)$ is $\mu + \alpha I(x > k)$ for some threshold k, with $\alpha > 0$ say, then B is positive but its numerical value fails to capture the nature of the dependence; the use of sampling weights should not be viewed as an alternative to careful model search and model criticism. The preceding discussion assumes implicitly that, for any given x, selection into the sample is independent of y. The opposite case, so-called *non-ignorable* sampling (after Rubin 1976), is less straightforward. In such situations sampling weights, where known, are used in order to provide protection against systematic bias even if the assumed model is correct (e.g. Pfeffermann 1993, Fienberg 1989, Hoem 1989). This is connected with the notion of *propensity score* (see Section 7.3.2).

7.3 Other topics

This section collects brief comments and references on a small selection of further topics of current interest.

7.3.1 Incomplete data

In a sense, all of statistical inference concerns incomplete data, but most usually the term refers to data that would have been observed under ideal conditions but that are, for some reason, missing. The mechanism for missingness plays a crucial role (Rubin 1976). For an overview see Little and Rubin (2002) or Schafer (1997). In social research, missingness most often arises through survey non-response: sample-selected cases might fail to provide some or all of the desired data; in panel studies a common problem is dropout. An important avenue for further research is the continued systematic study of practical methods for minimizing the extent of such missingness. However, some degree of missingness is inevitable, and so analytical methods have been developed that seek to overcome the resultant problems, which for standard statistical methods can be crudely characterized as systematic bias and misstatement of precision. The main general approaches are based on weighting and imputation. Weights calculated as reciprocals of estimated response probabilities can be used in some settings to reduce bias. Imputation methods are rather more general in scope, and in particular they can be used where the pattern of missingness is arbitrarily complex. Most attention has been paid to the method of *multiple* imputation (Rubin 1987, 1996), in which missing values are imputed at random, repeatedly in order to produce a set of m different 'completed' datasets; the completed datasets are then analysed by standard methods, and the observed between-imputed-datasets variation contributes an important component to the estimated imprecision of derived estimates. In this regard it is remarkable that in typical applications m is chosen to be a small number, such as 5. Motivation for this is given in Rubin (1987), where it is shown that efficiency loss in estimation of the parameters of interest is slight, and that interval estimates after a degrees-of-freedom correction are valid to a reasonable degree of approximation, with m as small as 5 in contexts with typical missing-data fractions. The estimation of a variance component using as few as 5 observations is, however, hugely ambitious: for example, a simple calculation based on the normal distribution shows that to be 90% sure of the first significant digit of σ^2 around 50 observations are needed. The use of m as small as 5 thus carries the risk of a poor estimate of precision in any particular application. A much greater value of m would, however, diminish the practical appeal of the method, especially for large datasets.

7.3.2   Policy evaluation and selection bias

The effectiveness of social policies, such as training schemes to help people out of long-term unemployment, is an important topic on which there is some methodological debate. In certain countries randomized experiments are routinely used, though care is needed in their design and interpretation; it may be, for example, that the measurable effect of a training scheme is different for those who would and those who would not have elected to follow the scheme if given a free choice. Where randomized study is not possible, evaluation typically proceeds by careful analysis of observational data, in which scheme participation is determined by the participants themselves. In order to minimize the potential impact of bias caused by such self-selection, analysis often compares outcomes for matched individuals, the matching being on attributes thought to be good predictors of scheme participation. This notion is distilled in *propensity score* matching (Rosenbaum and Rubin 1983, Rosenbaum 2002), where a logistic regression or other such model is constructed that estimates the probability of scheme participation—the propensity score—as a function of available predictor variables; participating and non-participating individuals with similar values of the estimated propensity score are then compared in order to estimate the effect of interest. Various aspects of this approach deserve further study, including properties in cases where true propensity is not fully independent of outcome given available predictor variables, and the impact of the detailed methods used in the estimation of propensity scores (e.g. Heckman *et al.* 1998). A useful collection of papers, giving a fairly broad view of current approaches in practice, resulted from a Royal Statistical Society meeting in July 2000 and is introduced by Dolton (2002).

7.3.3 Causality and graphical models

Questions of causal inference from observational data do, of course, extend far beyond the evaluation of policy interventions, and are at the core of much social science. A general consensus on interpretation of the word 'causal' is elusive. Common to most if not all social-scientific work, however, is that multiple causes operate, and this creates a prominent role for statistical analysis. Cox and Wermuth (2001) review three of the most prominent statistical interpretations of causality, and their relation to social research: causality as *stable association*, as *the effect of intervention*, and as *explanation of a process*. The last of these, in which subject-matter theory and statistical analysis combine to establish processes that are potentially causal, is common also in the natural and physical sciences; in social science the relevant substantive theory will most often derive from notions of rational choice. Where cause is taken to be the effect of an intervention, *counterfactual* reasoning is often used, in which a statistical comparison is made between responses in the presence and in the absence of the intervention (the difficulty in practice being that only one of those two responses is observed); this might involve the exclusion of certain types of variable, such as gender, as a potential cause, and it often demands great care in defining which other variables should be held fixed and which allowed to vary as the intervention is (hypothetically) introduced or removed. The deduction of causality from stable patterns of association requires temporal or subject-matter justification for the direction of any putative causal relationship, and—usually much more difficult—the elimination of alternative explanations. In statistical thinking on causality, graphical and structural-equation models based on conditional independence relations play a key role, and this has been an area of rapid development. Such models codify which variables might conceivably be related causally, and the direction of possible causal links. The strongest representation is a *directed acyclic graph* (DAG), from which conditional independence relationships may be read off using the central notion of *d-separation*. In many practical applications it is undesirable or impossible to assign a DAG-style direction to every possible association, and this has led to the development of chain-graph models in which variables are grouped as a partial ordering, with causal direction between groups specified. A good introduction to statistical aspects of causality is still Holland (1986); for more recent references see Sobel (2000) and Cox and Wermuth (2004). Pearl (2003), in a recent review paper with the same title as Holland's, emphasises the role of assumptions and the use of structural equation models; see also Pearl (2000). For graphical chain models and their interpretation, *inter alia*, Cox and Wermuth (1996) is essential reading.

7.3.4   Event histories

Event-history data is very common in some branches of social science, for example the study of individual careers. The dominant analytical approach, following the influential work of Cox (1972), is based on regression models—either parametric or semiparametric—for the log-hazard function of waiting-time distributions; see, for example, Blossfeld *et al.* (1989). Some common features of social-science applications are:

- recurrent events, for example repeated spells of unemployment;
- several states, for example {employed, unemployed, not in the labour force};
- discrete-time observation, e.g. data collected annually with no recall of precise dates of transition.

Other approaches aim to describe and explain 'whole history' aspects rather than transition intensities. One such is the 'optimal matching' analysis of Abbott and Hrycak (1990), where methods of alignment and distance estimation from DNA-sequence analysis are applied directly to suitably encoded career histories, a device that then allows the use of clustering and other such exploratory methods. The results are interesting, and it would be useful to consider whether a more tailored approach to the summary and comparison of such histories—which, for example, respects the direction of time, in contrast to the methods imported from molecular biology—might yield still stronger results.

7.3.5   Ecological inference

The topic of inference on individual-level relationships from aggregate data has a long history as well as various recent developments, which are critically reviewed in Wakefield (2004). The inherent difficulties are not ones that can readily be massaged away by clever modelling, and conclusions are often highly sensitive to detailed model assumptions. A promising general approach is to combine aggregate data with some additional information at the individual level. This raises some interesting methodological questions when, as seems likely often to be the case in practice, the individual-level data are in some way unreliable. For example, in the study of voter transitions over a pair of successive elections, the aggregate data—total votes cast at each election—are essentially perfect; but even an exit poll, probably the best type of individual-level data available in this context, will inevitably suffer from problems of differential refusal, false replies, faulty recall of previous vote, and so forth.

7.3.6   'Agent-based' models

Conventional statistical models typically establish and explain social regularities through associations between variables. The 'agent-based' approach aims instead to represent macro-level phenomena as consequences of the actions of interacting individuals who influence one another in response to the influences they themselves receive; for a variety of applications see, for example, the collection of papers summarized in Bankes (2002). Work with such models uses computer simulation, which often has no stochastic element, to study the evolution of large systems of individuals over long periods of time, perhaps many generations. Inference is made by 'tuning' micro-level behavioural rules to achieve an acceptable match between simulation results and established real-world regularities. This is largely unfamiliar territory for statisticians, but seemingly progress can be made at least under certain assumptions that relate system inputs and outputs through a Bayesian probability model (e.g. Poole and Raftery 2000). This is a topic on which further statistical input seems likely to be valuable.

7.3.7 Public-service 'league tables' and performance monitoring

The use of administrative and other data to rank institutions on aspects of their performance has become commonplace, at least in the UK, and has largely taken place without much statistically informed input. Recently, the Royal Statistical Society published a wide-ranging report by its *Working Party on Performance Monitoring in the Public Services* (Bird *et al.* 2005). The report stresses the importance of a clear protocol covering the design and implications of performance indicators, as well as methods for data collection, analysis, presentation and dissemination. Presentation, for example, should always reflect uncertainty due to measurement and/or haphazard variation, making use where appropriate of such devices as the 'funnel plot' (Spiegelhalter 2002). More work is needed both on the design of such presentational methods and on establishing their place in routine reporting of institutional-performance data, as well as on suitable methods of analysis for the clustered, multidimensional data that typifies this kind of work.

7.3.8 S (or R) is for social science

The uptake of new statistical methods by social scientists has for many years been patchy, and has been hampered to some extent by the dominance of one or two large commercial software packages which, although excellent for their core purposes, respond only rather slowly to novel statistical developments. Typically such packages do not encourage, or even necessarily allow, the programming of new methods by researchers themselves. By contrast, for statisticians the S language (Becker *et al.* 1988) has become the *lingua franca* for data analysis and for the rapid implementation of new models and methods. An important recent development is the open-source R project (Ihaka and Gentleman 1996), which provides both the S language and a rapidly growing array of fully documented 'packages' for specific methods. The availability of R, coupled with accessible introductory material for social scientists such as Fox (2002) and courses such as those given at the Inter-University Consortium for Political and Social Research Michigan Summer School and the Economic and Social Research Council Oxford

Spring School, has already had a marked effect on graduate research-methods teaching, and it suggests a future in which statistical methodology will play a still more important and immediate role in social research.

Acknowledgement

The author is very grateful for the careful reading and comments of a referee, and for the support of the UK Economic and Social Research Council through a Professorial Fellowhsip.

8
Biostatistics: the near future

Scott Zeger, Peter Diggle, and Kung-Yee Liang

8.1 Introduction

Over the last half-century, the influence of statistical ideas and methods has continued to expand from the foundation laid by Karl Pearson and Sir Ronald Fisher in the first half of the twentieth century. Statistical tools have advanced many areas of empirical science but perhaps none more than biology and medicine. The goal of this chapter is to look forward over the next decade to consider the future of biostatistics.

Biostatistics is a term with slightly different meanings in Europe and the United States. Here we take it to refer to statistical science as it is generated from and applied to studies of human health and disease. We distinguish it from the broader term *biometrics* that refers to the interface of statistics and all biological sciences including for example agricultural science and from *medical statistics*, which refers more narrowly to clinical than biomedical sciences.

In this chapter, we argue that a substantial component of the success of biostatistics derives not from the utility of specific models, even though there are many such successes, but rather to a model-based approach that Fisher, Sir David Cox and others have so forcefully advanced. We believe that this will be central to the continued advancement of biostatistics, at least for the near future. This approach:

- is built on a foundation of careful design and measurement;
- formulates scientific questions or quantities in terms of parameters θ in probability models $f(y; \theta)$ that represent in a parsimonious fashion, the underlying scientific mechanisms (Cox 1997);
- partition the parameters $\theta = (\psi, \lambda)$ into a subset of interest ψ and other 'nuisance parameters' λ necessary to complete the probability distribution (Cox and Hinkley 1974);
- develops methods of inference about the scientific quantities that depend as little as possible upon the nuisance parameters (Barndorff-Nielsen and Cox 1994); and
- thinks critically about the appropriate conditional distribution on which to base inferences.

This chapter starts with a brief review of what we believe are exciting biomedical and public health challenges capable of driving statistical developments in the next decade. We discuss how they may influence biostatistical research and practice. Section 8.3 then discusses the statistical models and inferences that are central to the model-based approach. We contrast the model-based approach with computationally intensive strategies for prediction and inference advocated by Breiman and others (e.g. Breiman 2001), and with more traditional methods of inference whose probabilistic basis is the randomization distribution induced by the particular experimental design (Fisher 1935), and that we might therefore call a design-based approach. We discuss the so-called *hierarchical* or *multilevel* specification of a statistical model as an example of the future challenges and opportunities for model-based inference. In Section 8.3, we also discuss the role of increasingly complex models in science and statistics. A key question is how to quantify the uncertainty in target parameters, incorporating more than the usual sampling variability. In Section 8.4, we discuss conditional inference. Recent examples from genetics are used to illustrate these issues. Sections 8.5 and 8.6 discuss causal inference and statistical computing, two other topics we believe will be central to biostatistics research and practice in the coming decade.

8.2   Biomedical and public health opportunities

8.2.1   The last 50 years

If we start our look forward with a brief look back at the past, it is clear that we live in a revolutionary period for biological science, and particularly for biomedicine and public health. It has been only 50 years since Watson, Crick, Franklin, Gosling, and others elucidated the basic mechanisms of molecular biology (Watson and Crick 1953, Franklin and Gosling 1953). Since then, the parallel advances in biotechnology and computer technology have produced a panoply of new measurements that are now driving statistical innovation.

In biomedical laboratories, we now routinely measure DNA sequences, gene and protein expression levels, protein structures and interactions. The new biotechnologies make it possible to control systems by breeding genetically designed laboratory animals, adding or removing putative genes, fabricating molecules, inserting human genes in other organisms, and so forth. Biomedical investigators have remarkable new tools to measure and to control systems permitting more informative studies of the mechanisms underlying health and disease.

The same fifty years has given rise to equally powerful technologies for population medical research. The modern randomized controlled trial has its roots in 1947 with the Medical Research Council Streptomycin in Tuberculosis Trial (MRC Streptomycin in Tuberculosis Trials Committee 1948). There are now generally accepted protocols for human experimentation such as the Nuremberg Code (Kious 2001) and the controlled clinical trial has become the single leading tool of clinical research. The advent of systematic testing of preventive and therapeutic therapies using controlled trials is arguably the most important development in clinical medicine during this period.

In 1952, Sir Richard Doll and Sir Austin Bradford Hill changed the practice of epidemiology with the development of the case-control study to investigate the association of smoking and lung cancer. Cornfield (1951) formulated the statistical framework for the analysis of case-control studies. Like the randomized trial in clinical research, the case-control study and related designs are now essential tools for every clinical and public-health researcher.

The progress in biostatistics over the last few decades has reflected the advances in medical and public-health research. The productivity in the subfields of statistical genetics, survival analysis, generalized linear models, epidemiological statistics, and longitudinal data analysis demonstrates the opportunities that medical and public-health breakthroughs of the last half-century have created for statistical science, and vice versa.

8.2.2 The next decade

A consideration of the near future of biostatistics best starts with consideration of the emerging opportunities in biomedicine and public health. The United States National Institutes of Health 'Roadmap' (Zerhouni 2003) and the United Kingdom Medical Research Council strategic plan (Medical Research Council 2001) present similar visions for the future. Collins *et al.* (2003) give an excellent review of how genomics will change biomedical research and practice.

There is consensus that, while the twentieth century has produced lists of genes and proteins, the priority for the next decade is to determine their functions. The goals are to identify and understand the gene–protein networks central to normal biology, aberrations and interactions with the environment that produce disease, and corrective interventions that can prevent or treat disease. The Medical Research Council plan states: 'The focus will shift to questions of protein structure, function and interactions, developments of physiological pathways and systems biology' (Medical Research Council 2001). The National Institutes of Health Roadmap identifies 'Building Blocks, Biological Pathways and Networks' as the first of its five priorities (Zerhouni 2003).

In the next decade, exhaustive gene lists will be refined for organisms at all levels. Once the sequence has been determined for a species, biologists will characterize the covariation across a population in key DNA markers and single nucleotide polymorphisms, differences among persons in the DNA bases that appear in particular locations. Parsimonious summaries of these patterns of association will be needed.

Associations—many of them false—of disease occurrence with single nucleotide polymorphisms, gene and protein expression levels will be discovered. Some associations will lead to the identification of biochemical pathways for normal biology and mistakes that cause disease. We should not be surprised if some aspect of the basic paradigm of molecular biology is supplanted or dramatically revised. For example, the field of epigenetics (e.g. Feinberg 2001) pre-supposes that molecular mechanisms other than DNA sequence contribute substantially to the control of cellular processes. Such a discovery could radically change the nature of biomedical research in the next decade.

The biotechnology industry will continue to produce more powerful laboratory methods to quantify genes and their RNA and protein expressions. There will be a key role for statistical design and evaluation of the measurement process itself. For example, the early gene expression measurements by Affymetrix, a leading manufacturer of microarrays, have been substantially improved by careful statistical analysis. Affymetrix quantified expression by comparing the binding of short sequences of DNA, roughly 25 bases to the target sequence RNA— so-called 'perfect match'—to a background rate of binding with a sequence whose middle base is changed. The idea was sensible: to correct for non-specific binding of RNAs with a sequence similar to but not exactly the same as the target. But Irizarry *et al.* (2003) have shown that there is an enormous price in variance to pay for the small improvement in bias when making this baseline correction. Similar opportunities certainly exist in quantifying protein expression, protein–protein interactions, and other fundamental measurements.

Measurement is also central to clinical research. For example, diagnosis and severity assessment for many psychiatric disorders such as depression and schizophrenia are based upon symptom checklists. Quality of life or health status is also measured via longer lists of items. Cox (1997) has discussed the opportunities for improved measurement in these situations where the basic datum is a multivariate vector of item responses.

Having quantified tens of thousands of gene or protein expression levels, the natural question is: which of these best predict the incidence or progression of disease, or the efficacy of a novel treatment? We should expect many more studies like the one by van de Vijver *et al.* (2002), who followed 295 patients with primary breast carcinoma for time to metastases. Gene expression levels for 25 000 genes were used to develop a 'risk profile' based upon roughly 70 genes. The profile was a strong discriminator of persons with early versus late progression. While crossvalidation was used to estimate the prediction error, the number of models with only main effects and two-gene interactions is too numerous to contemplate. The potential for false discoveries is enormous. Effective methods for searching model space, informed by current biological knowledge are needed. Ruczinski *et al.* (to appear) developed logic regression, a novel approach to address this problem of contending with high-dimensional predictor space when predicting protein structure from amino-acid sequences.

Similar problems are common in image analysis. See for example, Chetelat and Baron (2003), who predict onset of Alzheimer's disease using data from functional magnetic resonance images with activation intensities at 10^5 or so spatial locations or voxels in the brain.

The study of complex interactions or 'systems biology' is an emerging theme in biomedical research. Harvard University Medical School has recently created a new Department of Systems Biology intended to bring together a new community of researchers—biologists, physicists, engineers, computer scientists, mathematicians and chemists—to work towards a molecular description of complex cellular and multicellular systems, with an emphasis on quantitative measurement and mathematical and computational models.

There is no reference to statistics here! An illustration is the work by von Mering *et al.* (2003), who compare how proteins bind to one another for different species to identify critical subsets of interacting proteins that are preserved across species.

The next decade of clinical and public health research will also be driven by the twin genomics and computing revolutions. In epidemiology, the study of gene and environmental interactions will remain centre stage. For example, Caspi *et al.* (2003) reported results of a prospective study that identified two versions—'short' and 'long' alleles—of the serotonin transporter 5-HT T gene. Persons with one or two copies of the short allele exhibited more depression in response to stressful life events than individuals homozygous for the long allele. Similar research on gene–environment interactions will be central to cardiovascular and cancer epidemiology.

The amount of data and computing power will also make possible a new level of public-health surveillance (e.g. Dominici *et al.* 2003). For example, in their most recent work, Dominici, McDermott, and colleagues have assembled and crosslinked public databases for the USA over the period 1999–2002 comprising administrative medical records with diagnostic codes for the more than 40 million adults in the Medicare population, hourly air pollution time series for six major pollutants from thousands of Environmental Protection Agency monitors, hourly air and dew-point temperature from major weather centres, and census data at the postal-code level on socio-economic status, education, housing, and other factors that influence health. They are using these data to quantify the numbers of deaths, cases of disease and medical costs attributable to air pollution and its possible interaction with socio-economic status and other community-level factors.

In summary, biotechnology and modern computing have created new windows into the workings of biological systems in the form of new measurements that are inherently high-dimensional. We have now assembled long lists of genes, proteins and other biological constituents and are in the process of uncovering the interactions and networks that define normal and abnormal biology. These problems and similar ones in epidemiology and public health are likely to drive the development of statistical reasoning and methods for a decade or more.

8.3 Statistical models and inference

8.3.1 Statistical models

The word 'model' has many meanings in science and statistics. In toxicology, it may refer to an animal or cell system, in engineering, to a scaled physical approximation. Within statistical disciplines, the term model can cover any mathematical representation of the genesis of a set of data, y. From a scientific perspective, a useful distinction is between models that can be deduced from known or hypothesized scientific mechanisms underlying the generation of the data, and models that are simply empirical descriptions; we refer to these as *mechanistic* and *empirical* models, respectively. The dichotomy is an over-simplification but, we feel, a useful one. Consider for example the homogeneous Poisson process. As a model for the time sequence of positron emissions from a radioactive specimen, it has a well-accepted interpretation as a large-scale consequence of assumptions about fundamental physical processes operating at the submolecular scale; as a model for the time sequence of seizures experienced by an epileptic patient, its value rests on empirical validation against observed data. Besag (1974) argued, cogently in our view but ultimately unsuccessfully, for the use of the term 'scheme' rather than 'model' when referring to an empirical model.

From a statistical perspective, the key modelling idea is that data, y, constitute a realization of a random variable, Y: in other words our models are probabilistic models. We then need to ask where the randomness comes from. R. A. Fisher's answer, in the context of the analysis of agricultural field trials, was that the randomness was induced by the use of randomization in the allocation of experimental treatments to plots, leading to inferences that were essentially non-parametric. A second answer, which underlies most current mainstream statistics teaching and practice, is that the randomness is an inherent property of the process that produces each individual datum. Under this paradigm, the vector of plot-yields y in a field trial are modelled as a realization of a multivariate Gaussian random variable Y with appropriately structured mean vector and variance matrix, and inferences are based on calculated probabilities within this assumed distributional model.

These two approaches to inference can be termed *design-based* and *model-based*, respectively. In the context of the general linear model, the connection between the two was made by Kempthorne (1952). It is ironic that Fisher's analysis of variance methods are now taught almost exclusively under Gaussian distributional assumptions. The collection of papers published in the December 2003 issue of the *International Journal of Epidemiology* includes a fascinating discussion of the contrasting arguments advanced by Fisher and by Bradford Hill for the benefits of randomization in the contexts of agricultural field experiments and clinical trials, respectively (Chalmers 2003, Armitage 2003, Doll 2003, Marks 2003, Bodmer 2003, Milton 2003).

Most statisticians would accept that model-based inference is appropriate in conjunction with a model that, to some degree at least, is mechanistic. It is less clear how one should make inferences when using an empirical model. In practice, design-based inference copes more easily with problems whose structure is relatively simple. The development of progressively more complex data structures involving multidimensional measurements, together with temporal and/or spatial dependence, has therefore led to an increased focus on model-based inference, which then raises the awkward question of how sensitive our inferences might be to the assumed model. To quote Coombs (1964), 'We buy information with assumptions', with the unspoken corollary that what is bought may or may not give good value.

The model-based approach lends itself equally well to classical likelihood-based or Bayesian modes of inference. The last decade has seen a strong shift towards the use of Bayesian inference in many branches of statistics, including biostatistics. Whether this is because statisticians have been convinced by the philosophical arguments, or because the computing revolution and the development of Monte Carlo methods of inference have apparently reversed the historical computational advantage of classical methods over their Bayesian rivals, is less clear. For a thought-provoking discussion of likelihood-based inference, including a critique of Bayesian inference, see Royall (1997).

In recent years, a third broad approach to inference has developed, which we might call algorithm-based. This sits naturally at the interface between statistics and computing, and can be considered as a computationally intensive version of John Tukey's early ideas concerning exploratory data analysis (Tukey 1977), in which the traditional emphasis on formal inference is replaced by a less formal search for structure in large, multidimensional datasets. Examples include a range of non-parametric smoothing methods, many originating at Stanford University and the University of California at Berkeley (e.g. Breiman *et al.* 1984, Friedman 1991, Breiman and Friedman 1997), and classification methods based on neural networks (Ripley 1994). The cited examples illustrate the key contributions that statisticians continue to make to algorithm-based inference. Nevertheless, its emergence also represents a challenge to our discipline in that algorithm-based methods for data analysis are often presented under the computer-science-oriented banner of 'informatics,' rather than as an integral part of modern statistical methodology.

8.3.2 Hierarchically specified models and model-based inference

A recurrent theme during the last fifty years of statistical research has been the unification of previously separate methods designed for particular problems. Perhaps the most important to date has been Nelder & Wedderburn's (1972) unification of numerous methods including analysis of variance, linear regression, log-linear models for contingency tables, and logit and probit models for binary data, within the class of generalized linear models; for a comprehensive account, see McCullagh and Nelder (1989). This class represents the state-of-the-art for modelling the relationship between a set of mutually independent univariate responses and associated vectors of explanatory variables.

Hierarchically specified models represent a comparable unification of models for dependent responses. Their defining feature is that observed responses $Y = (Y_1, \ldots, Y_n)$ are modelled conditionally on one or more layers of latent random variables that are usually of scientific interest but are not directly observable. Using the notation $[\cdot]$ to mean 'the distribution of', the simplest hierarchical model specification would take the form $[Y, S] = [Y \mid S][S]$, where S denotes a latent random variable. The emphasis here is on the hierarchical nature of the specification, rather than the model itself. For example, consider the hierarchical specification for an observeable quantity Y as $Y_i \mid S \sim N(S, \tau^2)$, with the Y_i conditionally independent given S, and $S \sim N(\mu, \nu^2)$. This is identical to the non-hierarchical specification $Y \sim N_n(\mu 1_n, \nu^2 1_n 1_n^{\mathrm{T}} + \tau^2 I_n)$, where 1_n and I_n are, respectively, the $n \times 1$ unit vector and $n \times n$ identity matrix. The model arises in the context of longitudinal data analysis where, Y represents a sequence of measurements on a single subject. In that context, a possible advantage of the hierarchical specification is that it invites a mechanistic interpretation for the assumed covariance structure of Y, namely that it arises as a consequence of random variation between the average response obtained from different subjects, as represented by the latent random variable S. More elaborate examples are discussed below.

In contrast to classical generalized linear models, it is hard to identify a single, seminal paper on hierarchically specified models. Rather, the essential idea was proposed independently in different settings, and these separate contributions were only later brought together under a common framework.

The Kalman filter (Kalman 1960) is an early example of a hierarchical model for time-series data. In its basic form, it models an observed univariate time-series Y_t with $t = 1, 2, \ldots$ in relation to an unobserved 'state variable' S_t by combining a linear stochastic process model for S_t—the state equation—with a linear regression for Y_t conditional on S_t—the observation equation—and uses the model to derive a recursive algorithm for computation of the minimum mean square error predictor for S_t, given the observed data Y_s for $s \le t$. This results in the standard hierarchical factorization $[Y, S] = [Y|S][S]$. Within the time-series literature, models of this kind have become known as state-space or dynamic time-series models (Harvey 1989, West and Harrison 1997).

Cox (1955a) had earlier introduced a class of hierarchically specified models, now known as Cox processes, for point process data. A Cox process is an inhomogeneous Poisson process whose intensity $\Lambda(x)$ is itself the realization of an unobserved, non-negative valued stochastic process. For example, a log-Gaussian Cox process (Møller *et al.* 1998) takes $\Lambda(x) = \exp S(x)$, where $S(x)$ is a Gaussian process. Models of this kind are likely to become increasingly important for modelling spatial point process data in which the local intensity of points is modulated by a combination of observed and unobserved environmental factors, and will therefore play a key role in environmental epidemiology. Although the Gaussian assumption is a strong one, it is attractive as an empirical model because of its relative tractability and the ease with which it can incorporate adjustments for observed, spatially referenced explanatory variables. Incidentally, Cox processes bear the same relationship to Poisson processes as do frailty models for survival data to the classic Cox proportional hazards model (Cox 1972a). In each case, an unobserved stochastic effect is introduced into the model to account for unexplained variation over and above the variation compatible with a Poisson point process model for the locations of events in space or time, respectively.

Lindley and Smith (1972) were concerned not so much with developing new models as with reinterpreting the classical linear model from a Bayesian perspective. Hence, their use of the hierarchical specification was to consider the linear regression relationships as applying conditional on parameter values that were themselves modelled as random variables by the imposition of a prior distribution. The same formalism is used in classical random effects linear models.

Motivated primarily by social science applications, Goldstein (1986) used hierarchically specified models to explain how components of random variation at different levels in a hierarchy contribute to the total variation in an observed response. For example, variation in educational attainment between children might be decomposed into contributions arising from components of variation between schools, between classes within schools, and between children within classes.

Finally, note that one way to extend the generalized linear model class for independent responses is to introduce a latent stochastic process into the linear predictor, so defining a generalized linear mixed model. Breslow and Clayton (1993) review work on such models up to that date, and discuss approximate methods of inference that have been largely superceded by subsequent developments in classical or Bayesian Monte Carlo methods (Clayton 1996).

Hierarchically specified models of considerably greater complexity than the examples given above are now used routinely in a wide range of applications, but especially as models for spatial or longitudinal data. Spatial examples include models for smoothing maps of empirical measures of disease risk (Clayton and Kaldor 1987, Besag *et al.* 1991), or for geostatistical interpolation under non-Gaussian distributional assumptions (Diggle *et al.* 1998). Longitudinal examples include joint modelling of longitudinal measurements and time-to-event data in clinical trials (Wulfsohn and Tsiatis 1997), in which stochastic dependence between the measurement process and the hazard function for the time-to-event process are induced by their shared link to a latent Gaussian process.

Despite the widespread, and widely accepted, application of hierarchically specified models, their routine use is not without its attendant dangers. More research is needed on some fundamental issues concerning model formulation, computation and inference.

With respect to model formulation, a key issue is the balance between simplicity and complexity. Simple models tend to be well identified. Complex models promise greater realism, but only if they are scientifically well founded. When models are empirical, simplicity, consistent with achievement of scientific goals, is a virtue. A challenge for the next decade is to develop a theory of model choice that appropriately combines empirical and subject-matter knowledge. In our opinion, it is doubtful whether such a theory could ever be reduced to a mathematical formalism, and we are skeptical about the role of automatic model-selection algorithms in scientific research.

With respect to computation, we need a better understanding of when off-the-shelf Markov chain Monte Carlo algorithms are and are not reliable, and better tools for judging their convergence. At least for the near future, we also need to recognize that many applications will continue to need mechanistic models and tailored fitting algorithms, using the combined skills of subject-matter scientists, applied statisticians, and probabilists.

Perhaps the most challenging questions of all concern inference for hierarchically specified models. For Bayesian inference, how should we choose joint priors in multiparameter models? How can we recognize poorly identified subspaces of the parameter space? And, whether or not we choose to be Bayesian, when might unverifiable modelling assumptions unduly influence our conclusions? In a sense, all of these questions have straightforward formal answers, but dealing with them thoroughly in practice is impossible because of the multiplicity of cases that need to be examined. Furthermore, there seems to be an increasing divergence between Bayesian philosophy, in which the notion of a 'good' or 'bad' prior has no meaning, and Bayesian practice, in which there appears to be a decreasing emphasis on topics such as prior elicitation through expert opinion, and a *de facto* shift to using computationally convenient families of prior within which location and/or scale parameters act as 'tuning constants' somewhat analogous to multidimensional bandwidth choices in non-parametric smoothing. The coming decade will provide better tools for the Bayesian data analyst to quantify the relative amounts of information provided by data and by prior specification, whether at the level of model choice or parameter values within a chosen model, and a better understanding of how these influence estimates and substantive findings.

8.3.3 Sources of uncertainty

The endproduct of a piece of statistical work is the result of a long sequence of decision-making in the face of uncertainty. An oversimplified representation of this process is that we iteratively address the following questions:

1. What is the *scientific* question?
2. What experimental *design* will best enable the question to be answered?
3. What statistical *model* will incorporate an answer to the question whilst dealing adequately with extraneous factors?
4. How should we *analyse* the resulting data using this model and interpret the estimates and measures of their uncertainty?

Currently, statistical theory deals explicitly and very thoroughly with the last of these, through the theory of statistical inference. In essence, inference allows us to say: 'this is what we have observed, but we might have observed something different', and to moderate our conclusions accordingly. To a limited extent, techniques such as Bayesian model averaging, or classical nesting of the model of interest within a richer class of possible models, allow us to take account of uncertainty at the level of model formulation, but here the methodology is less well developed, and less widely accepted. Rather, it is generally accepted that model formulation is, at least in part, a subjective process and as such not amenable to formal quantification. All statisticians would surely agree that careful attention to design is of vital importance, yet our impression is that in the formal training of statistical graduates, courses on design typically occupy a very small fraction of the syllabus by comparison with courses on inference, modelling and, increasingly, computation. We predict that some relatively old ideas in experimental design, such as the construction of efficient incomplete block designs, will soon enjoy a revival in their importance under the perhaps surprising stimulus of bioinformatics, specifically gene expression microarray data. As discussed above, microarray data are in one sense spectacularly high-dimensional, but a typical microarray experiment is cost limited to only a small number of arrays and these, rather than the many thousands of individual gene expression levels, are the fundamental experimental units at which design questions must be addressed. See, for example, Kerr and Churchill (2001), Yang and Speed (2002), and Glonek and Solomon (2004).

Finally, the best comment we can offer on the importance of addressing the right question is to quote Sir David Cox in an interview published in volume 28, issue 1, page 9 of the newsletter of the *International Statistical Institute*. In answer to the question of what advice he would offer to the head of a new university department of statistics, he included 'the importance of making contact with the best research workers in other subjects and aiming over a period of time to establish genuine involvement and collaboration in their activities.'

8.3.4 Model complexity

The new measurements enabled by biotechnologies will certainly lead some to increasingly complex descriptions of biological systems. For example, Figure 1 of Jeong *et al.* (2001) represents protein–protein interactions for 1870 proteins shown as nodes, connected by 2240 direct physical interactions for the yeast *Saccharomyces cerevisiae*. They represent by the colour of a node, the phenotypic effect of deleting the gene that encodes its corresponding protein, and argue that proteins with more connections are more essential to yeast survival.

Biostatisticians have long recognized the trade-off of biological realism against parsimony in their models estimated from a finite set of observations. As the richness of datasets increases, more complex models will surely be contemplated. An important question for the next decade will be where to find the new balance between realism and parsimony.

Cox (1997) offered a hierarchy of probability models according to their scientific basis that extends the two categories we discussed above:

1. purely empirical models;
2. 'toy' models;
3. intermediate models; and
4. quasi-realistic models.

Types 1 and 4 of this list correspond essentially to what we have called 'empirical' and 'mechanistic', models. Moving down the list, the scientific complexities are represented to an increasing degree. It seems plausible for at least two reasons that biostatistical models will strive for increasing biological content over the coming decade.

First, parameters in more realistic models are inherently more useful and interesting to scientists. The enormous success of the proportional hazards model (Cox 1972) is partly because its parameters are relative risks with simple biological interpretations. The hierarchical models discussed above are especially popular among social scientists who study the effects of covariates at the individual and community levels because their regression parameters can be given causal interpretations.

Second, imposing scientific structure is a partial solution to the dimensionality of our outcome measures. For example, in early statistical analysis with DNA microarrays, gene expressions were most often treated as exchangeable (e.g. Hastie *et al.* 2000). But there is substantial scientific knowledge that can be used to arrange genes into prior clusters, for example because they encode proteins involved in a common metabolic pathway. Bouton *et al.* (2003) have developed DRAGON, a software tool that queries genomic databases and creates such gene groupings. Biological knowledge can be used to reduce the number of variables from the order of 10^4 genes to 10^2 classes of genes.

As is well known to statisticians, there is a price to pay for increasing 'realism' in the face of limited data. Often, key parameters are difficult to identify. For example, Heagerty and Zeger (2000) have shown how parameter estimates in hierarchically specified conditional random effects models depend quite strongly on difficult-to-verify assumptions about the variances of the unobserved random effects.

8.4 Conditional inference

8.4.1 Introduction

Sir David Cox published two seminal papers in 1958. The first (Cox 1958*b*) addressed the statistical principle of *conditioning*. The second (Cox 1958*c*) applied this idea to the analysis of binary responses, in particular using the logistic regression model. In the fifty years since their publication, these and subsequent articles have established conditional inference as a cornerstone of model-based applied statistics. In this section, we discuss three reasons why conditioning has had so much impact to date and consider its role in the future of biostatistics. As above, we will represent the full set of parameters by $\theta = (\psi, \lambda)$ where ψ represents the subset of scientific interest and λ represents the nuisance parameters.

8.4.2 Inducing the proper probability distribution

One fundamental argument in favour of conditioning is to ensure that the probability distribution used for inference about ψ is consistent with the data in hand (Cox 1958b). Specifically, it implies use of a probability distribution for the observed data conditional on ancillary statistics defined as those functions of sufficient statistics whose distributions are independent of ψ. The idea is to focus on models for random variables that are directly relevant to ψ without investing time in modelling less relevant ones. A simple example of an ancillary statistic is the sample size because, in most situations, the probability mechanism for an observation does not depend on the choice of sample size. Because the likelihood function and statistical uncertainty about ψ do depend upon the sample size, this dependence must be made explicit, as happens through conditioning.

The most compelling argument for conditional inference was a toy example given in Cox (1958b). Here, the parameter of interest is the mean value for a biological measurement. A coin is tossed to determine which one of two competing laboratories will be selected to run the biological assay. One laboratory has high precision, that is, low variance, and the other has low precision. The conditioning principle demands that inferences about the unknown mean be drawn conditional on the lab that actually made the measurement. That another laboratory, with very different precision, might have made the measurement, is ancillary information.

8.4.3 Elimination of nuisance parameters

The closely related second objective of conditioning is to focus inferences on the parameters of interest ψ, minimizing the impact of assumptions about the parts of the probability mechanism indexed by nuisance parameters λ. The parameters λ are necessary to jointly specify the mechanism but are not of intrinsic interest to investigators. Examples are common in biomedical studies. In case-control studies of disease etiology, ψ is the odds ratio relating a risk factor to disease, and λ is the probability of exposure for the control group. In the proportional hazards model (Cox 1972), ψ is the log of relative hazards comparing individuals with covariate value $X = x$ versus those with $X = 0$. The nuisance parameters comprise the hazard function for the reference group, which is left unspecified.

A key issue mentioned in Section 8.1 is how to develop methods of inference about the parameters of interest that depend as little as possible on the nuisance parameters. This is important because it is well documented that the quality of inference for ψ depends critically on how these two sets of parameters are intertwined with each others (e.g. Cox and Reid 1987, Liang and Zeger 1995). Neyman and Scott (1948) demonstrated that the maximum likelihood estimates for ψ can be inconsistent when the dimension of λ increases with the sample size. This is the situation epidemiologists face in the matched case-control design, which is commonly used when confounding variables are difficult to measure. Examples include controlling for a case's neighborhood or family environment or genes. In this situation, the number of nuisance parameters is roughly equal to the number of cases. Cox (1958c) showed how to eliminate the nuisance parameters by conditioning on sufficient statistics for the nuisance parameters. The modern version of this approach, conditional logistic regression, is in routine use.

Conditional logistic regression has had important application in genetic linkage where the primary goal is to find the chromosomal region of susceptibility genes for a disease using phenotypic and genetic data from family members. One design that is attractive because of its simplicity is known as the case–parent trio design (Spielman *et al.* 1993). Blood samples are drawn from diseased offspring and both of their parents. Each parent–child dyad forms a matched pair. The 'case' is defined to be the target allele of a candidate gene that was transmitted from the parent to the offspring. The 'control' is the allele that was not transmitted to the offspring. Thus for each trio, two matched pairs are created and one can test the no-linkage hypothesis by testing whether the target allele is preferentially transmitted to diseased individuals. The connection between the conventional one-to-one matched case-control design and the case–parent trio design enables investigators to address important questions that the new technologies for genotyping large numbers of markers have made possible. A key question for the coming decade is how to quantify evidence that the candidate gene interacts with environmental variables to cause disease. Conditional inference is likely to be essential here.

8.4.4   Increasing efficiency

In the coming decade, genetic research will continue to focus on complex diseases that involve many genes and more complicated gene–environment interactions. Traditional genetic linkage methods are designed to locate one susceptibility gene at a time. Unless one has large numbers of homogeneous families, the power to detect the putative gene is small. Recently, Nancy Cox and colleagues (Cox *et al.* 1999) applied statistical conditioning to address this issue in a search for genes associated with Type-I diabetes. Taking advantage of a previous linkage finding on chromosome 2, they carried out a genome-wide scan on the remaining chromosomes by conditioning on each family's linkage score from chromosome 2. Through this process, strong linkage evidence on chromosome 15 was discovered. This constitutes a sharp contrast to the previous analysis without conditioning, which found no evidence of linkage on chromosome 15. This approach was further developed for studies that use identical-by-descent sharing by Liang *et al.* (2001). They have similar findings in data from an asthma linkage study reported by Wjst *et al.* (1999).

The examples above illustrate that proper conditioning can help to address the challenging issue of gene–gene interaction by increasing the statistical power in locating susceptibility genes. A limitation of this approach is that it is conditional on one region at a time. With thousands of single nucleotide polymorphisms, an important research question is how best to extend the conditional approach to higher-dimensional cases. Similar opportunities exist when searching for associations between gene or protein expression levels and disease outcomes or with voxel-specific image intensities and disease outcome or progression. The model-based approach provides a way forward.

8.5 Determining cause

Most biostatisticians would agree that quantification of the evidence relevant to determining cause and effect is an essential responsibility of statistical science. For example, the practice of medicine is predicated upon the discovery of treatments that cause an improvement in health by reversing a disease process or by alleviating its symptoms.

The meaning of the term 'cause' is generally accepted among statisticians to be close to the Oxford English Dictionary definition: 'That which produces an effect' (Simpson and Weiner 1989). But how this idea is best implemented in empirical research has been, is today, and will be for the next decade, the subject of important work.

In the biomedical sciences, Koch's postulates were among the earliest attempts to implement causal inference in the modern era. Robert Koch, a nineteenth-century physician who discovered the anthrax and tubercle bacilli, proposed four criteria by which an infectious agent would be established as the cause of an illness: the organism must be present in all cases; it must be isolated from the host and grown in culture; it must cause the disease when administered to a susceptible host; and it must be isolated from the newly infected host. The Koch postulates remain an important example of the integration of biological reasoning with experimental data to establish cause (Munch 2003). In addition to establishing the modern randomized clinical trial and case-control study, Sir Austin Bradford Hill posited a number of desirable factors to determine whether an association is causal (Bradford Hill 1965): strength of evidence, consistency of association across studies, specificity of the association, that the causal agent precedes its effect, a dose–response relationship, biological plausibility, and coherence with other relevant knowledge, and the ability to draw analogy from other similar problems (Rothman and Greenland 1998). Like Koch, Hill envisaged a qualitative, interdisciplinary process to establish cause. His ideas grew out of the debate in the 1950s about whether smoking caused lung cancer, and his attributes are relied upon today in text books (e.g. Rothman and Greenland 1998) and court rooms.

Biological reasoning about mechanisms is central to the Koch and Bradford Hill approaches to causal inference. Is there a biological mechanism that explains the statistical association? Are there alternative mechanisms that would give rise to the same observations? Statistical evidence from scientific studies is central. But the evidence is assembled through a qualitative process in which current biological and medical knowledge plays a prominent role. The importance of mechanisms in determining cause has been stressed by Cox (1992) and by McCullagh in Chapter 4 of this volume.

A second line of causal reasoning is based upon the formal use of counterfactuals. The causal effect of a treatment for a person is defined as the difference in outcome between two otherwise identical worlds: one in which the treatment was taken, the other where it was not. The counterfactual definition of 'cause' dates back to the eighteenth century or earlier (Rothman and Greenland 1998) and is prevalent in economics and more recently statistical sciences. Formal counter-factual analysis involves direct modelling of the pair of responses $Y_i(t = 1)$ and $Y_i(t = 0)$ for unit i with $(t = 1)$ and without $(t = 0)$ the treatment or risk factor.

Statistical methods for causal inference have frequented the leading journals for at least three decades (e.g. Rubin 1974, Holland 1986). The formal use of counterfactuals in probability models is particularly effective in randomized clin-ical trials and other similar studies where the assignment of treatments occurs according to a known or well-approximated mechanism. Counterfactual models allow us to study the effects of departures from this leading case, for example to study the effects of dropouts (e.g. Scharfstein *et al.* 1999) or failure to comply (Frangakis and Rubin 1999).

Counterfactual thinking has also become central to epidemiological research, where randomized trials are not possible for the study of most risk factors. See, for example, Greenland *et al.* (1999) and Kaufman *et al.* (2003) and references therein.

Formal counterfactual analysis is now commonplace in problems far from the randomized trial. For example, Donald Rubin, an early exponent and key researcher on formal causal inference, is also the statistical expert for the tobacco industry in their civil suits brought by the states and the United States Justice Department (Rubin 2002). He has testified that the medical costs caused by smoking in a population are properly determined by comparing two worlds: one in which smoking occurred, and the other in which it did not. To quantify the effects of the alleged fraudulent behaviour of the tobacco companies, he calls for a comparison of two worlds: one with and the other without behaviour that took place over several decades in thousands of discrete acts (Rubin 2001).

It is hard to argue in the abstract with these causal targets for inference whether in a randomized controlled trial, an epidemiologic study or an assess-ment of a complex industrial behaviour. But an important issue is the role of statistical models in causal inference. Should statistical models be used to esti-mate key components, for example the relative risk of disease given exposure, which are then combined through a qualitative process with background knowl-edge and theory to determine cause? The process of determining that smoking causes lung cancer, cardiovascular disease and premature death is an example of one such scientific process. Or should formal causal models be relied upon to or-ganize the evidence, prior knowledge and beliefs about mechanisms, alternative explanations and other relevant biological and medical factors?

Cox (1986, 1992) pointed to the multilayer process of establishing cause in his discussion of Holland (1986), saying 'Is not the reason that one expects turning a light switch to have the result it does not just direct empirical observation but a subtle and deep web of observations and ideas—the practice of electrical engineering, the theory of electrical engineering, various ideas in classical physics, summarized in particular, in Maxwell's equations, and underneath that even ideas of unified field theory?'

The statistical approach to causal inference will continue to be refined in the coming decade. We will learn how to more accurately quantify the contributions to causal inference from statistical evidence versus from assumptions, particularly when making inference about a complex world that did not occur. We can envision the evolution of a hierarchy of causal inferences. The term *causal estimation* might be reserved for the randomized studies and for modest departures from them. Inferences about causal targets that critically depend on unverifiable assumptions might more accurately be termed *causal extrapolations* or in some cases, for example about the causal effects of the behaviours of the tobacco industry, *causal speculations*. These qualitative distinctions represent points along a continuum. In our opinion the question how best to quantify the respective roles of statistical evidence and model assumptions in causal inference remains an important topic for further research.

8.6 Statistical computing

The explosion of information technologies during the last few decades has changed radically the way in which empirical science is conducted. The collection, management and analysis of enormous datasets has become routine. The standard responses in biostatistical research have radically changed. Journal articles twenty years ago dealt mainly with binary, count or continuous univariate response variables. Generalized linear modelling (Nelder and Wedderburn 1972, McCullagh and Nelder 1989) was a breakthrough because it unified regression methods for the most common univariate outcomes.

In current studies, the response is commonly of very high dimension—an image with a million discrete pixels, a microarray with a continuous measure of messenger RNA binding for each of 30 000 genes, or a schizophrenia symptoms questionnaire with 30 discrete items. Computing power has made possible the routine gathering of such intrinsically multivariate data. The emerging fields of computational biology, bioinformatics and data mining are attempts to take advantage of the huge growth in digitally recorded information and in the computing power to deal with it.

It is safe to predict that biostatistics research in the coming decades will become increasingly interwoven with computer science and information technologies. We have already discussed the role of computationally intensive algorithms and resampling-based inference. A second major change is the near-instantaneous interconnection among scientists that the Internet makes possible. International research groups now work closely together on 'big science' projects, such as the development of the Linux operating system and the Human Genome Project.

In statistics, Internet groups, operating in the Linux model, have created R, a statistical computing and graphics language that now dominates most centres of statistical teaching and research. It is organized to permit continuous expansion as unconnected researchers produce R-packages for specialized methods.

In bioinformatics, the `Bioconductor` package is a leading tool for managing and analyzing gene-expression array data (Gentleman and Carey 2003). It is created and managed by a group of investigators from around the world, who have implemented protocols for extending the package themselves and for accepting extensions offered by others. In just three years, it has grown to have substantial influence on the practice of biostatistics in molecular genetics laboratories everywhere. `Bioconductor` demonstrates how the Internet can reshape biostatistical research to involve larger teams of statisticians, computer scientists, and biologists, loosely organized to achieve a common goal. A cautionary note is that an understandable focus on the computational challenges of bioinformatics brings with it a danger that the continuing importance of fundamental statistical ideas such as efficient design of incomplete block experiments may be forgotten.

The specialty of bioinformatics also reveals another trend. Because genetic information is collected and shared through web-based data systems, methods of analysis are also shared in the same way. There is a strong contrast with the past dissemination of statistical procedures: when a new method was developed, the researcher would make available an often difficult-to-use program for others to try; subsequently the method might became established and eventually added to commercial software. For example, generalized additive models (Hastie and Tibshirani 1990) took about a decade to appear in the commercial package `Splus` and to this day remain unavailable in other commonly used packages.

Statistical methods that are easy to access and use will be more influential among biologists, whose standard is to share data and software via publicly available websites (see, for example, Bouton *et al.* 2003, Colantuoni *et al.* 2003). The success of biostatistical research centres will depend increasingly on their ability to disseminate their methods and papers to the endusers more rapidly. The risks of quickly adopting innovative methods should not be ignored, but given the appropriate cautions, we expect the trend toward more rapid dissemination of biostatistical methods to continue.

8.7   Summary

The next decade will likely be a time of continued expansion for biomedical science and for biostatisticians. New biotechnologies are producing novel, high-dimensional measurements that will sustain the demand for innovative statistical tools. The focus on biological systems and networks will make the modelling of interactions particularly important (Cox 1984a). The model-based approach to statistical science advanced by Sir David Cox and colleagues can continue to bear scientific fruit in the decades ahead.

9

The Early Breast Cancer Trialists' Collaborative Group: a brief history of results to date

Sarah Darby, Christina Davies, and Paul McGale

9.1 Introduction

Breast cancer is the commonest type of cancer among women in most developed countries. There are about a million new cases diagnosed each year worldwide, and around 35,000 new cases annually in the United Kingdom alone (Quinn *et al.* 2001). For such a common disease, widely practicable treatments that produce only moderate effects on long-term survival, such as increasing the number of women surviving for more than ten years after diagnosis from 50% to 55%, could result in the avoidance of many thousands of deaths each year. Therefore, it is important to be able to distinguish such treatments from those that have no effect, or even a deleterious effect, on overall survival.

In developed countries, most women who are diagnosed with breast cancer are diagnosed when the disease is at an early stage and is detected only in the breast itself—'node-negative' women—or in the breast and the lymph nodes near the affected breast—'node-positive' women. The primary treatment for most such women is surgery. However, the extent of the surgery considered necessary has varied substantially at different times and in different countries. There is also a wide variety of 'adjuvant' treatments that can be given in addition to surgery, and many hundreds of trials comparing the different treatments and combinations of treatments have been undertaken. The design of most of these trials is that women who satisfy a pre-specified set of entry criteria, for example in terms of age or extent of disease at diagnosis, are allocated at random to one of two possible treatment 'arms' that differ only with respect to the treatment being evaluated. For example the extent of surgery might be identical in both trial arms, but the women allocated to one of the trial arms might also be offered radiotherapy.

Where there are several trials that address similar questions it is in principle possible to obtain estimates of the differences between treatments by combining the data from them. This approach leads to much more precise estimates than those given by any individual trial, and such results will also be more stable than those for individual trials. Inevitably, trials with extreme results tend to receive more attention than those with more moderate results. This produces a natural tendency for unduly selective emphasis on those trials or subcategories of patients where, by chance alone, the results are misleadingly positive or mis-leadingly negative. Most such biases can be avoided by appropriate combination of the results of all trials that address similar questions. This combination cannot be done satisfactorily from published data alone (Stewart and Parmar 1993), and the inclusion of unpublished as well as published data is necessary to avoid bias. Furthermore, the information available from the published trials is not sufficient to allow a uniform analysis of all the available data with appropriate stratifi-cation for factors that will affect survival such as age, time since diagnosis, or nodal status. Thus, analysis based on individual patient data is necessary.

With these issues in mind, collaboration was sought in 1983–1984 between the co-ordinators of all randomized trials of the treatment of early breast cancer that satisfied certain criteria, and in 1985 the Early Breast Cancer Trialists' Collaborative Group (EBCTCG) was initiated. It has continued since then in five-yearly cycles. At the time of writing, the analyses resulting from the fourth cycle of the EBCTCG have been finalized, and initial preparations are being made for the fifth cycle, which will include data up to 2005. The fourth-cycle analyses are now available via the University of Oxford Clinical Trial Service Unit website, `www.ctsu.ox.ac.uk`, and in the published literature; see the note added in proof on page 196. The present chapter summarizes the main results of the earlier cycles of the EBCTCG, both in Table 9.1 and in the text, and it also comments on recent trends in breast cancer mortality. The results presented here are mainly those for overall survival. However, the original publications also consider other endpoints, including mortality from breast cancer and mortality from other specific causes of death, as well as breast cancer recurrence, that is, the return of the original cancer after a period of remission. In almost all cases, where a treatment has a beneficial effect on overall survival, this occurs following an earlier and larger effect on breast cancer recurrence.

9.2 The first cycle of the EBCTCG (1984 and 1985)

The first meeting of the nascent EBCTCG took place in London in October 1984 and preliminary analyses of data from two categories of trial were pre-sented. Both were concerned with the evaluation of 'systemic' treatments—those involving drugs that would reach not just the breast and local tissues, but all parts of the body to which microscopic deposits of the cancer might already have spread. The first category included trials in which women in one treat-ment arm received only the standard treatment schedule at the centre involved in terms of the extent of surgery and whether or not radiotherapy was given, while women in the other treatment arm received the same standard treatment plus treatment with the anti-oestrogen agent tamoxifen, which can inhibit the growth of tumours with appreciable oestrogen receptor expression—so-called 'ER-positive' tumours. The second category of trials included patients in which the two trial arms differed only by the addition of some form of long-term cytotoxic chemotherapy, consisting of the administration of one or more drugs intended to kill cancer cells.

The data were analysed using standard statistical methods comprising significance tests and appropriately weighted estimates of treatment effects, based on log-rank analyses, together with life-table estimates of survival, which have been described elsewhere (EBCTCG 1990). Before these analyses, no generally agreed conclusions had emerged about the effects on mortality of either of these types of systemic therapy in early breast cancer, but it was apparent even from the preliminary results presented at the 1984 meeting that there were clearly significant reductions in short-term mortality among those receiving either tamoxifen or chemotherapy (Anonymous 1984). The final published data confirmed these preliminary findings (EBCTCG 1990, 1988). Individual data were available in 1985, when the EBCTCG was formally established, from a total of 28 trials of the effects of tamoxifen in which a total of 16,500 women had been randomized, of whom nearly 4000 had died by the end of the available follow-up. When all ages at diagnosis were combined, there was a highly significant reduction ($p < 0.0001$) in mortality for the women who had been allocated to tamoxifen compared with the women who had not. More detailed analysis indicated—misleadingly, as became apparent from later analyses of larger datasets—that the beneficial effect of tamoxifen was concentrated in women who were aged 50 or more at diagnosis, in whom the annual death rate from all causes combined was reduced by about one fifth. Furthermore, since not all patients complied with the treatment assigned, the true beneficial effect of tamoxifen is likely to be somewhat greater than this.

As regards the trials of cytotoxic chemotherapy, data were available from a total of 40 trials involving over 13,000 randomized women, of whom just over 4000 had died. There was a highly significant reduction in the overall death rate in trials of any chemotherapy versus no chemotherapy ($p = 0.003$) and also in trials of polychemotherapy—that is, chemotherapy with more than one drug for more than one month—versus single-agent chemotherapy ($p = 0.001$). In contrast to the tamoxifen trials, the beneficial effect of the chemotherapy appeared to be concentrated in women aged under 50 at diagnosis, in whom the annual death rate was reduced by about one quarter. The chemotherapy trials also suggested that administration of chemotherapy for 12–24 months might offer little survival advantage over administration of the same chemotherapy for about six months.

9.3   The second cycle (1990)

Breast cancer is unusual in that survival without apparent recurrence for the first five years after diagnosis is by no means a guarantee that cure has been achieved. Indeed, among women diagnosed with early breast cancer, mortality rates from the disease remain substantially elevated for at least the next ten to twenty years (Quinn *et al.* 2001). In the first cycle of the EBCTCG the 8000 or so deaths in the randomized women were approximately evenly distributed over years 1, 2, 3, 4, and 5+ of follow-up, but there was little useful information

TABLE 9.1. The main results for overall survival in the randomized controlled trials considered by the Early Breast Cancer Trialists' Collaborative Group. Standard errors (se) in percent are given in parentheses.

Cycle	Data included	Main results for overall survival and references
First	Trials started before 1985. Follow-up to 1985.	Tamoxifen (28 trials involving 16,500 randomized women): Highly significant ($p < 0.0001$) reduction in mortality in trials of 'tamoxifen versus no tamoxifen' over about five years of follow-up (EBCTCG 1988).
		Chemotherapy (40 trials involving over 13,000 randomized women): Highly significant ($p = 0.003$) reduction in mortality in trials of 'any chemotherapy versus no chemotherapy', and also ($p = 0.001$) in trials of 'polychemotherapy versus single-agent chemotherapy', both over about five years of follow-up. Chemotherapy for 12–24 months may offer little survival advantage over 6 months (EBCTCG 1988).
Second	Trials started before 1985. Follow-up to 1990.	Tamoxifen (40 trials involving 30,000 randomized women): Highly significant 17% (se 2; $p < 0.00001$) reduction in overall mortality rate in trials of 'tamoxifen versus no tamoxifen' over about 10 years of follow-up. Indirect comparisons suggest longer term treatment (~2–5 years) better than shorter. Polychemotherapy plus tamoxifen clearly better than polychemotherapy alone at ages 50–69 (EBCTCG 1992).
		Chemotherapy (31 trials involving 11,000 randomized women): Highly significant 16% (se 3; $p < 0.00001$) reduction in overall mortality rate in trials of 'polychemotherapy versus no chemotherapy' over about 10 years of follow-up. No advantage in long-term treatment (~12 months) over shorter term (~6 months). Tamoxifen and polychemotherapy may be better than tamoxifen alone at ages 50–69 (EBCTCG 1992).
		Ovarian ablation (10 trials involving 3000 randomized women): Highly significant 25% (se 7; $p = 0.0004$) reduction in overall mortality rate for women treated at age < 50 over at least 10 years of follow-up. No significant effect for those aged 50+ when treated. (EBCTCG 1992).
		Immunotherapy (24 trials involving 6300 randomized women): No significantly favourable effects of immunotherapy found (EBCTCG 1992).
		Local therapies (64 trials involving 28,500 randomized women): Radiotherapy reduced rate of local recurrence by factor of three and breast conserving surgery involved some risk of recurrence in remaining tissue, but no definite differences in overall survival at 10 years (EBCTCG 1995).

Third	Trials started before 1990. Follow-up to 1995.	Tamoxifen: (55 trials involving 37,000 women): No significantly favourable effect in women with ER-negative tumours, but for 30,000 women with ER-positive or untested tumours highly significant effects with 12% (se 3), 17% (se 3) and 26% (se 4) reductions in the overall mortality rate in trials of 1, 2, and 5 years respectively of tamoxifen versus no tamoxifen, over about 10 years of follow-up. Proportionate benefit applies regardless of nodal status, age, menopausal status, daily tamoxifen dose, and whether or not chemotherapy was given (EBCTCG 1998b). Chemotherapy (69 trials involving 30,000 women): Highly significant reductions in overall mortality rate of 25% (se 5; $p < 0.00001$) in women aged < 50 and 11% (se 3; $p = 0.0001$) in women aged 50–69 in trials of "polychemotherapy versus no chemotherapy". Proportionate benefit similar regardless of nodal status, menopausal status (given age), ER-status and whether or not tamoxifen had been given. No advantage of more than about 6 months of treatment (EBCTCG 1998a).
		Ovarian ablation (12 trials involving 3500 women): No benefit for women aged over 50, but highly significant improvement in 15-year survival among those aged 50 or under when treated (15-year survival 52% versus 46%; $p = 0.001$) in trials of "ovarian ablation versus no ablation". Further evidence needed on effect of ablation in the presence of other adjuvant treatments (EBCTCG 1996).
		Radiotherapy (40 trials involving 20,000 women): Substantial reduction in breast cancer mortality largely offset by an increased risk of mortality from cardiovascular disease, so that overall 20-year survival was 37% with radiotherapy and 36% without ($p = 0.06$) in trials of "radiotherapy versus no radiotherapy". The ratio of benefit (from reduced mortality from breast cancer) to harm (from increased mortality from cardiovascular disease) was strongly affected by nodal status, age and decade of follow-up (EBCTCG 2000).
Fourth	Trials started before 1995. Follow-up to 2000.	Data on ~80,000 women randomised in trials of tamoxifen, ~50,000 women in trials of chemotherapy, 10 000 women in trials of ovarian ablation and ~38,000 women in trials of local therapies have recently been published (EBCTCG 2005); see the note added in proof on page 196.

on the effects of the treatments being compared after year 5. Therefore, the second cycle of the EBCTCG included trials that began before 1985, as in the first cycle, but the follow-up was now extended for a further five years, to 1990. This enabled the effects of treatment on mortality to be evaluated not just to five, but also to ten years after the diagnosis of breast cancer (EBCTCG 1992). Data were available for a total of 40 trials of the effect of tamoxifen versus the same standard treatments but without tamoxifen. These trials included nearly 30,000 women, comprising approximately 98% of all those randomized into eligible trials, and just over 8000 of them had died. A highly significant effect of tamoxifen was once again apparent ($p < 0.00001$), and women who had received the drug had 17% (standard error 2%—here and below we give standard errors in parentheses after the corresponding estimates) lower mortality rate over the entire period of follow-up than those who did not. It was also clear that tamoxifen substantially reduced—by 39% (9%), giving $p < 0.00001$—the risk of development of 'contralateral' breast cancer, that is, of a completely new cancer arising in the previously unaffected breast.

In the second cycle of the EBCTCG, data were available for 11,000 women randomized in a total of 31 trials of adjuvant polychemotherapy versus no chemotherapy—79% of all women randomized into eligible trials—of whom over 3500 had died. A highly significant effect of polychemotherapy was demonstrated ($p < 0.00001$) and women who had received polychemotherapy had mortality rate 16% lower than those who did not.

Treatment of pre-menopausal women with ovarian ablation, which destroys ovarian function, thus altering sex hormone levels and inducing an artificial menopause, may affect the progression of breast cancer and also survival. Data were available for just over 3000 women in ten randomized trials of the effect of ovarian ablation and these demonstrated a highly significant ($p = 0.0004$) reduction in mortality of 25% (7%) for the 1800 or so women treated under the age of 50. Curiously, this is the one EBCTCG result that did not prove durable. For the 1326 women randomized to ovarian ablation when aged over 50, ovarian ablation had no significant effect either on overall mortality or on recurrence-free survival. This is possibly due to the fact that the majority of such women would have been post-menopausal at the time they were randomized.

In the second cycle of the EBCTCG, data were also available on 6300 women randomized in 24 trials of immunotherapy given to increase the immune response of the woman's body to the tumour; eight trials of bacillus Calmette–Guérin, nine of levamisole, and seven of other agents. Neither in total, nor in any of the three subgroups, nor in any of the 24 separate trials was there any significantly favourable difference between treatment or control in either recurrence-free or overall survival—indeed, women given immunotherapy had 3% (4%) higher rates for both endpoints. Perhaps by chance, for bacillus Calmette–Guérin the adverse effect of 20% (8%) just reached a conventional level of statistical significance ($p = 0.02$). These results show that an overview can yield a strongly null result and, in particular, the results for bacillus Calmette–Guérin in the overview contrasted with previous claims of a benefit from it in 'historically controlled' comparisons (EBCTCG 1992, Hortobagyi *et al.* 1978).

In the second cycle of the EBCTCG, the reduction in mortality following either tamoxifen or the polychemotherapy regimens tested in the trials available at that time was highly significant both during and after years 0–4, so the cumulative differences in survival are larger at ten than at five years after initial treatment. Both direct and indirect randomized comparisons indicate long-term polychemotherapy, over 12 months, for example, to be no better than shorter, for example, of six months. However, indirect comparisons suggested that longer-term tamoxifen, such as daily for two or for five years, is significantly more effective than shorter tamoxifen regimens, such as daily for only 1 year. This observation, together with promising preliminary findings from several trials, which randomly assign women to different durations of adjuvant tamoxifen therapy—such as two versus five years or five versus ten years (Peto 1996)—led in the 1990s to the establishment of the ATLAS and aTTom trials (`www.ctsu.ox.ac.uk/atlas/`) that are seeking to establish reliably the effects of prolonging adjuvant tamoxifen by an extra five years among women who have already received a few years—usually five or so—of tamoxifen prior to being enrolled in the trials. These two trials address the question of the appropriate duration of adjuvant hormonal therapy in general. Therefore, their results should be relevant not only to tamoxifen itself but also to other hormonal treatments as they become available.

In women aged over 70 at diagnosis, the second cycle of the EBCTCG showed the efficacy of tamoxifen, but few women in this age range had been included in the trials of polychemotherapy. Among women aged 50–69 not only were tamoxifen and polychemotherapy both demonstrated to have a beneficial effect when given individually, but a directly randomized comparison also demonstrated polychemotherapy plus tamoxifen to be better than polychemotherapy alone; a 20% (4%) reduction in mortality rate was seen ($p < 0.00001$). Directly randomized comparisons also suggested that polychemotherapy plus tamoxifen may be better than tamoxifen alone, but at that time the reduction in mortality of 10% (4%) was not highly significant. At ages below 50, although both polychemotherapy and ovarian ablation were demonstrated to have a beneficial effect when given individually, the numbers available for the examination of their effects when given in combination were small and the associated standard errors correspondingly large.

In addition to examining data on the effects of systemic treatments, the second cycle of the EBCTCG also considered trials of the effects of different local therapies (EBCTCG 1995). Data were available from 36 trials comparing radiotherapy plus surgery with the same type of surgery alone, ten comparing more-extensive surgery with less extensive surgery and 18 comparing more-extensive surgery with less-extensive surgery plus radiotherapy. Information was available on mortality for 28,500 women, over 97% of the women randomized to eligible trials. Some of these local therapies had substantially different effects on the rates of local recurrence of the breast cancer: in particular, the addition of radiotherapy to surgery reduced the rate of local recurrence by a factor of three, and breast-conserving surgery involved some risk of recurrence in the remaining tissue. However, at least in the 1990 overview, these trials did not demonstrate any differences in overall survival at ten years.

9.4   The third cycle (1995)

In the third cycle of the EBCTCG, the number of eligible trials was expanded to include all those that began before 1990. Information was obtained on approximately 37,000 women who had been randomized in 55 trials of adjuvant tamoxifen versus no tamoxifen, comprising about 87% of all the women randomized into eligible trials (EBCTCG 1998b). Compared with the second cycle, the amount of evidence on events occurring more than five years after randomization was substantially increased and, in terms of numbers of deaths, the amount of evidence from trials of about five years of tamoxifen was doubled. Nearly 8000 of the randomized women had a low, or zero, level of oestrogen-receptor protein measured in their primary tumour; these are 'ER-negative' tumours. Among these women the overall effects of tamoxifen were small, with the annual mortality rate reduced by only 6% (4%), and there was no suggestion of any trend towards greater benefit with longer treatment. In contrast, among the remaining women, of whom there were 18 000 with ER-positive tumours and nearly 12,000 women with untested tumours, the results were striking: for trials of one year, two years, and five years of adjuvant tamoxifen, mortality was reduced by 12% (3%), 17% (3%), and 26% (4%), respectively, during ten years of follow-up. Not only were all these reductions highly significant individually, but there was some evidence of a greater effect with longer treatment ($p = 0.003$). The proportionate mortality reductions were similar for women with node-positive and node-negative disease, but in absolute terms the reductions were greater in node-positive women, in whom survival is generally poorer. In trials with the longest course of treatment, of about five years of tamoxifen, the absolute improvements in ten-year survival were 11% (2%) for node-positive (61% versus 50%; $p < 0.00001$) and 6% (1%) for node-negative (79% versus 73%; $p < 0.00001$). These benefits were largely irrespective of age, menopausal status, daily tamoxifen dose, and of whether or not the women in both arms of the trial had received chemotherapy. In terms of other outcomes, about five years of tamoxifen approximately quadrupled the incidence of endometrial cancer ($p = 0.0008$), and halved the incidence of cancer in the contralateral breast ($p < 0.00001$): in absolute terms, however, the decrease in the incidence of contralateral breast cancer was about twice as large as the increase in the incidence of endometrial cancer. Tamoxifen had no apparent effect on the incidence of colorectal cancer or, after exclusion of deaths from breast or endometrial cancer, on any of the other main categories of cause of death. About one extra death per 5000 woman-years of tamoxifen was attributed to pulmonary embolus but, based on the evidence available at that time, the excess was not statistically significant.

Information was available in the third cycle of the EBCTCG on about 18,000 women randomized in 47 trials of prolonged polychemotherapy versus no chemotherapy, about 6000 women randomized in 11 trials of longer versus shorter polychemotherapy, and about 6000 women in 11 trials of anthracycline-containing regimens versus cyclophosphamide, methotrexate, and fluorouracil (EBCTCG 1998a). In trials of polychemotherapy versus no chemotherapy there were highly significant reductions in mortality both for women aged under 50, for whom there was a 27% (5%) reduction ($p < 0.00001$), and for women aged 50–69 at randomization, who showed an 11% (3%) reduction ($p = 0.0001$), while few women aged 70 or over had been studied. After taking both age and time since randomization into account, the proportionate reductions in mortality were similar in women with node-negative and node-positive disease. These proportionate reductions suggest that, for women aged under 50 at randomization, the effect of polychemotherapy would be to increase a typical ten-year survival of 71% for those with node-negative disease to 78%, giving an absolute benefit of 7%, while for women with node-positive disease it would be to increase a typical ten-year survival of 42% to 53%, an absolute benefit of 11%. For women aged 50–69 at randomization, the corresponding increases would be from 67% to 69% for those with node-negative disease, an absolute benefit of only 2%, and 46% to 49% for those with node-positive disease, giving an absolute benefit of only 3%. At a given age, the benefits of polychemotherapy appeared largely independent of menopausal status at presentation, ER status of the primary tumour, and of whether or not adjuvant tamoxifen had been given. In addition, the directly randomized comparisons of polychemotherapy did not indicate any survival advantage with the use of more than about six months of polychemotherapy. In contrast, the directly randomised comparisons provided some evidence that, compared with cyclophosphamide, methotrexate, and fluorouracil alone, the anthracycline-containing regimens reduced mortality slightly, to a five-year survival of 69% versus 72% ($p = 0.02$). In terms of other endpoints, polychemotherapy reduced the incidence of cancer in the contralateral breast by about one-fifth and had no apparent adverse effect on deaths from causes other than breast cancer. Although the EBCTCG data show that polychemotherapy can improve long-term survival, it does have considerable short-term side effects, including leukopenia, nausea and vomiting, thromboembolic events, thrombocytopenia, anaemia, infection, mucositis, diarrhoea, and neurological toxicity (Pritchard *et al.* 1997, Fisher *et al.* 1990). The incidence of such events has not been reviewed by the EBCTCG, but clearly needs to be taken into account in treatment decisions.

Trials of the effect of ovarian ablation were also considered in the third cycle of the EBCTCG (EBCTCG 1996). Many of these trials began before 1980, and so 15 years of follow-up were available for analysis. Among the 1354 women aged 50 or over when randomized, there was no significant improvement of ovarian ablation on survival, as in the results obtained in the second cycle. However, among the 2102 women aged 50 or under when randomized, 15-year survival was highly significantly improved among those allocated ovarian ablation—52.4%

versus 46.1%, so 6.3% (2.3%) fewer deaths per 100 women treated—and the benefit was significant both for node-positive and for node-negative disease. In the trials of ablation plus cytotoxic chemotherapy versus the same chemotherapy alone, the benefit of ablation appeared smaller than in the trials in the absence of chemotherapy, and although there was no significant heterogeneity between these two subgroups, the benefit of ovarian ablation on survival was not significant when the trials in which both arms had received chemotherapy were considered on their own. It was concluded that further randomized evidence was needed on the additional effect of ovarian ablation in the presence of other adjuvant treatments and also on the relevance of hormone receptor measurements, which were only available for four of the 12 trials.

Many of the trials comparing radiotherapy plus surgery with the same type of surgery alone started even earlier than the trials of ovarian ablation. Therefore, in the third cycle of the EBCTCG, it was possible to examine the impact of radiotherapy at both ten and at twenty years (EBCTCG 2000). Data were available for a total of 40 trials, involving 20,000 women, half with node-positive disease. In these trials the radiotherapy fields included not only chest wall or residual breast, but also the axillary, supraclavicular, and internal mammary chain lymph nodes. Breast cancer mortality was reduced in the women who had received radiotherapy ($p = 0.0001$) but mortality from certain other causes, in particular cardiovascular disease, was increased, and overall 20-year survival was 37.1% with radiotherapy versus 35.9% without ($p = 0.06$). Nodal status, age and decade of follow-up strongly affected the ratio of breast cancer mortality to other mortality, and hence affected the ratio of absolute benefit to absolute hazard. It was estimated that, without the long-term hazard, radiotherapy would have produced an absolute increase in 20-year survival of about 2–4%, except for women at particularly low risk of local recurrence. The average hazard seen in these trials would, however, reduce this 20-year survival benefit in young women and reverse it in older women.

Until recently, it had been thought that an increased risk of cardiovascular disease occurred only following substantial doses of radiation, and improvements in radiotherapy techniques for breast cancer in recent years have tended to reduce radiation doses to the heart. However, there is now mounting evidence that doses that have traditionally been regarded by clinical oncologists as unimportant in terms of cardiovascular risk may, in fact, carry an appreciable long-term risk. The main evidence for this comes from the follow-up studies of the survivors of the atomic bombings of Hiroshima and Nagasaki, where significant dose-response relationships for heart disease and for stroke have been reported following whole body uniform doses (Preston *et al.* 2003). These data also suggest that if there is a threshold dose for non-cancer disease mortality then it cannot be greater than about 0.5 Gy and this suggests that, even with modern radiotherapy techniques, some cardiovascular risk may well remain. Further research is now being carried out, using data from cancer registries (Darby *et al.* 2003) together with detailed dosimetry data, to characterize more precisely the cardiovascular risk from radiotherapy for breast cancer.

9.5 The fourth cycle (2000)

The fourth cycle of the EBCTCG, whose results are currently being prepared for publication, has included even more data than the third round, with about 38,000 randomized women in trials of local therapies—including comparisons of radiotherapy plus surgery with the same type of surgery alone, more-extensive surgery with less extensive surgery, and more-extensive surgery with less-extensive surgery plus radiotherapy—and about 50,000 randomized women in trials of tamoxifen versus no tamoxifen, and 30,000 in the more interesting question of longer versus shorter durations of tamoxifen therapy, about 50,000 women involved in chemotherapy trials, and 10,000 in trials of ovarian ablation. In addition, for many of the women included in previous EBCTCG cycles, longer follow-up is available.

9.6 Recent trends in breast cancer mortality

During the period 1950–1960, the annual mortality rate for breast cancer in the age-range 35–69 years, calculated as the mean of the seven age-specific rates 35–39, 40–44, ... , 65–69 in the United Kingdom was around 58 per 100,000 women; see the left part of Figure 9.1. From about 1960, the rate started to rise, as in many other European countries, and this rise continued steadily until the late 1980s, by which time it was over 70 per 100,000 women. In 1990, however, it suddenly started to fall and it has fallen continuously since then, so that in 2000 it was only just over 50 deaths per 100,000, well below its value in the 1950s.

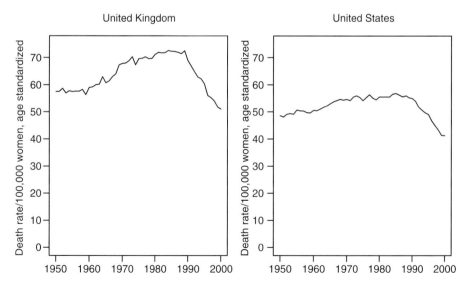

FIG. 9.1. Breast cancer mortality for women at ages 35–69 in the United Kingdom and United States, 1950–2000. The data are averaged annual rates in component five-year age groups, obtained using World Health Organization mortality and United Nations population estimates.

During the 1990s, lesser decreases have also occurred in several other countries, including the United States (Peto *et al.* 2000); see the right part of Figure 9.1. Recent data on breast cancer incidence, as opposed to mortality, are difficult to interpret as they are affected by screening. For example, in the UK the national screening program was introduced in 1987 for women aged 50–64, and has undoubtedly resulted in many cancers in this age range being diagnosed earlier than they would otherwise have been. However, there is no suggestion from the incidence rates at ages below 50 or above 65 of a sudden decrease in the underlying trend of breast cancer incidence rates that might be responsible for the recent decrease in mortality (Quinn *et al.* 2001). Some of the downward trends in mortality are likely to be the effect of the screening program, and detailed analyses have estimated that in the UK screening is responsible for 30–40% of the fall in mortality in age-groups 55–69 (Blanks *et al.* 2000, Sasieni 2003). The remaining 60–70% is likely to be explained partly by the tendency towards earlier presentation outside the screening program, but mostly by the increasing use of tamoxifen and chemotherapy from the mid-1980s, which has been helped by the EBCTCG analyses. In 2000, a United States National Institutes of Health Consensus Statement on adjuvant therapy for breast cancer (National Institutes of Health 2000) drew heavily on the findings of the EBCTCG for their recommendations on the use of tamoxifen, ovarian ablation and polychemotherapy. As a result of this and of similar treatment guidelines in other countries, further falls in breast cancer mortality worldwide can be expected in the future. At the same time new treatments for breast cancer that require rigorous assessment continue to be developed and provide further questions for assessment in future cycles of the EBCTCG.

9.7   Note Added in Proof

Since this chapter was written, the report on the effects of chemotherapy and hormonal therapy of the fourth cycle of the EBCTCG has been published (EBCTCG, 2005). This shows that some of the widely practicable adjuvant drug treatments that were being tested in the 1980s, which substantially reduced 5-year recurrence rates, also substantially reduce 15-year mortality rates. In particular, for middle-aged women with ER-positive disease, the breast cancer mortality rate throughout the next 15 years would be approximately halved by 6 months of anthracycline-based chemotherapy followed by 5 years of adjuvant tamoxifen. For, if mortality reductions of 38% (age < 50 years) and 20% (age 50–69 years) from such chemotherapy were followed by a further reduction of 31% from tamoxifen in the risks that remain, the final mortality reductions would be 57% and 45%, respectively. Overall survival would be comparably improved, since these treatments have relatively small effects on mortality from the aggregate of all other causes. Further improvements in long-term survival could well be available from newer drugs, or better use of older drugs.

10
How computing has changed statistics

Brian D. Ripley

10.1 Introduction

When I became a graduate student in 1973 I was pleased not to have to do any computing: although I had done some programming and some statistical consultancy during vacation employment, the experience had impressed upon me how tedious the implementation of statistical methods could be. I happily worked on stochastic point processes on abstract spaces, but after a couple of years people started suggesting to me practical applications of things I had been working on, and of course they wanted numerical answers or, even better, pretty pictures. Our computing resource was one IBM mainframe shared by the whole university, and statistics graduate students were so far down the pecking order that we each were allowed only a few seconds CPU time per day. A reasonably well-organized student could work out in less than an hour enough things to try to use up the day's allocation, and spend the rest of the working day trying to prove theorems.

Thirty years later the resources available are much less of a constraint. Often graduate students have computers on their own desktops that are much more powerful than the mainframe to which we had rationed access, and many have their own machines at home. Professors tend to have several machines, and for those who can convince funding agencies they need it, clusters of hundreds of CPUs are available. What are not available are single machines very much faster than those we provide to incoming research students.

The pace of change in available resources is obvious, and we attempt to quantify it in the next section. I believe though that attitudes to statistical computing have moved much less rapidly, and the extent to which useful statistics has become critically dependent on statistical computing is underappreciated, together with the effort that is being put into developing software and computing environments for modern statistics.

My story has in one sense little to do with David Cox: as his friends know, David sometimes likes to dabble in computing with interesting data but has had the great good sense to find collaborators who have managed the computing. That is not to say that he is either uninterested or disinterested in statistical computing: he is one of those who has recognized our critical dependence on computing and who has encouraged many people in the field.

10.2   The pace of change

Part of our brief for this volume was to give some history. That is daunting in the field of statistical computing: so much has changed in my working life, and that is less than half the length of David Cox's. My predecessor at the University of Strathclyde—Rupert Leslie who retired in 1983—still used 'computer' to refer to a human being, in the mould of the famous picture of R. A. Fisher caressing his electric calculator; see the frontispiece of Box (1978). One of the most famous of all failures of foresight was the quoted remark of Thomas Watson, then Chairman of IBM, in 1943 that

'I think there is a world market for maybe five computers'

—*he* certainly meant machines. If you want to follow that up, make use of one of the revolutionary changes wrought by computers: *Google* it.

The pace of change in computing has been relentless. The famous *Moore's Law* states that the number of transistors on an integrated circuit will increase exponentially. Moore (1965) projected to 1975, but his prediction has been followed for nearly 40 years, and Moore himself in 1997 thought it would hold good for another 20 years, and then hit physics-based limits. As always, simple things are never quite as simple as they appear, but Moore's law has been a good model for the costs of affordable storage, both RAM and disc space. Originally Moore stated a doubling every year but in 1975 he amended the prediction to a doubling every two years, and in the late 1980s an Intel executive observed in a glossy flyer that

'Moore's Law was driving a doubling of computing performance every 18 months.'

which is the form in which Moore's law is most widely known.

The folk version of Moore's law *is* a reasonable indicator of the pace of change. When I moved to Oxford in 1990 my home computer was a 25-MHz Sun workstation with 12 Mb of RAM and 400 Mb of disc space: its successor bought in late 2003 has a 2.6 GHz processor with 1 Gb RAM and 160 Gb of disc space, and was a quarter of the price. The folk version of Moore's Law predicts a 400-fold increase, and although the processor speed has increased by less than that, what the processor can do per clock cycle has increased—and my office machine has two such processors, while my department recently acquired a machine with over 100 faster 64-bit processors. So computers have got relentlessly faster and cheaper, so much so that we can each have several of them, and we can expect that pace of change to continue for the foreseeable future.

How have statisticians adapted to such changes? Like all computer users we have become more impatient and expect an instant answer, and with the possibility of a near-instant response we try out many more ideas when faced with a set of data. There is a tendency to replace thinking by trying, though, and sensible conclusions are not reached all that much faster.

Some evidence comes from the way we assess the practical skills of Master's students. Twenty-five years ago David Cox and I were both at Imperial College London and concerned with how to teach applied statistics, in the sense of Cox and Snell (1981) and Venables and Ripley (2002). We noticed when we stopped teaching hand calculations and just presented regression as a black box whose output needed to be interpreted that there was a jump in the quality of students' reports. However, my recollection is that we had three-hour practical exams—down from seven hours in the days of electronic calculators—and expected the students to do two or three pieces of data analysis. Nowadays in Oxford we set problems of similar size, with four or five to be done in a five-day week in open-book conditions. Some students must spend close to 100 hours in front of a computer during that week. These days the reports are elegantly typeset and the figures are beautiful and often in colour, but it is not clear that the analyses are any more penetrating.

One big difference is that we now assess parts of statistics, notably multivariate analysis, that were inaccessible then for lack of software and computing power.

Storage is another aspect. John Tukey was an early advocate of the use of computers in statistics, yet *Exploratory Data Analysis* (Tukey 1977) was developed for hand calculation. I recall asking him why, and being told that he only had his HP calculator with him on plane trips for consulting. As all plane travellers know, businessmen have for a decade or so been crunching their spreadsheets *en route* to sales opportunities, and I am sure that if John were still with us he would have embraced the new technologies. I think it was in 1998 I first encountered people who carried their 'life's work' around in a small pack of CD-ROMs. Today we have DVDs holding 4.5 Gb, and my career's statistics work, including all the data, fits on a couple of those. We think of writing a book as a large amount of work but each of mine amount to just 1–2 Mb. Before very long a 'life's work' will fit on a key-fob memory device.

10.3   What can we do with all that power?

One sense of 'computer-intensive' statistics is just statistical methodology that makes use of a large amount of computer time. Examples include the bootstrap, smoothing, image analysis, and many uses of the EM algorithm. However, the term is usually used for methods that go beyond the minimum of calculations needed for an illuminating analysis, for example by replacing analytical approximations by computational ones, or requiring numerical optimization or integration over high-dimensional spaces. I wrote ten years ago about neural networks (Ripley 1993):

> Their pervasiveness means that they cannot be ignored. In one way their success is a warning to statisticians who have worked in a simply-structured linear world for too long.

The extent of non-linearity that is not just being contemplated but is being built into consumer hardware goes far beyond non-linear regression in the sense of, for example, Bates and Watts (1988). So here are several ways to use 'all that power':

- working with (much) larger datasets;
- using more realistic models and better ways to fit models;
- exploring a (much) larger class of models;
- attempting a more realistic analysis of existing simple models; and
- better visualization of data or fitted models or their combination.

Not everything has changed at the same rate, if at all. Computers have helped collect data, but we have already reached some limits. In 1993 Patty Solomon and I worked on a dataset of all 2843 pre-1990 AIDs patients in Australia. This was published as an example in Venables and Ripley (1994), and I had to work quite hard to analyse it within my then computing resources: it took around an hour. I don't know how long it would take now—nor do I care as the whole of the survival analysis chapter of what has become Venables and Ripley (2002) can be run in 7 seconds. That was a large dataset in survival analysis then, and it still is.

Data mining is a currently popular term for exploring large datasets, although one of my favourite quotes is (Witten and Franke 2000, p. 26)

> Cynics, looking wryly at the explosion of commercial interest (and hype) in this area, equate data mining to statistics plus marketing.

Many datasets are already close to the maximal possible size. Over the last three years my D. Phil. student Fei Chen has been looking at data mining in the insurance industry—motor insurance actuaries already have databases of 66 million motor insurance proposals, some one third of all drivers in the USA. There are around 30 items for each driver, and that is not going to increase much as potential customers will not be prepared to answer more questions—and the more questions they are asked the less reliable their answers will become.

There *are* fields in which there is the potential to collect substantially more data on common activities. So-called *Customer Relationship Management* uses loyalty cards to track the shopping habits by individual customer in, for example, supermarkets. Fei is now employed to do data mining for fraud detection, looking for unusual patterns of activity in, say, credit-card transactions. But even these fields have limits that are not so far off given the changes predicted by Moore's law, and it seems that working with all the available data will be the norm in almost all fields within a few years.

In the theory of computational complexity an exponential growth is regarded as very bad indeed, and most lauded algorithms run in at most polynomial time in the size of the problem, in some suitable sense. But when the available resources are growing exponentially the constants do matter in determining for polynomial-time algorithms when they will become feasible, and for exponential algorithms if they ever will. Consider linear regression, with n cases and p regressors. The time taken for the least-squares fitting problem is $O(np^2)$—this is the number of operations required to form the normal equations; to solve a $p \times p$ system needs $O(p^3)$ where we know $p \leq n$, and more accurate methods based on the QR decomposition have the same complexity. For a fixed set of explanatory variables this is linear in n. We know that we need $n \geq p$ to determine the coefficients, but if we have access to a large amount of data, how large should we take n? About 1997 some software packages started to boast of their ability to fit regressions with at least 10,000 cases, and Bill Venables and I discussed one evening if we had ever seen a regression that large (no) and if we ever would. Although we 'know' that theory suggests that the uncertainty in the estimated coefficients goes down at rate $O(1/\sqrt{n})$, it is easy to overlook the conditions attached. The most important are probably:

- the data are collected by a process close to independent identically distributed sampling; and
- the data were actually generated by the linear regression being fitted for some unknown set of coefficients.

Neither of these is likely to be realistic. Large datasets are rarely homogeneous and often include identifiable sub groups that might be expected to behave differently. Over the years I have known him I have heard David Cox stress quite often the benefits of applying simple analyses to such subgroups and then analysing the results for the subgroups as a new dataset. A more formal analysis might make use of *mixed models*, which include random effects for different subgroups (Diggle *et al.* 2002, Pinheiro and Bates 2000). These require orders of magnitude more computation and under reasonable assumptions may be quadratic in the number of groups. There are examples from educational testing that probably exceed the limits of current software.

Suppose the homogeneity were satisfied. A failure of the second assumption will lead to systematic errors in prediction from the model, and it is very likely that systematic errors will dwarf random errors before n reaches 10,000. As another famous quotation goes (G. E. P. Box, 1976):

All models are false, but some are useful,

and as n increases the less falsehood we will tolerate for a model to be useful—at least for prediction. So for large n we find ourselves adding more explanatory variables, adding interactions, non-linear terms or even applying non-linear regression models such as neural networks. This assumes, of course, that there is a payoff from an accurate model, but if many data points have been collected it usually is the case that the model will be applied to even larger numbers of cases. It seems that in practice p increases at roughly the same rate as n so we really have a $O(n^3)$ methodology. On the other hand, the number of possible submodels of a regression model with p regressors is $2^p - 1$, so exhaustive search remains prohibitive as a method of model selection. For some selection criteria we can short-circuit some of the fits: in 1974 Furnival and Wilson were able to select the 10 best subsets of 30 or so regressors in a couple of minutes and the practical limit today is only about treble that. The S-PLUS implementation of their procedure is still limited to $p \leq 30$, and for $n = 1000, p = 30$ takes around

0.1 seconds on my home machine. The R implementation allows $p < 50$ and even more regressors if a safety catch is released. However, $p = 50$ took around 1 minute and $p = 70$ took 25 minutes, in all cases on an example where almost all the regressors were actually unrelated to the response.

Just because we can now apply simple methods to large datasets should not of itself encourage doing so.

Other areas of statistics have algorithms that scale much less well. Even today, running a multidimensional scaling algorithm (Borg and Groenen 1997, Cox and Cox 2001, Venables and Ripley 2002) on a dataset of 200 points is only just about fast enough for a demonstration during a lecture. Kruskal (1964*b*) pointed out that this was a problem that was hard to optimize, and one thing we can do with lots of computing power is to try to optimize better by starting from a range of initial configurations. This is something that should be common practice, but when looking for a good example of factor analysis for Venables and Ripley (2002) and to test the factor analysis code I wrote for R, I found that the majority of published examples I tried to recompute were only displaying a suboptimal local extremum and often the best solutions I could find were degenerate ones, so-called *Heywood cases.*

Multidimensional scaling represents a dissimilarity matrix through distances between points in a low-dimensional configuration space. Most often Euclidean distance is used, but psychometricians have been interested in determining if the best L_p metric occurs for $p = 2$. They call the L_p metric *Minkoswki distance*, and the idea goes back to Kruskal (1964*a*), one of the earliest papers on multidimensional scaling; for a more modern account see Borg and Groenen (1997, §16.3). So we need to apply the multidimensional scaling optimizer a number of times for each of say ten values of p, when a single run for 200 points currently takes a few seconds.

This example raised an issue that is normally ignored. It transpired that when run under the Windows version of R, the code for $p \neq 2$ was several times slower than that for $p = 2$, but this was not the case for other platforms. Some quick calculations showed that for each two-dimensional configuration some 40 000 calculations of x^p were being done, and tens of configurations were being assessed. A few months earlier Phillippe Grosjean had pointed out that x^p was rather slow in the run-time library used by R under Windows and I had asked if there was a practical problem that used so many such calculations that one would notice the difference. The fates had found me one! Some digging showed that a fast but inaccurate x^p function had been replaced by a slow but accurate one. This application did not need the accuracy, but it is hard to know when you do need it, and general-purpose software needs to be accurate first and fast second.

We do rely on the basic software we use unknowingly to be fast, accurate, and reliable. It is not always deserving of our blind faith.

Given that we are worrying about computational speed today, how was Kruskal able to work on such problems in 1964? Part of the answer must be that he had an unusually rich computing environment at what was then Bell

Telephone Laboratories, but the main answer is problem size. We were looking at a multidimensional scaling plot of cases, whereas psychologists are often interested in a plot of variables, of which there are normally many fewer. Kruskal (1964*a*) said

When n is large, say $n = 50$ or 60 there are a great many dissimilarities ...

and his example using Minkowski distances has $n = 14$ cases. It is not easy to estimate well the computational complexity of multidimensional scaling algorithms. Given n cases (or variables) to be represented in d (usually 2) dimensions there are nd parameters to be optimized over and $n(n-1)/2$ observed and fitted dissimilarities to be compared. For each evaluation of the objective we must perform an isotonic regression of the fitted dissimilarities on the observed ones, and although that is at least quadratic in n, its complexity depends on the data. And how complex the optimization task is appears to depend on n. What we can do is collect data from some experiments. Figure 10.1 shows the results of such an experiment. A power law with a slope of just over two is a reasonable fit to these data. What is worth remarking is that this experiment, including all the programming and drawing the graph, took about 15 minutes, with a little

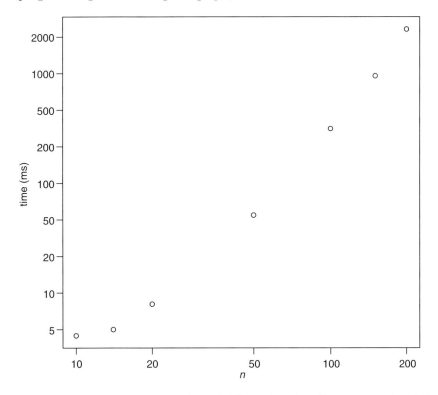

FIG. 10.1. Times to run non-metric multidimensional scaling on samples of size n from a dataset of 213 cases, including $n = 14$. Note the log–log axes.

help from R. Yet I had been wondering for a decade if $O(n^2)$ was a reasonable estimate of the computational complexity of non-metric multidimensional scaling.

What can we do if n is too large for the available computational resources? One idea is to be less ambitious and give up the finely tuned fit criterion used in non-metric multidimensional scaling. Teuvo Kohonen, an electrical engineer, described his motivation in developing what he called *self-organizing maps* as (Kohonen 1995, p. VI)

> I just wanted an algorithm that would effectively map similar patterns (pattern vectors close to each other in the input signal space) onto contiguous locations in the output space.

This sounds very like multidimensional scaling, and the only real difference is that his 'output space' is discrete—a hexagonal lattice, say—rather than continuous. Kohonen really was much less ambitious, and wrote down an algorithm that involves repeated passes over the dataset and has been proved not to actually optimize anything (Erwin *et al.* 1992, Ritter *et al.* 1992). Its advantages are that it can be run for as long as is feasible, and that additional cases can be plotted on the self-organizing map, in contrast to multidimensional scaling. There are corresponding disadvantages: different runs of the algorithm give different displays, often radically different ones, and without a fit criterion we cannot even say which gives a better fit.

10.4   Statistical software

It is not really the change in computational speed as predicted by Moore's Law that has affected the way we do things. If computers were still accessed by the interfaces of the 1960s they would be the preserve of a specialist cadre of operators/programmers, and it is the software that they run that has made computers so widely acceptable. Point-and-click interfaces are now so commonplace that we encounter graduate students who have never seen a command line.

It is statistical software that has revolutionized the way we approach data analysis, replacing the calculators—mechanical, electrical, electronic—used by earlier generations. We should remember that data analysis is not done only, or even mainly, by statisticians: spreadsheets, notably Excel, are almost certainly the dominant tools in data analysis.

Software has often been an access barrier to statistical methods. Many times over the years I would have liked to try something out or give my students a chance to try a method out, and have been frustrated by the inaccessibility of software—for reasons of expense, usage conditions, or machine requirements. There are different amounts of choice available to users of Windows, Unix in its many variants, Linux, and MacOS X. An old adage is

> one should choose one's hardware and operating system to run the software one needs

but many users do not have that degree of choice, and for those who do, having multiple platforms is costly, inconvenient, and demands a lot more knowledge. The standardization in the Unix-alike world on remote access protocols means that one can often access Unix/Linux software on a server from any of these systems, but that is not so for the Windows and MacOS X interfaces.

Software choice is important, with far-reaching implications. Perhaps not as far-reaching as an advertisement I saw recently that contained the quotation

'I'm a LECTURER IN STATISTICS—responsible for ensuring that good statistical practise [sic] becomes the norm with new generations of analysts. **Which is why I chose Xxxxx.**'

This does seem an ambitious goal for one lecturer or one piece of software. Package 'Xxxxx' describes itself as 'a cutting-edge statistical software package'. One of the most difficult tasks in training the data analysts of the future is predicting what it will be important for them to know. Having software available biases that choice, but teaching the right mix of methodology and how to use it well is far more important. However, package 'Xxxxx' is not available for the operating system I use daily, and so I know little of its capabilities.

It seems remarkable to me how little effort statisticians put into getting their favourite statistical methods used, in sharp contrast to many other communities who also do data analysis. Some are very entrepreneurial, and after developing a novel method rush off to patent it and found a company to sell software to implement it. It has become recognized that as the developer of a method some protection is needed that when people use 'your' method it really is what you intended. One solution is to expect a *reference implementation*, some code that is warranted to give the authors' intended answers in a moderately sized problem. It need not be efficient, but it should be available to anyone and everyone.

Open-Source software

This freedom of choice has exercised software developers for a long time, under headings such as 'Open Source' and 'Free Software'. These are emotive terms, coined by proselytizing zealots but with sound arguments. The Free Software Foundation is 'free as in speech, not free as in beer'. The term 'Open Source' was coined more recently, with a precise definition due to Bruce Perens (DiBona *et al.* 1999). The aim is to make software

available to anyone and everyone

and in particular to be freely redistributable. Other freedoms, such as the freedom to know how things work, may be equally important.

The R project is an Open Source statistics project. It may not be nirvana, and it may not be suitable for everyone, but it is a conscious attempt to provide a high-quality environment for leading-edge statistics that is available to everyone. It is free even 'as in beer'. You can download the source code (at `www.r-project.org`) that compiles on almost all current Unix and Linux systems, as well as binary versions for the major Linux distributions, FreeBSD, MacOS X and 32-bit Windows. This makes it a good environment for a reference implementation of a statistical method as the only barrier to understanding how it works, precisely, is skill.

To the end-user R looks like a dialect of the S language developed by John Chambers and his colleagues over 25 years, and documented in a series of books (Becker *et al.* 1988, Chambers and Hastie 1992, Chambers 1998).

R was originally written in the early 1990s by Ross Ihaka and Robert Gentleman, then of the University of Auckland: Ihaka and Gentleman (1996) is an early report on their work. Early versions were used at Auckland for elementary classes, on Macintoshes with 2 Mb of memory. By 1997 other people had become involved, and a *core team* had been set up with write access to the source code. There was a Windows version, and Linux users pushed development forward, there being no S version available for Linux at the time.

The R core team aim to provide a basic system and facilities for commonly used statistical methods. It can be extended by *packages*, with over 500 being available from CRAN, a central repository run by the core team, another 50 at a computational biology repository at `www.bioconductor.org`, and others less widely available. R's stronghold is in the academic world but it is now increasingly being used in commercial companies and even in their products.

Software quality

For me, one of the principal attractions of R is that with enough diligence I can find out exactly what is the algorithm implemented and come to a view if it has been correctly implemented. The core team are very good at fixing bugs, but there is a public repository at `bugs.r-project.org` so you can look up what others users think is wrong, and judge their credibility.

Kurt Hornik and other members of the R community have put a lot of thought into how to maintain a large statistical software system. A rather formal quality assurance scheme has been designed, and there are quality control checks on both core R and the packages that go into the repositories. Note that this is mainly on form and not content: we can check that software does what its designer expected it to do, but not that the formulae he implemented are correct nor that the algorithm is stable enough, and so on. There is a danger in regression testing that the software works in unusual cases where it used to fail but no longer gives the correct answer in the common cases.

It probably is the case that by far most incorrect statistical analyses result from user error rather than incorrect software, but the latter is often not even considered. The issue has been highlighted recently by a controversy in June 2002 over the link between 'sooty air pollution' and higher death rates (Dominici *et al.* 2002). The *New York Times* reported on June 5, and the Johns Hopkins' press release stated

> Their work appeared in the *New England Journal of Medicine* and other peer-reviewed publications. While updating and expanding their work, the investigators recognized a limitation of the S-plus statistical program used to analyze data for the study. The S-plus program is standard statistical software used

by many researchers around the world for time-series and other analyses. The Hopkins investigators determined that the default criteria in one regression program used for examining patterns and fitting the statistical model (referred to as *convergence criteria*) were too lenient for this application, resulting in an upward bias in the estimated effect of air pollution on mortality.

To be blunt, this was a case of users blaming their tools with only a little cause—the need to change this default is in a certain well-known book (Venables and Ripley 2002). But all credit to them for actually checking. Bert Gunter, a senior statistician at Merck, commented

> Data analysis is a tricky business—a trickier business than even tricky data analysts sometimes think.

The message is that for all but the simplest data analyses we rely on lots of decisions made by software designers just as much as we rely on those theorems about distributions of test statistics. However, we expect those theorems to have published proofs we can consult, and to have been refereed by experts in the field. We accept software with no access to the workings and no review procedure. If the convergence criteria had not been a user-settable parameter the Hopkins team might never have discovered the sensitivity.

This is just a larger-scale version of the comment we made about x^p: accuracy first, speed second. Nowadays we rely on the ability to fit a Poisson log-linear model accurately as much as we rely on our calculators' ability to multiply. I suspect few readers will have consulted a book of statistical tables in the last year, instead using the equivalents built into statistical packages. Beware: they are found wanting alarmingly frequently.

Graphical software

For many years I was frustrated by the inability to draw elegant graphs, and to explore graphically data and models. We do perhaps now have the hardware to do what I have been seeking, but we do not yet have the software. One problem is that statistical packages, including R, are stuck in a 1970's model for graphics, putting ink down on two-dimensional paper. The only difference is that we may now be allowed a limited choice of coloured inks.

One of the things I noticed about S when first teaching with it in the 1980s was how much better its default plots looked than those of the statistical packages we had been using previously, although this also had much to do with the availability of laser printers. Not very much has changed since. Graphical presentation is a skill that we teach, often by bad examples, in first statistics courses and then ignore. A lot of research has been done on visual perception, good colour schemes and so on, but it has not yet reached our software. Two thought-provoking monographs are Spence (2001) and Wilkinson (1999).

10.5 Case studies

I selected two examples to illustrate several of the points made so far.

Classification Trees—CART

Classification trees is one area that illustrates the importance of software. They have been somewhat independently developed in machine learning, electrical engineering, and statistics, from the mid-1970s to the end of the 1980s: Ripley (1996) reviews the development.

The 1984 book *Classification and Regression Trees* (Breiman *et al.* 1984) was a seminal account. Unusually for statisticians, they marketed their software, CART®.

The other communities also marketed their software. Ross Quinlan even wrote a book (Quinlan 1993) about his, C4.5, containing the source code *but not allowing* readers to use it. The C code could be bought separately, but it 'may not be used for commercial purposes or gain'.

The net effect is that classification trees did not enter the mainstream of statistical methodology. Neither CART nor C4.5 had a user-friendly interface. I was frustrated—this was appealing methodology from which I was locked out in the 1980s by non-availability of software. We did manage to procure a teaching licence for some machine-learning software, but it was very slow.

The advent of classification and regression trees in S in 1991 made the technique much more accessible. Unfortunately the implementation was bug-ridden. Eventually, I decided to write my own implementation to try to find out what the correct answers were. I wish I had realized how easy it would be: all the commercial vendors had a vested interest in making it seem obscure and hard. That software is now the `tree` package in R, and Terry Therneau had reimplemented CART during his Ph.D. and his code formed the basis of the package `rpart`.

Some lessons I learnt were:

- having the source code available makes it *much* easier to find out what is actually done;

- having independent open implementations increases confidence in each;

- people keep on reporting discrepancies between the implementations. Almost inevitably these are not using comparable 'tuning' parameters, and people never appreciate how important these are; and

- classification trees are not as easy to interpret as people in the machine-learning community make out.

Figure 10.2 supports the last two claims. This is a dataset of 214 fragments of glass found at scenes of crime, and used as a running example in Ripley (1996) and Venables and Ripley (2002). There are six classes of glass, window float glass, window non-float glass, vehicle window glass, containers, tableware, and vehicle headlamps. A classification tree divides the nine-dimensional space of observations into cuboid regions with faces parallel to the axes. The trees shown have eight and nine regions, and were produced in package `rpart` by selecting one of the two supported fitting criteria. They fit equally well and they give classification rules that is easy to interpret—they just give rather different rules.

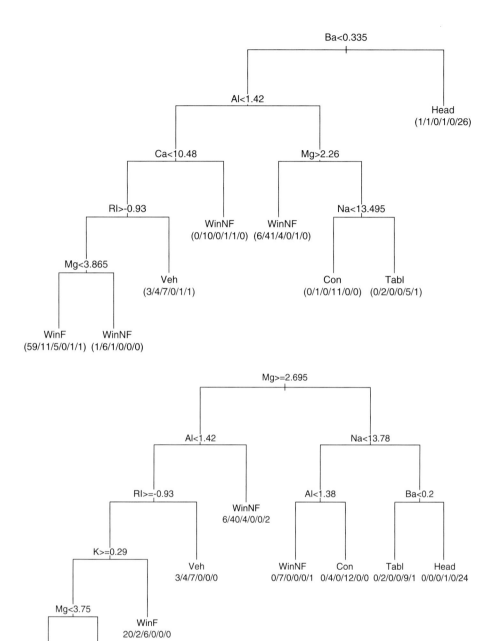

FIG. 10.2. Two classification trees of the same dataset, using the 'information' index (top) and the Gini index. The numbers below each node are the numbers of cases in each of the six classes reaching that node.

Serial brain scans

Bradley *et al.* (2002) reported on a serial magnetic resonance imaging study on the early detection of physical brain changes in Alzheimer's patients. There were 39 subjects in three groups, marked on the strips in Figure 10.3, each being brain-scanned 2–4 times over up to 15 months. Everyone's brain shrinks with age at about 0.4% per year, and not uniformly. Diseases such as Alzheimer's change both the overall rate and the differences in rates in different parts of the brain.

This is an example with massive amounts of raw data, for these scans are measured in megabytes. However, much of the data is not useful information, and pairs of scans were used to delimit brain regions in the same way for each of the pair and then to measure volumes of the regions, reducing megabytes of data to tens of bytes of information.

Figure 10.3 shows the data and for each patient a dashed line showing the mean rate of change for the alloted group. Two patients whose panels are marked with a dot were later shown to have been incorrectly allocated to the 'normals' group. As is typical of medical studies, the real shortage of data is in the small number of diseased patients, especially as one expects diseased groups to be more variable than the control group.

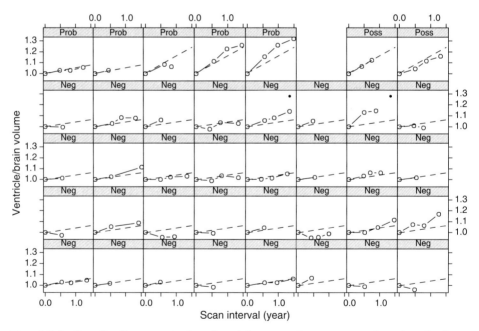

FIG. 10.3. Results from a study of serial magnetic resonance imaging brain scans of 39 subjects. The open circles joined by lines are the data for that subject, whereas the dashed lines are 'fitted' lines from the model. The labels refer to the *a priori* groups of *negative, possible* and *probable*.

The results are promising and have been used to estimate how large the time spacing between pairs of measurements and how large an experimental group would need to be to detect the effect of an anti-Alzheimer's drug.

How I came to be involved with this is an interesting comment on the sociology of science. The data were collected by the OPTIMA project run by Oxford University's Professor of Pharmacology, David Smith. He had talked about this study at a meeting in Washington University, St. Louis, and someone in the audience had advised him to use generalized estimating equations (Diggle *et al.* 2002) and furthermore that Brian Ripley in his own University was *the* expert— a rare case of getting too much credit for software development. I did not use generalized estimating equations, but fitted a linear mixed-effects model with a random slope for each subject. The dashed lines shown in Figure 10.3 are the best linear unbiased predictors (Robinson 1991) for each subject. From the estimated variance components we calculated power analyses for various hypothetical studies. With such small sample sizes it did prove to be important to use a formal model to obtain as precise as possible estimates of the crucial variance components.

The availability of software has been crucial to the surge of interest in mixed models. It is not that we see many more problems needing random effects than we did a decade ago but that we can now proffer software solutions. In this example I used the `nlme3` software for S-PLUS of José Pinheiro and Douglas Bates.

10.6 The future

Thomas Watson's failure of foresight quoted at the start of this chapter discourages too much crystal-ball gazing! Statistical computing is a field in which we stand on the shoulders of others, and issues that filled the books on statistical computing a generation ago are now taken so much for granted that they are completely unknown to the present generation of students.

It is clear that quality of the tools we use is already becoming a major issue. Statistical computing environments are now far too complex for empirical testing to be effective, and we need to build quality in by documenting exactly the algorithms used and allowing subject-matter experts to 'referee' the code.

We may be coming to the end of the era of user interfaces that look like high-level computer programs. Indeed, many commercial products have all but abandoned that form of interface in favour of systems based on menus and dialogue boxes, often restricted to Windows or MacOS platforms. No one seems as yet to have found a way to build a graphical interface in an easily extensible style. My hunch is that this will come, and the way we interact with computers to do statistics by the time of David Cox's 100th birthday will be something as yet unforeseen.

11

Are there discontinuities in financial prices?

Neil Shephard

11.1 Introduction

11.1.1 Main questions

In this chapter I will try to discover if there are discontinuities in financial price processes. Such discontinuities or jumps are important as they are hard to hedge and so increase the risk associated with financial instruments. My approach will be based around two radically different methods: one parametric, and the other semiparametric. I will argue that in order to find such discontinuities it is important to use relatively high-frequency data and that the easiest way of doing this is through the semiparametric approach. When this is implemented empirically, the evidence suggests that jumps are frequent—much more frequent than is commonly believed. Parametric analysis based on a novel particle filter, using standard low-frequency daily returns data, is unable to pick up price discontinuities that are plainly present in the price process, because the daily data blur the impact of discontinuities to such a degree that jumps are of only moderate importance at the daily level.

The parametric and semiparametric approaches to inference will be compared using a dataset on the exchange rate between the Japanese yen and United States dollar. This time series covers the ten-year period from 1 December 1986 until 30 November 1996. The original dataset records prices every five minutes during the working week and was kindly supplied by Olsen and Associates in Zurich (Dacorogna *et al.* 2001). Over this period the rate drops from around 200 to around 110 yen to the dollar, roughly where it stood in mid-2004. A declining rate indicates a weakening dollar.

The upper left panel of Figure 11.1 shows the evolution of the daily percentage change in the rate from 1 December 1986 over the first two years of the dataset. There is a 20% fall during this period. The upper right panel records the daily percentage changes, which are often called returns in financial econometrics. It shows some dramatic daily changes, which tend to cluster together. More detailed information is available from the higher-frequency data. The lower left panel is like the panel above, but shows the results only for the first five days, while the lower right panel shows five-minute returns over this shorter period. I will use these high-frequency data below.

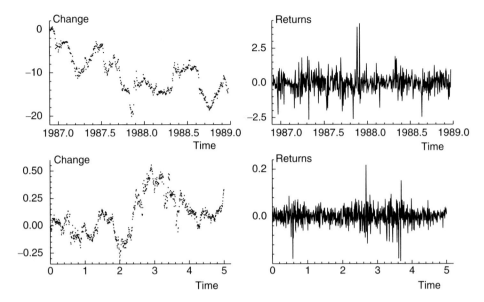

FIG. 11.1. Yen/dollar exchange rate, based on the Olsen dataset. Upper left: percentage change since 1 December 1986 over first two years of sample. Upper right: daily percentage changes. Lower left: percentage change over first five days of sample. Lower right: five-minute returns over first five days in sample.

11.1.2 Notation and models

I will let one unit of time represent a day throughout, and log-prices Y_t for time $t \geq 0$ will be defined on a filtered probability space, $(\Omega, \mathcal{F}, (\mathcal{F}_t)_{t \geq 0}, P)$; typically $\mathcal{F}_t$ will be the natural filtration. In the subset of financial economics that rules out arbitrage—that is, riskless money making—opportunities, Y must follow a semimartingale process, written $Y \in \mathcal{SM}$, with respect to $\mathcal{F}_t$. Such a no-arbitrage assumption seems sensible over moderate time intervals but over very small intervals of time market frictions play a large role even in thickly traded markets such as the market for the dollar and yen. Frictions such as discreteness of trading, bid/ask spread effects—the difference between the price one pays when one instigates the buying of a good and what one obtains when one tries to sell it—or finiteness of liquidity over small time intervals, are called market microstructure effects. They limit the realism of the semimartingale models of prices. For the moment I will ignore them, returning to this issue in Section 11.4.1.

If $Y \in \mathcal{SM}$ then it is possible to write

$$Y = A + M,$$

although this decomposition is not unique when Y exhibits jumps. Here A is a finite variation process—written $A \in \mathcal{FV}$—which means that the sum of the absolute values of the increments of this process measured over any set of time intervals is finite. Further, M is a local martingale—written $M \in \mathcal{M}_{\text{loc}}$—which is a convenient generalization of a martingale in financial economics. For an excellent discussion of probabilistic aspects of semimartingales see Protter (1992), while their attraction from an economic viewpoint is discussed by Back (1991). The A process can be thought of informally as the reward to the investor for being exposed to the risk M. This will become clearer when we get to eqn (11.2). The Y process will often be divided into the piece that has continuous sample paths and the piece with jumps. I generically denote this with superscripts c and d respectively; thus M^c denotes the continuous component of M.

A key role in the econometrics of price processes is played by the quadratic variation process. This always exists if the multivariate price process $Y \in \mathcal{SM}$ and is defined as (e.g. Protter 1992, Section II.6)

$$[Y] = \underset{n \to \infty}{\text{p-lim}} \sum_{j=1}^{n} \left(Y_{t_j} - Y_{t_j - 1} \right) \left(Y_{t_j} - Y_{t_j - 1} \right)^{\mathrm{T}}, \tag{11.1}$$

where ^{T} denotes a transpose and p-lim denotes the probability limit. The limit can be shown to hold for any sequence of partitions $t_0 = 0 < t_1 < \cdots < t_n = t$ with $\sup_j \{t_j - t_{j-1}\} \to 0$ for $n \to \infty$. It sums up outer-products of vectors of returns measured over infinitesimal time intervals. It will be useful later to note that

$$[Y]_t = [Y^c]_t + \sum_{0 \le s \le t} \Delta Y_s \Delta Y_s^{\mathrm{T}},$$

where $\Delta Y_t = Y_t - Y_{t-}$ are the jumps in the process. This decomposition is always unique even though the decomposition of Y into Y^c and Y^d is not.

In the special case where $[Y]$ has absolutely continuous sample paths, then the martingale representation theorem (e.g. Karatzas and Shreve 1991, p. 170) implies that

$$M_t = \int_0^t \Theta_u \mathrm{d}W_u,$$

where W is a vector of Brownian motion and Θ is a matrix whose elements are measurable and adapted. This model for M is called a stochastic volatility process; see for example Ghysels *et al.* (1996) and Shephard (2004). I will sometimes write this integral in the minimalist notation used in stochastic analysis,

$$M = \Theta \bullet W,$$

where $\bullet$ denotes stochastic integration. Under these conditions the lack of arbitrage implies that A must have absolutely continuous sample paths; see Back (1991, p. 380) and Karatzas and Shreve (1998, p. 3). Taken together this means, under these conditions, that log-prices must be of the form of a Brownian semimartingale, written $Y \in \mathcal{BSM}$, that is,

$$Y_t = \int_0^t a_u \mathrm{d}u + \int_0^t \Theta_u \mathrm{d}W_u.$$

Such a model can be called semiparametric from a statistical viewpoint as a and Θ are infinite-dimensional measurable random functions. When $Y \in \mathcal{BSM}$ then $[Y]_t = \int_0^t \Theta_s \Theta_s^T ds$, which means that conditional on $\mathcal{F}_t$, dY_t is Gaussian with

$$\mathrm{E}\left(dY_t \mid \mathcal{F}_t\right) = dA_t = a_t dt, \quad \mathrm{Cov}\left(dY_t \mid \mathcal{F}_t\right) = d[Y]_t = \Theta_t \Theta_t^T dt. \qquad (11.2)$$

Thus the A process is the integral of the infinitesimal conditional mean process while $[Y]$ is the integral of the infinitesimal conditional covariance process.

11.1.3   Outline

In this chapter I will simply ask whether Y has a continuous sample path. This is not straightforward when the volatility of prices is allowed to change through time, for non-Gaussian returns are then not evidence for jumps. Broadly there are two approaches to answering the question, parametric and semiparametric. I will briefly review them, emphasizing where I think problems with our current knowledge lie.

Section 11.2 focuses on semiparametric methods, while Section 11.3 looks in detail at likelihood based procedures. I compare these methods in the context of a dataset in Section 11.3.3. Additional issues are presented in Section 11.4. Conclusions are presented in Section 11.5.

11.2   Semiparametric analysis

11.2.1   Realized quadratic variation

In principle, high-frequency financial data can be used to estimate $[Y]$ so long as the usual no-arbitrage assumptions hold. Let $\delta > 0$ denote a time period between high-frequency observations. Then I compute the associated vector returns as

$$y_j = Y_{\delta j} - Y_{\delta(j-1)}, \quad j = 1, 2, 3, \ldots,$$

and calculate the *realized quadratic variation process* as

$$[Y_\delta]_t = \sum_{j=1}^{\lfloor t/\delta \rfloor} y_j y_j^T.$$

By the definition of the quadratic variation process $[Y_\delta]_t \overset{p}{\to} [Y]_t$ as $\delta \downarrow 0$, provided that $Y \in \mathcal{SM}$; here and below $\overset{p}{\to}$ denotes convergence in probability. I will first discuss quantifying $[Y_\delta]_t - [Y]_t$ in the context of no jumps, as the style of analysis will continue in the case where there are discontinuities. If $Y \in \mathcal{BSM}$ then $[Y_\delta]_t \overset{p}{\to} \int_0^t \Theta_s \Theta_s^T ds$. In this case a central limit theory for $\delta^{-1/2} \left([Y_\delta]_t - [Y]_t\right)$ can be derived under the single additional assumption that $\int_0^t a_{(l)u}^2 du < \infty$ for every l, where $a_{(l)}$ is the l-element of the a process. Barndorff-Nielsen and Shephard (2004b) detail this result, building on earlier work by Jacod (1994), Jacod and Protter (1998), and Barndorff-Nielsen and Shephard

(2002, 2004*a*). Here I will only discuss the univariate result. Barndorff-Nielsen and Shephard (2004*b*) show that

$$\delta^{-1/2}\left([Y_\delta]_t - [Y]_t\right) \rightarrow \sqrt{2}\int_0^t \Theta_u^2 dB_u, \qquad (11.3)$$

where B is Brownian motion independent of Y and the convergence is in law stably as a process. This type of convergence, introduced by Rényi (1963) and studied by Aldous and Eagleson (1978), is stronger than convergence in law; see Jacod and Protter (1998, pp. 269–270). Expression (11.3) means that as $\delta \downarrow 0$ for a fixed t,

$$\frac{\delta^{-1/2}\left([Y_\delta]_t - [Y]_t\right)}{\sqrt{2\int_0^t \Theta_u^4 du}} \xrightarrow{d} N(0,1),$$

where $\xrightarrow{d}$ denotes convergence in distribution. The integrated quarticity $\int_0^t \Theta_u^4 du$ can be estimated from the returns. In particular, under the $Y \in \mathcal{BSM}$ assumption, $\delta^{-1}\sum_{j=1}^{\lfloor t/\delta\rfloor} y_j^4 \xrightarrow{p} 3\int_0^t \Theta_u^4 du$, so a use of the delta method yields that as $\delta \downarrow 0$,

$$\frac{\log[Y_\delta]_t - \log[Y]_t}{\sqrt{\frac{2}{3}\sum_{j=1}^{\lfloor t/\delta\rfloor} y_j^4 / \left([Y_\delta]_t\right)^2}} \xrightarrow{d} N(0,1). \qquad (11.4)$$

This result is feasible: except for the unknown $[Y]_t$ it can be computed directly from the data and so can be used to construct confidence intervals, for example for $[Y]_t$. This allows us to perform inference on $[Y]$ without specifying a detailed model for a or for Θ. Hence inference based on this limit theory is semiparametric.

The central limit theory holds, in particular, for the increments of the quadratic variation process. There are rather practical reasons for looking at daily increments

$$V(Y)_i = [Y]_i - [Y]_{i-1} = \int_{i-1}^i \Theta_u \Theta_u^{\mathrm{T}} du.$$

The spot volatility Θ has strong intraday patterns caused by social norms and timetabled macroeconomic and financial announcements. These regularities do not contradict the semimartingale assumption on Y, but they make it hard to exploit high-frequency data by building temporally stable intraday models; see, for example, the careful work of Andersen and Bollerslev (1997). However, daily quadratic variation is somewhat robust to these types of intraday patterns, in the same way as yearly inflation is somewhat insensitive to seasonal fluctuations in the price level. If I write the jth high-frequency observation on the ith day as

$$y_{i,j} = Y_{i+\delta j} - Y_{i+\delta(j-1)}, \quad j = 1, \ldots, \lfloor 1/\delta\rfloor,$$

then the ith daily realized-variation is defined as

$$V(Y_\delta)_i = \sum_{j=1}^{\lfloor 1/\delta \rfloor} y_{i,j} y_{i,j}^{\mathrm{T}}.$$

Clearly $V(Y_\delta)_i \overset{p}{\to} V(Y)_i$ as $\delta \downarrow 0$, while in the univariate case (11.4) implies that

$$\frac{\log\left(\sum_{j=1}^{\lfloor t/\delta \rfloor} y_{i,j}^2\right) - \log V(Y)_i}{\sqrt{\frac{2}{3} \sum_{j=1}^{\lfloor t/\delta \rfloor} y_{i,j}^4 / \left(\sum_{j=1}^{\lfloor t/\delta \rfloor} y_{i,j}^2\right)^2}} \overset{d}{\to} N(0,1).$$

The central limit theorem for the logarithm of daily realized variation can be used to compute daily confidence intervals for $V(Y)_i^{1/2}$ in the univariate case, based on the so-called realized volatilities $V(Y_\delta)_i^{1/2}$. Figure 11.2 implements this method for the yen/dollar rates, together with daily 95% confidence intervals. The figure shows significantly time-varying and partially predictable levels of volatility through time. Similar results could be displayed for realized regression- and correlation-type quantities, which are discussed in detail in Barndorff-Nielsen and Shephard (2004a). Again these would show significant movements through time, although correlations tend to be more stable over time than do volatilities.

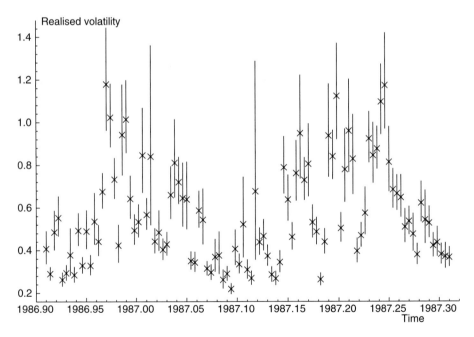

FIG. 11.2. Daily realized volatility (crosses) computed using 10-minute returns for the yen/dollar exchange rate, starting on 1 December 1986. Also shown are 95% confidence intervals computed using the log-based feasible central limit theorem.

This way of looking at the historical evolution of market volatility has a long history in financial economics. Examples include Merton (1980), Poterba and Summers (1986), Schwert (1989), Richardson and Stock (1989), Schwert (1990), Taylor and Xu (1997), and Christensen and Prabhala (1998). The new element here is the formal $\mathcal{BSM}$-based analysis and its distribution theory.

11.2.2 Forecasting volatility

Before discussing our central topic of testing for discontinuities, we take a slight detour to mention some recent literature on forecasting volatility that may be of interest to statisticians.

Having constructed a time series of estimated $V(Y)_i$, it is possible to try to extrapolate these estimates into the future. A systematic study of this is given in Andersen *et al.* (2001) and Andersen *et al.* (2003*a*), whose work is based on the following observations. Ito's formula implies that if $Y \in \mathcal{SM}^c$, then

$$YY^{\mathrm{T}} = [Y] + Y^{\mathrm{T}} \bullet Y + Y \bullet Y^{\mathrm{T}}$$
$$= [Y] + Y \bullet A^{\mathrm{T}} + A \bullet Y^{\mathrm{T}} + Y \bullet M^{\mathrm{T}} + Y^{\mathrm{T}} \bullet M.$$

So if $M \in \mathcal{M}$, then

$$\mathrm{E}(Y_t Y_t^{\mathrm{T}} \mid \mathcal{F}_0) = \mathrm{E}\left([Y]_t \mid \mathcal{F}_0\right) + \mathrm{E}\left(\int_0^t Y_s \mathrm{d}A_s^{\mathrm{T}} \mid \mathcal{F}_0\right) + \mathrm{E}\left(\int_0^t A_s \mathrm{d}Y_s^{\mathrm{T}} \mid \mathcal{F}_0\right).$$

Over small intervals of time, the second and third of these terms will in practice be small, so

$$\mathrm{E}(Y_t Y_t^{\mathrm{T}} \mid \mathcal{F}_0) \simeq \mathrm{E}\left([Y]_t \mid \mathcal{F}_0\right).$$

In terms of realized covariations this means that when we want to forecast the outer product of future returns given current data, we find

$$\mathrm{E}(y_i y_i^{\mathrm{T}} \mid \mathcal{F}_{i-1}) \simeq \mathrm{E}\left\{V(Y)_i \mid \mathcal{F}_{i-1}\right\}.$$

Given the sequence of estimators $V(Y_\delta)_1, \ldots, V(Y_\delta)_{i-1}$ of $V(Y)_1, \ldots, V(Y)_{i-1}$, Andersen and co-workers have built models of future covariation by projecting past realised covariations forward. Andersen *et al.* (2003*a*) and Andersen *et al.* (2005*b*) are fine examples of this work. An extensive review of some of this material, placed in a wider context, is given by Andersen *et al.* (2005*a*).

11.2.3 Realized bipower variation

So far I have assumed that $Y \in \mathcal{BSM}$, so prices have continuous sample paths *a priori*. I now consider testing this assumption.

Barndorff-Nielsen and Shephard (2003, 2004*c*) have studied an alternative measure of volatility that is robust to discontinuities in the sample path. Their

idea, which has so far only been stated in the univariate case, is to move away
from taking squared returns. Instead they work with realized bipower variation

$$\{Y_\delta\}_t = \frac{\lfloor t/\delta \rfloor}{\lfloor t/\delta \rfloor - 2} \sum_{j=3}^{\lfloor t/\delta \rfloor} |y_{j-2}| \, |y_j| , \qquad (11.5)$$

where I have chosen to look at the second-order lag version, which has some
interesting robustness properties against market microstructure effects, though
their theory carries over to any finite non-zero lag. It is possible to show that if
$Y \in \mathcal{BSM}$ then

$$\{Y_\delta\}_t \xrightarrow{p} \mu_1^2[Y]_t = \mu_1^2 \int_0^t \Theta_u^2 du, \qquad (11.6)$$

with $\mu_r = \mathrm{E}\,|U|^r$ where $r > 0$ and U is a standard normal variate. It will be
convenient to write this probability limit as

$$\{Y\}_t = \text{p-lim}_{\delta \downarrow 0} \{Y_\delta\}_t .$$

See Barndorff-Nielsen *et al.* (2004*a*) for this general result and the above-mentioned
papers for earlier special cases.

The result, eqn (11.6), which seems simply to deliver a scaled version of
quadratic variation, is more interesting than first appears. The most stimulating
feature of realized bipower variation is its behaviour under a Brownian semimartingale plus jump process

$$Y_t = \int_0^t a_u du + \int_0^t \Theta_u dW_u + \sum_{j=1}^{N_t} c_j,$$

where N is a simple counting process. Then

$$\{Y\}_t = \mu_1^2 \int_0^t \Theta_u^2 du \quad \text{and} \quad [Y]_t = \int_0^t \Theta_u^2 du + \sum_{j=1}^{N_t} c_j^2,$$

so $\{Y\}$ is unaffected by jumps, while $[Y]$ increases. This implies that it is theoretically possible to identify $\int_0^t \Theta_u^2 du$ and $\sum_{j=1}^{N_t} c_j^2$. The reason for the robustness
of $\{Y_\delta\}$ to jumps is that only a finite number of terms in eqn (11.5) are affected
by them. When there are no jumps then $|y_j| \downarrow 0$ as $\delta \downarrow 0$ and so, in the limit,
only terms with contiguous jumps can contribute to the probability limit. But
the probability of having contiguous jumps goes to zero as $\delta \downarrow 0$ owing to the
assumption that N is a simple counting process. The detailed behaviour of N
or c does not affect this argument.

Under some stronger assumptions including that $Y \in \mathcal{BSM}$, Barndorff-
Nielsen and Shephard (2003) showed that

$$\sqrt{\frac{\delta^{-1}}{\int_0^t \Theta_u^4 du / \left(\int_0^t \Theta_u^2 du \right)^2}} \left(\frac{\mu_1^{-2} \{Y_\delta\}_t}{[Y_\delta]_t} - 1 \right) \xrightarrow{d} N(0, \vartheta),$$

where $\vartheta = \pi^2/4 + \pi - 5 \simeq 0.609$. This can be used as the null distribution of a test for discontinuities in price processes in the time interval $(0, t)$, for I will indicate shortly that $\int_0^t \Theta_u^4 du$ can be robustly estimated in the presence of jumps. Clearly one can compute the daily realized bipower variation

$$B(Y_\delta)_i = \frac{\lfloor 1/\delta \rfloor}{\lfloor 1/\delta \rfloor - 2} \sum_{j=2}^{\lfloor 1/\delta \rfloor} |y_{i,j-2}| \, |y_{i,j}|$$

and compare $\mu_1^{-2} B(Y_\delta)_i$ to $V(Y_\delta)_i$ to see if there are jumps on the ith day. Such t-tests are independent over days. Huang and Tauchen (2003) have shown that such tests have quite good finite sample behaviour under modelling assumptions thought realistic in financial economics.

11.2.4 Empirical results

The upper panel of Figure 11.3 contrasts the daily realized continuous sample path volatilities based on 10-minute returns, $\{\mu_1^{-2} B(Y_\delta)_i\}^{1/2}$, with the daily realized volatilities standard $V(Y_\delta)_i^{1/2}$. They are similar on most days. The lower

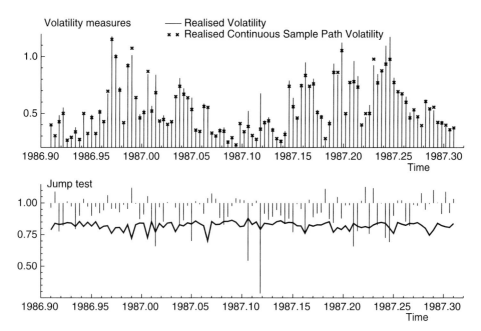

FIG. 11.3. Volatility measures for the first 100 days in the sample, using 10-minute returns for the yen/dollar exchange rate. Upper panel: daily realized continuous sample path volatility and realized volatility. Lower panel: ratio jump test statistic. Numbers below one indicate evidence for jumps. The solid line shows the 99% critical value for a null hypothesis of no jumps in the price process. Values of the test statistic below this line show statistically significant evidence for jumps.

panel shows the daily jump ratio test statistics $\mu_1^{-2}B(Y_\delta)_i/V(Y_\delta)_i$ together with 99% one-sided critical values for the null hypothesis of no jumps in the price process on day i. Here the asymptotic standard error of the statistic is estimated by

$$\frac{\lfloor 1/\delta \rfloor}{\mu_{4/3}^3(\lfloor 1/\delta \rfloor - 4)} \sum_{j=5}^{\lfloor 1/\delta \rfloor} |y_{i,j-4}|^{4/3}\,|y_{i,j-2}|^{4/3}\,|y_{i,j}|^{4/3}\,/\,\left(\mu_1^{-2}B(Y_\delta)_i\right)^2.$$

Some moderate volatility days in the middle of the dataset show fairly significant signs of discontinuities.

Figure 11.4 shows the yen/dollar exchange rate every ten minutes on the 52nd day of the sample, the day with the lowest jump ratio shown in the lower panel of Figure 11.3. It suggests a very large sustained change in the price. The jump is not dramatic in comparison with some price moves seen on the foreign exchange market, but as it happens on a low-volatility day, the jump ratio test picks it up.

Over the ten years of data, around 17% of days have jump test statistics statistically significant at the 99% level, and around 8% of the days are significant at the 99.9% level. This is appreciable evidence of price discontinuities in exchange rate data. Andersen *et al.* (2003c) have studied interest rate and stock

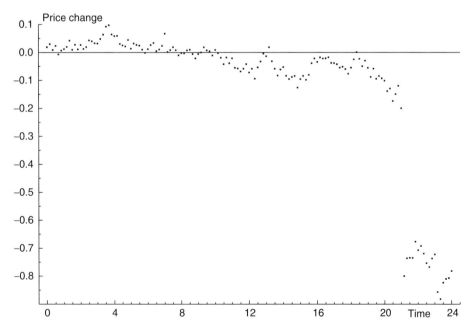

FIG. 11.4. Percentage change in the yen/dollar exchange rate every ten minutes during the 52nd day in the sample, relative to the start of the trading day (time 0). The x-axis has units of hours.

return data and have found even higher fractions of days on which there were jumps in the price process. Detailed case studies by Barndorff-Nielsen and Shephard (2003) suggest that most discontinuities are associated with public news announcements, such as macroeconomic statistical statements by governments on inflation, balance of payments and unemployment—see also the related literature on macroeconomic news shocks, including Ederington and Lee (1993), Andersen and Bollerslev (1998) and Andersen *et al.* (2003*b*).

11.3 Parametric analysis

11.3.1 Model building

The realized measures of volatility and jumps are semiparametric and exploit high-frequency data rather than relying on the temporal dependence between days in the return data. In the financial economics literature there is a stream of papers that tries to model jumps parametrically. Examples of this work include Carr *et al.* (2003), Carr and Wu (2004), Eberlein *et al.* (2001), Eraker *et al.* (2003) and Li *et al.* (2004). These models are then typically estimated using daily data. In this section I will follow this approach, comparing the conclusions I find using this inference method with those reported in the previous section.

In this section I will ignore the effect of drift, which is small for exchange rates, and simply model the local martingale component of prices. There are broadly three ways of carrying this out, modelling the price process as

$$Y = \sigma \bullet W + Z, \quad Y = \sigma \bullet Z, \quad Y_t = Z_{\int_0^t \sigma_u^2 \mathrm{d}u},$$

where Z includes jumps, W denotes Brownian motion, and σ is the spot volatility. If Z is Brownian motion then all of these models are equivalent. For statisticians, the most familiar special case of this is perhaps where Z is a Poisson process; then the last of these processes is a doubly stochastic Poisson process (Cox 1955*a*).

Here I will focus entirely on the $\sigma \bullet W + Z$ case, modelling the Z as a simple compound Poisson process

$$Z_t = \sum_{j=1}^{N_t} c_j, \quad c_j \overset{iid}{\sim} N(0, \omega_c^2),$$

which has Gaussian jumps and $N_0 = 0$. The literature consistently reports results with ω_c being quite large, while N_t is typically assumed to be a homogeneous Poisson process with modest intensity, λ_N, indicating that there are a handful of large jumps each year. Throughout, I will make the simplifying assumption that W, σ, N, c are totally independent processes—which is somewhat unreasonable empirically but is a potentially useful starting point for exchange-rate data that have straightforward dynamics relative to stock market indices. In this section I assume the simple diffusion for σ,

$$\mathrm{d}\log \sigma_t^2 = -\rho^* (\log \sigma_t^2 - \mu)\mathrm{d}t + \omega_\sigma \mathrm{d}B_t. \tag{11.7}$$

It is important to notice that these highly specific assumptions strongly contrast with the methods discussed in the previous section.

My data will be daily returns,

$$y_i = Y_i - Y_{i-1}, \quad i = 1, \ldots, T.$$

This is absolutely key in the results I will report, for I am now using a coarser information source than the one in the previous section. I will try to make up for this loss of information by using fully parametric methods. The daily returns are conditionally independent, given the path of (σ, Z) with

$$y_i \mid z_i, \sigma_i \sim N(z_i, \sigma_i^2), \tag{11.8}$$

where

$$\sigma_i^2 = \int_{i-1}^{i} \sigma_u^2 \mathrm{d}u, \quad z_i = Z_i - Z_{i-1}. \tag{11.9}$$

In order to simplify my exposition I will use an Euler approximation over the day to eqn (11.7) and eqn (11.9), that is

$$\log \sigma_{i+1}^2 \mid \sigma_i^2 \sim N(\mu + \rho \log \sigma_i^2, \omega_\sigma^2), \tag{11.10}$$

while the z_i are independent and identically distributed Poisson variables with mean λ_N. If the time series of z_i and σ_i^2 were observed then inference on the parameters of this process would be straightforward, but these processes are latent. One way of carrying out inference on this type of non-Gaussian state space model, where z_i, σ_i are the states, is by using Markov chain Monte Carlo methods. Papers about this for stochastic volatility models include Kim *et al.* (1998) and Yu (2004), as well as the references they provide.

Here I have chosen to tackle the inference problem using a particle filter, which will allow me to learn about the model online, updating my views on unknown parameters and the position of the volatility as new information arrives. Papers on particle filters include Gordon *et al.* (1993) and Pitt and Shephard (1999), while Doucet *et al.* (2001) is an excellent book of essays about them. The first uses of particle filters in this context were Kim *et al.* (1998) and Pitt and Shephard (1999), while more general papers on non-linear filtering using particle filters include Jacod *et al.* (2001), Jacod and Moral (2001) and the references contained within. Johannes *et al.* (2002) focuses on these issues in the context of financial econometrics. Here I will exploit an extension of standard particle filters suggested by Storvik (2002) that uses the property of sufficient statistics. I will first briefly review this idea in a general state-space framework, before returning to the special features of our model.

11.3.2 Online learning through a particle filter

The model (11.8) and (11.10) is a Markov random field, which I write abstractly in terms of densities

$$f(y_i \mid \alpha_i, \theta), \quad f(\alpha_{i+1} \mid \alpha_i, \theta),$$

where $\theta = (\mu, \rho, \omega_\sigma, \lambda_N, \omega_c)^{\mathrm{T}}$, y_i are observations and $\alpha_i = (z_i, \sigma_i^2)^{\mathrm{T}}$ is the vector of states. The choice of state vector is somewhat arbitrary, for one could write

$$Z_t = \sum_{j=1}^{t} \delta_j c_j, \quad c_j \overset{iid}{\sim} N(0, \omega_c^2), \quad \delta_j \overset{iid}{\sim} \text{Bernoulli}(\lambda_N),$$

so $N_t = \sum_{j=1}^{t} \delta_j$, and then define the state vector as $\alpha_i^* = (\delta_i, c_i, \sigma_i^2)$. It is clear that α_i^* is a richer state vector, for it implies knowledge of α_i but not vice versa. Typically it is a bad idea to perform online parameter learning by particle filtering using α_i^* rather than α_i, as it increases the degree of complete information and makes data imputation methods computationally more challanging. This is the case here, for the methods I outline below would simply collapse if I had used α_i^* instead of α_i.

The task is to compute iteratively

$$f(\alpha_i, \theta \mid \mathcal{F}_i),$$

where $\mathcal{F}_i$ is $y_1, \ldots, y_i$. In order to carry out online learning, I assume that I can:

- calculate a sufficient statistic, S_i, for θ given $\mathcal{F}_i$ and $\alpha_1, \ldots, \alpha_i$, the complete data;
- calculate a sufficient statistic, $S_{i+1|i}$, given $\mathcal{F}_i$ and $\alpha_1, \ldots, \alpha_{i+1}$;
- simulate from $\alpha_{i+1} \mid \alpha_i, \theta$;
- simulate from $\theta \mid S_i$; and
- compute $f(y_i \mid \alpha_i, \theta)$.

Storvik (2002) performs particle filtering on $f(\alpha_i, S_i \mid \mathcal{F}_i)$ using θ as an auxiliary variable, discarding this additional variable at each step. His method is based on sampling from

$$f(\alpha_{i+1}, S_{i+1|i}, \theta \mid \mathcal{F}_{i+1}) \propto f(y_{i+1} \mid \alpha_{i+1}, \theta) f(\alpha_{i+1}, S_{i+1|i}, \theta \mid \mathcal{F}_i)$$
$$= f(y_{i+1} \mid \alpha_{i+1}, \theta) \int f(\alpha_{i+1}, S_{i+1|i} \mid \alpha_i, S_i, \theta)$$
$$f(\theta, \alpha_i, S_i \mid \mathcal{F}_i) d\alpha_i dS_i$$
$$= f(y_{i+1} \mid \alpha_{i+1}, \theta) \int f(\alpha_{i+1}, S_{i+1|i} \mid \alpha_i, S_i, \theta)$$
$$f(\theta \mid S_i) f(\alpha_i, S_i \mid \mathcal{F}_i) d\alpha_i dS_i. \tag{11.11}$$

This is carried out by first assuming a large sample from $f(\alpha_i, S_i \mid \mathcal{F}_i)$,

$$\alpha_i^j, S_i^j, \quad j = 1, \ldots, J, \tag{11.12}$$

which is used to approximate eqn (11.11), yielding the empirical updating density

$$\widehat{f}(\alpha_{i+1}, S_{i+1|i}, \theta \mid \mathcal{F}_{i+1}) \propto f(y_{i+1} \mid \alpha_{i+1}, \theta) \sum_{j=1}^{J} f(\alpha_{i+1}, S_{i+1|i} \mid \alpha_i^j, S_i^j, \theta) f(\theta \mid S_i^j).$$

In order to complete the particle filter, samples must be drawn from this density, simply discarding the θ draws to produce a sample of particles $\alpha_{i+1}^j, S_{i+1}^j, j = 1, \ldots, J$. One can think of these as approximately coming from $f(\alpha_{i+1}, S_{i+1} \mid \mathcal{F}_{i+1})$. Some of the many ways in which to carry this out are discussed by Pitt and Shephard (1999). Here is a very simple algorithm:

1. for each j, simulate

$$\theta^{j,k} \sim \theta \mid S_t^j, \quad k = 1, \ldots, K;$$

2. for each j, simulate from

$$\alpha_{i+1}^{j,k} \sim \alpha_{i+1} \mid \alpha_i^k, \theta^{j,k}, \quad k = 1, \ldots, K,$$

which, in turn, is used to calculate $S_{i+1|i}^{j,k}$. Use y_{i+1} to calculate $S_{i+1}^{j,k}$, yielding the pairs $\left(\alpha_{i+1}^{j,k}, S_{i+1}^{j,k} \right)$;

3. compute

$$w_i^{j,k} \propto f(y_{i+1} \mid \alpha_{i+1}^{j,k}, \theta^{j,k});$$

then resample $\left(\alpha_{i+1}^{j,k}, S_{i+1}^{j,k} \right)$ with probability proportional to $w_i^{j,k}$, giving a sample of size J. Label the sample

$$\left(\alpha_{i+1}^j, S_{i+1}^j \right), \quad j = 1, \ldots, J.$$

The computational load of this is linear in the number of observations and costs $O(JK)$ floating-point operations per observation.

A key feature is that the above strategy updates particles one observation at a time. In principle, one could use a fixed-lag strategy, updating using $k > 0$ observations at a time, as introduced concurrently and independently in this context by Clapp and Godsill (1999) and Pitt and Shephard (2001). Subsequent implementations of this type of approach include the so-called practical filter (Johannes *et al.* 2004).

The above approach requires the specification of some priors for θ. To simplify my work I have assumed independence across the elements of θ, using Gaussian priors on μ and ρ, centred at -0.02 and 0.96 respectively with corresponding variances of $1/40$ and $1/4$; a $\Gamma(2 \times 0.13^2, 2)$ prior, having mean $1/0.13^2$, for ω_σ^{-2}; a $\mathrm{Be}(0.2, 1.0)$ prior for λ_N; and a $\Gamma(2 \times 3^2, 2)$ prior for ω_c^{-2}. This is not entirely satisfactory as it postulates prior independence for parameters such as μ and ρ, but I shall skirt the computational complexity of using more sensible priors.

11.3.3 Empirical results

The upper left panel of Figure 11.1 shows the daily log-prices for the yen/dollar rate, while the upper right panel shows the corresponding daily returns. Figure 11.5 plots the corresponding simulation based estimator of the posterior median of θ given the daily data $y_1, \ldots, y_i$ for $i = 50, 51, \ldots, 550$, together with the corresponding 0.01 and 0.99 quantiles. These graphs are given for a variety

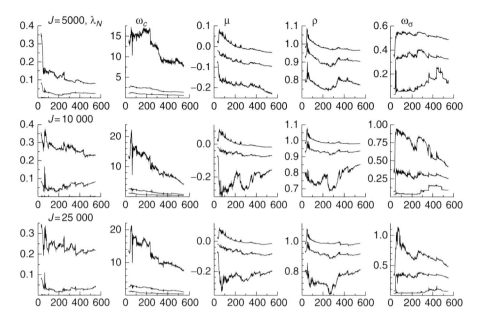

FIG. 11.5. Posterior 0.01, 0.5, and 0.99 quantiles for the parameters λ_N, ω_c, μ, ρ and ω_σ, conditional on $y_1, \ldots, y_i$, for $i = 50, \ldots, 550$. The quantiles are computed using $J \times K$ particles with $K = 20$ and $J = 5000$ (upper panels) $J = 10\,000$ (middle panels), and $J = 25\,000$ (lower panels).

of values of J, with $K = 20$ throughout, and show large ranges of uncertainty for most parameters, with the range falling somewhat during this period. The results are quite stable as J varies, with only small changes as J increases. The most sensitive parameter is ω_σ. Overall this suggests that conclusions from the particle filter output should be useful.

Table 11.1 shows the posterior median and 0.01 and 0.99 quantiles at the end of the sample, $i = 550$. Most of the mass of the posterior on λ_N is around small values, indicating that jumps are rare according to this model. When the process does jump, the standard deviation of the jump is around 1, similar to the average daily realized volatility during the sample, 0.709. The persistence of the stochastic volatility component is modest, although that parameter is poorly estimated. Corresponding results for the prior are also shown in Table 11.1; for most parameters they are consistent with the posterior. The distribution for μ is somewhat misplaced, although not dramatically.

The overall importance of jumps and stochastic volatility components in this model can be assessed by computing the marginal likelihoods, via the prediction decomposition

$$f(y_1, \ldots, y_T) = f(y_1) \prod_{i=2}^{T} f(y_i \mid \mathcal{F}_{i-1}),$$

TABLE 11.1. Posterior inference on the parameters in the model given all 550 observations in the sample. Shown are the 0.01, 0.5 and 0.99 quantiles, computed using the particle filter using $J = 25,000$ and $K = 20$. Also given are the corresponding prior quantiles.

	Quantiles of prior			Quantiles of posterior		
	0.01	0.5	0.99	0.01	0.5	0.99
λ_N	0.0113	0.345	0.911	0.000657	0.0412	0.219
ω_c	1.39	3.60	29.5	0.407	1.04	7.62
μ	-0.0957	-0.0200	0.0557	-0.198	-0.0834	-0.0143
ρ	0.719	0.960	1.20	0.807	0.915	0.986
ω_σ	0.0545	0.119	0.541	0.0696	0.296	0.486

where $\mathcal{F}_{i-1} = (y_1, \ldots, y_{i-1})$. This integrates out the effect of parameter uncertainty at each time step. A simple estimator of this given the output of the particle filter may be computed using

$$\widehat{f}(y_i \mid \mathcal{F}_{i-1}) = \frac{1}{JK} \sum_{j,k} w_i^{j,k}.$$

If this estimator is insensitive to J, then it provides a viable alternative to the standard Markov chain Monte Carlo method for performing this computation; see for example Chib (1995) and Chib and Jeliazkov (2001).

Table 11.2 gives the corresponding marginal log-likelihoods, $\log f(y_1, \ldots, y_T)$, calculated for the full model, the model that removes the jumps, a Brownian motion plus jump model, and a pure Brownian motion model. Each marginal likelihood automatically penalizes the model for its complexity, by integrating out the prior belief on the unknown parameters, so they can be compared directly across the different models. The results are given for varying values of J, to see whether conclusions drawn are sensitive to this. The stochastic volatility model dominates the Brownian motion and Brownian motion plus jump models, which

TABLE 11.2. Estimated log marginal likelihood for a variety of models, using a particle filter for various values of J and $K = 20$ throughout. BP denotes the Box–Pierce statistic using 20 lags computed using reflected predictive distribution functions, which should have a χ_{20}^2 distribution for a well-specified model.

	Estimated log marginal likelihood			BP
	$J = 5,000$	$J = 10,000$	$J = 25,000$	
Stochastic volatility, jump	-557.8	-557.2	-557.5	11.9
Stochastic volatility	-555.4	-555.8	-555.5	12.7
Brownian motion, jump	-576.9	-576.7	-576.9	50.4
Brownian motion	-583.9	-583.9	-583.9	50.1

both imply independent identically distributed returns. The stochastic volatility model fits better than the stochastic volatility plus jump model, because of the penalty on the complexity of the added jump model. On longer time series or using stock return data rather than exchange rates, I would expect to see more evidence of jumps.

Table 11.2 shows Box–Pierce statistics computed on the time series of so-called reflected predictive distribution functions, $2 |pr(y_{i+1} | \mathcal{F}_i) - 0.5|$, approximated by the particle filter. If the model and prior are correctly specified, then $u_{i+1} = pr(y_{i+1} | \mathcal{F}_i)$ should be a sequence of independent identically distributed uniform random variables; see, for example, Rosenblatt (1952), Smith (1985), Shephard (1994) and, in the context of stochastic volatility, Kim *et al.* (1998). The reflected uniform variables, introduced by Kim *et al.* (1998), should have the same properties under the correct model, but have the advantage that they have some power to detect misspecification in the modelling of the volatility process: if there is unmodelled volatility clustering then the sequence of u_i should be close to 0.5 for low-volatility periods and around 0 and 1 for high-volatility periods. Such u_i would be uncorrelated, but their reflected versions $2 |u_i - 0.5|$ will be serially correlated. The table shows statistically significant volatility patterns being picked up for the Brownian motion and the Brownian motion plus jump models, as expected, but the stochastic volatility models with and without jumps have quite good diagnostics.

11.3.4 Comparison of parametric and semiparametric methods

To see if there has been a jump on day i I used the output of the particle filter to estimate $1 - pr(z_i = 0 | \mathcal{F}_i)$, the posterior probability that there has been a jump on day i given the information up to time i. This is shown in the lower left panel of Figure 11.6, and indicates some quite high probabilities, although the evidence for jumps on particular days is not overwhelming. I also calculated $E(z_i | \mathcal{F}_i)$, the average size of the jump on that day. This is drawn in the lower right panel, together with the time-varying conditional volatility $E(\sigma_i | \mathcal{F}_i)$. This is much more stable through time than the corresponding graph, which is not shown here, when jumps are excluded from the analysis.

It is interesting to compare the times at which there is appreciable posterior probability of there having been a jump and the times at which the semiparametric jump test was significant. One might expect these two sets of times to be close to one another if the parametric model is modelling jumps effectively. Suppose, for instance, that the parametric model has detected a possible jump if the posterior probability of a jump is above 0.2. In the sample there are 11 days out of 550 with such evidence. Of these four have statistically significant semiparametric jump tests at the 0.999 level, out of a total of 51 such days. Thus the days that the parametric test flags as potentially having a jump often seem not to have jumps in them and, more strikingly, the parametric approach misses almost all the real discontinuities in the process.

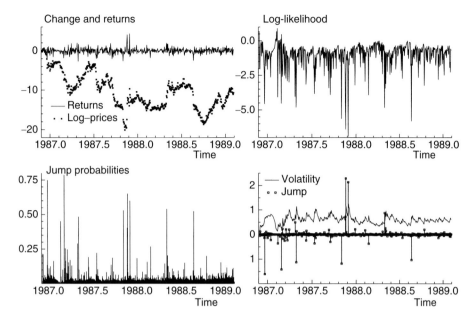

FIG. 11.6. Assessment of jumps on particular days. Upper left: daily returns and log-prices. Upper right: daily predictive likelihood $\log f(y_i \mid \mathcal{F}_{i-1})$. Lower left: posterior probability of jumps, $1 - \mathrm{pr}(z_i = 0 \mid \mathcal{F}_i)$. Lower right: posterior expected value of volatility $\mathrm{E}(\sigma_i \mid \mathcal{F}_i)$ and jumps $\mathrm{E}(z_i \mid \mathcal{F}_i)$.

11.4   Open problems

11.4.1   Market microstructure noise and realized measures

A major potential problem with the approach used in Section 11.2 is that the price process may not be well described by a $\mathcal{BSM}$ plus jump process due to market microstructure noise. Such noise induces discreteness of prices, irregular trading, and bid/ask bounce, and is likely to be modest for returns measured over large time intervals of 10–20 minutes, but becomes central over very small intervals. Interesting work on such issues include Delattre and Jacod (1997), Corsi *et al.* (2001), Bandi and Russell (2003), Hansen and Lunde (2003) and Zhang *et al.* (2003).

11.4.2   General bipower variation results

Realized bipower variation measures integrated variance by computing the quantity $\sum_{j=3}^{n} |y_{j-2}| \, |y_j|$, where $n = \lfloor 1/\delta \rfloor$, which is somewhat robust to jumps. A natural question is whether this is a good way to obtain robustness. Recently, Barndorff-Nielsen *et al.* (2004a) have studied general statistics of the form

$$\frac{1}{n} \sum_{j=3}^{n} g\left(\sqrt{n}\, y_{j-2}\right) h\left(\sqrt{n}\, y_j\right),$$

as $n \to \infty$ when $Y \in \mathcal{BSM}$. Here g and h are known general functions possessing at most polynomial growth. It would be interesting to know what choice of functions yields statistics that are robust to jumps and that maximize power. In particular, one may be inspired by the robustness literature to look at functions that behave like squares for small values of absolute returns, but are truncated at some point to downweight large observations, though choosing the truncation point may be hard. Earlier papers on this type of approach include Mancini (2001, 2003).

11.4.3 Particle filters

I have used a rather simple particle filter to learn about the parameters and states of the parametric model. There is a substantial literature that suggests the computational load of these algorithms can be reduced by carrying out the sampling more efficiently. For example, one could use fixed-lag particle filtering or better-chosen proposal densities. Johannes *et al.* (2004) give a particular implementation of such thoughts on some relatively simple models, combining an elaborate Markov chain Monte Carlo algorithm with fixed-lag filtering to produce a fairly effective fitting algorithm.

11.4.4 Conditional expectations and martingales

The ability to iteratively compute the distribution of θ given $\mathcal{F}_i$ recursively offers a number of inferential possibilities. I have already shown that it allows us to compute the marginal likelihood for the model. It seems to be less well known in statistics that the time series of the $\mathrm{E}\left(\theta \mid \mathcal{F}_i\right)$ is a martingale if the prior and likelihood are correctly specified. This property can be used to test for the correct specification of the model and prior. The assumption that the prior is correct is, of course, somewhat artificial.

11.5 Conclusion

I have looked at a very active area of financial econometrics, the study of discontinuities in the price process. I have shown that parametric and semiparametric methods yield rather different results. The semiparametric analysis is based on high-frequency data, while the parametric approach is based on the coarser daily returns. There are some strong reasons to suggest that the semiparametric analysis may be preferable in this case.

Acknowledgements

My research is supported by the UK Economic and Social Research Council. All the calculations here are based on software written by the author using the Ox language (Doornik 2001). I thank an editor and three referees for their detailed comments on a previous draft. Comments of Ole Barndorff-Nielsen, Tim Bollerslev, Clive Bowsher, Nour Meddahi, Omiros Papaspiliopoulos, and Michael Pitt helped improve the exposition.

12

On some concepts of infinite divisibility and their roles in turbulence, finance and quantum stochastics

Ole Eiler Barndorff-Nielsen

12.1 Introduction

Lévy theory—taken to mean the theory of infinite divisibility, Lévy processes, Lévy bases, and so forth—and its applications has, over the last decade, been a very active and wide-ranging area of research. The present chapter surveys a few aspects and ramifications of these developments. The choice of topics is determined by the author's own interests, and the chapter in no way pretends to be a balanced or comprehensive account of the subject areas concerned.

It is a great pleasure to note here that my own interest in some of these topics arose from seminal comments in two of Sir David Cox's papers (Cox 1981, 1984b), as referred to in Barndorff-Nielsen et al. (1990).

Section 12.2 recalls a number of basic concepts and results from Lévy theory. Various extensions and applications of these are then considered in Sections 12.3–12.6, which discuss tempo-spatial modelling, time-change, and some applications to finance and to turbulence. Section 12.7 discusses some, newly introduced, regularising mappings of Lévy measures and infinitely divisible distributions. The ideas for these mappings arose out of quantum stochastics, more specifically free probability, but the theory summarised in Section 12.7 falls purely within classical infinite divisibility. Recent developments in quantum stochastics have led to the discovery of several fascinating new concepts of 'independence', with associated concepts of 'infinite divisibility', 'Lévy processes', and so forth, the most prominent being free independence. These are briefly discussed in Section 12.8.1, and the relation to the Upsilon mappings, which form the subject of Section 12.7, is indicated. Section 12.8.2 concerns infinitely divisible quantum instruments; these play an important role in the modern theory of continuous-time quantum measurements and related stochastic process models.

12.2 Classical infinite divisibility, Lévy processes and Lévy bases

12.2.1 Introduction

The classical concept of infinite divisibility was introduced and studied by de Finetti, Kolmogorov, Lévy, and Khintchine, during the second quarter of the twentieth century. It defines a random variable X to be *infinitely divisible* if for all $n \in \mathbb{N}$ there exist independent and identically distributed random variables $X_1, \ldots, X_n$ such that $X \overset{L}{=} X_1 + \cdots + X_n$, where $\overset{L}{=}$ means equality in distribution. This condition may equivalently be expressed in terms of the probability laws μ and μ_n of X and X_1 as $\mu = \mu_n^{*n}$, where $*$ indicates ordinary convolution of measures.

In the second half of the twentieth century there has been steady progress in the development of the theory of infinite divisibility and of Lévy processes. A *Lévy process* is a stochastic process $Z = \{Z_t\}_{t \geq 0}$ having independent and stationary increments; it is usually assumed that $Z_0 = 0$ and that Z_t is continuous in probability. Thus the concept of Lévy processes is intimately related to that of infinite divisibility. To a major extent these more recent developments are due to Japanese mathematicians (see the bibliography in Sato 1999); in the early stages particularly to Kiyoshi Ito, who showed the importance of looking at Lévy processes from the sample path point of view and gave mathematical rigour to what is now called the *Lévy–Ito representation*, which is a stochastic process version of the *Lévy–Khintchine formula* for the cumulant transform of infinitely divisible distributions. The monographs by Bertoin (1996) and Sato (1999) represent culminations of many of these theoretical achievements.

Until around 1980 infinite divisibility and Lévy processes were widely conceived as being of quite limited applied interest. Since then the picture has changed dramatically; see for instance the collection of state-of-the-art surveys in Barndorff-Nielsen *et al.* (2001*a*). To some extent this is related to the fact that many of the concrete probability distributions used in applications were shown, often by quite difficult mathematical analysis, to be infinitely divisible or even self-decomposable, and also to the increase in computing power that has allowed a better understanding of the features of Lévy processes, in particular by simulation.

Although the definitions of infinite divisibility and Lévy processes may sound innocent these concepts entail many intriguing, subtle and difficult mathematical properties and aspects.

12.2.2 Infinite divisibility

The key result in the theory of infinite divisibility is the celebrated *Lévy–Khintchine formula* for the cumulant function of any random variable X with infinitely divisible law μ. Writing

$$C_\mu(\zeta) = \mathrm{C}\{\zeta \ddagger X\} = \log \mathrm{E}(e^{i\zeta X})$$

for the *cumulant function*, the formula states that

$$C_\mu(\zeta) = ia\zeta - \frac{1}{2}b\zeta^2 + \int_{\mathbb{R}} \left\{ e^{i\zeta x} - 1 - i\zeta x \mathbf{1}_{[-1,1]}(x) \right\} \rho(\mathrm{d}x), \qquad (12.1)$$

where $\mathbf{1}$ denotes the indicator function of a set, i is the imaginary unit, the *characteristic triplet* (a, b, ρ) of μ or X satisfies $a \in \mathbb{R}$, $b \geq 0$, and the *Lévy measure* ρ is characterized by the properties of having no atom at zero and of satisfying

$$\int_{\mathbb{R}} \max\{1, x^2\} \rho(\mathrm{d}x) < \infty. \tag{12.2}$$

Formulae (12.1) and (12.2) have natural extensions to vector random variates and to random measures; see eqn (12.6).

For distributions μ on the positive half-line $\mathbb{R}_{\geq 0}$ it is convenient to work instead with the *kumulant transform*, that is, the log Laplace transform

$$K_\mu(\theta) = -a\theta - \int_0^\infty (1 - \mathrm{e}^{-\theta x}) \rho(\mathrm{d}x),$$

where we also write

$$K_\mu(\theta) = \mathrm{K}\{\theta \ddagger X\} = \log \mathrm{E}(\mathrm{e}^{-\theta X}).$$

In this case

$$\int_0^\infty \max\{1, |x|\} \rho(\mathrm{d}x) < \infty;$$

in other words, a Lévy measure ρ such that $\int_0^\infty 1 \wedge |x| \rho(\mathrm{d}x) = \infty$ cannot correspond to a probability distribution that is concentrated on $\mathbb{R}_{\geq 0}$.

In many examples of interest, the Lévy measure ρ has a density with respect to Lebesgue measure. We denote the *Lévy density* by $u(x) = \mathrm{d}\rho(x)/\mathrm{d}x$, and we write $\bar{u}(x) = xu(x)$.

Example 1 (Generalized inverse Gaussian distribution)   The probability density of the generalized inverse Gaussian distribution $\mathrm{GIG}(\lambda, \delta, \gamma)$ is

$$p(x) = \frac{(\gamma/\delta)^\lambda}{2K_\lambda(\delta\gamma)} x^{\lambda-1} \exp\left\{-\frac{1}{2}\left(\delta^2 x^{-1} + \gamma^2 x\right)\right\}, \qquad x > 0,$$

where K_λ denotes a modified Bessel function of the third kind. Important special cases are the gamma distributions that occur for $\delta = 0$ and the inverse Gaussian laws obtained by taking $\lambda = -\frac{1}{2}$. The generalized inverse Gaussian laws are all infinitely divisible with characteristic triplet $(0, 0, \rho)$ and absolutely continuous Lévy measure ρ.

For the gamma distribution $\Gamma(\lambda, \alpha)$ with probability density function

$$p(x) = \frac{\alpha^\lambda}{\Gamma(\lambda)} x^{\lambda-1} \mathrm{e}^{-\alpha x}, \qquad x > 0, \quad \alpha, \lambda > 0,$$

where $\Gamma(\alpha)$ represents the gamma function, the Lévy density $u(x)$ equals $x^{-1}\mathrm{e}^{-\alpha x}$, $x > 0$. For the inverse Gaussian density function

$$p(x) = (2\pi)^{-1/2} \delta \mathrm{e}^{-\delta\gamma} x^{-3/2} \exp\{-(\delta^2 x^{-1} + \gamma^2 x)/2\}, \qquad x > 0, \quad \gamma, \delta > 0,$$

the Lévy density is

$$u(x) = (2\pi)^{-1/2}\delta x^{-3/2}\exp(-\gamma^2 x/2), \qquad x > 0.$$

The Lévy density for the general GIG(λ, δ, γ) law is, however, far from simple (Halgreen 1979):

$$u(x) = x^{-1}\left[\delta^2\int_0^\infty e^{-x\xi}g_\lambda(2\delta^2\xi)\mathrm{d}\xi + \max\{0,\lambda\}\right]\exp\left(-\gamma^2 x/2\right), \qquad x > 0,$$

where

$$g_\lambda(x) = \left[(\pi^2/2)x\left\{J_{|\lambda|}^2(\sqrt{x}) + N_{|\lambda|}^2(\sqrt{x})\right\}\right]^{-1}, \qquad x > 0,$$

and J_ν and N_ν are Bessel functions.

Example 2: (Normal inverse Gaussian distribution)   The normal inverse Gaussian distribution with parameters α, β, μ and δ has probability density function

$$p(x; \alpha, \beta, \mu, \delta) = a(\alpha, \beta, \mu, \delta)q\left(\frac{x-\mu}{\delta}\right)^{-1}K_1\left\{\delta\alpha q\left(\frac{x-\mu}{\delta}\right)\right\}e^{\beta x}, \quad x \in \mathbb{R},$$

where $q(x) = \sqrt{(1+x^2)}$ and

$$a(\alpha, \beta, \mu, \delta) = \pi^{-1}\alpha\exp\left\{\delta\sqrt{(\alpha^2-\beta^2)} - \beta\mu\right\},$$

and where K_1 is the modified Bessel function of the third kind and index 1. The domain of variation of the parameters is $\mu \in \mathbb{R}$, $\delta \in \mathbb{R}_{>0}$, and $0 \le |\beta| < \alpha$. The distribution is denoted by NIG($\alpha, \beta, \mu, \delta$). It follows immediately from the expression for $a(\alpha, \beta, \mu, \delta)$ that its kumulant-generating function has the simple form

$$\delta[\sqrt{(\alpha^2-\beta^2)} - \sqrt{\{\alpha^2-(\beta+u)^2\}}] + \mu u.$$

The characteristic triplet of NIG($\alpha, \beta, \mu, \delta$) is $(a, 0, \rho)$, where the Lévy measure ρ has density (Barndorff-Nielsen 1998b)

$$u(x) = \pi^{-1}\delta\alpha|x|^{-1}K_1(\alpha|x|)e^{\beta x}, \tag{12.3}$$

while

$$a = \mu + 2\pi^{-1}\delta\alpha\int_0^1\sinh(\beta x)K_1(\alpha x)\mathrm{d}x.$$

For more information on properties of the normal inverse Gaussian law and on its use in modelling and inference, particularly in finance and turbulence, see Barndorff-Nielsen and Shephard (2005) and references therein. A recent application to turbulence is briefly discussed in Section 12.6.1.

12.2.3 Classes of infinitely divisible distributions

If we denote the class of infinitely divisible laws on the real line $\mathbb{R}$ by $\mathcal{ID}(*)$, where $*$ refers to the convolution of probability measures, there is a hierarchy

$$\mathcal{G}(*) \subset \mathcal{S}(*) \subset \mathcal{T}(*) \subset \mathcal{L}(*), \ \mathcal{B}(*) \subset \mathcal{ID}(*), \tag{12.4}$$

where $\mathcal{G}(*)$ is the class of Gaussian laws, $\mathcal{S}(*)$ the class of stable laws, $\mathcal{T}(*)$ the extended *Thorin class*, $\mathcal{B}(*)$ the extended *Bondesson class*, and $\mathcal{L}(*)$ the *Lévy class* of *self-decomposable* laws. For distributions on $\mathbb{R}_{>0}$ with lower extremity 0 we write the hierarchy as

$$\mathcal{S}_+(*) \subset \mathcal{T}_+(*) \subset \mathcal{L}_+(*), \ \mathcal{B}_+(*) \subset \mathcal{ID}_+(*).$$

We now sketch the properties of these classes.

Thorin classes $\mathcal{T}_+(*)$ and $\mathcal{T}(*)$ The class $\mathcal{T}_+(*)$ is also known as the family of generalized gamma convolutions. It was introduced by Thorin (1977, see also Thorin 1978) as a tool to prove infinite divisibility of the log-normal distribution, and $\mathcal{T}_+(*)$ and its ramifications have since been studied extensively, particularly by Bondesson (1992).

By definition, $\mathcal{T}(*)$ is the smallest class of distributions that is closed under convolution and weak limits and contains the laws of all random variables of the form $\varepsilon X + a$ where X follows a gamma distribution, $\varepsilon = \pm 1$, and $a \in \mathbb{R}$. In terms of the Lévy–Khintchine representation, $\mathcal{T}(*)$ is characterised as the subfamily of $\mathcal{ID}(*)$ having characteristic triplet $(a, 0, \rho)$ where $a \in \mathbb{R}$ and ρ has a density u such that $\bar{u}$ is completely monotone, that is,

$$\bar{u}(x) = xu(x) = \begin{cases} \int_0^\infty e^{-x\xi} \nu(\mathrm{d}x), & x > 0, \\ \int_{-\infty}^0 e^{x\xi} \nu(\mathrm{d}x), & x < 0, \end{cases}$$

where ν is a measure on the real line (Barndorff-Nielsen and Thorbjørnsen 2004*b*).

Bondesson classes $\mathcal{B}_+(*)$ and $\mathcal{B}(*)$ The class $\mathcal{B}_+(*)$ was studied extensively by Bondesson (1981, 1992); see also Sato (1999, p. 389) and Barndorff-Nielsen *et al.* (2004*d*).

By definition (Barndorff-Nielsen *et al.* 2004*d*) $\mathcal{B}(*)$ is the smallest class of distributions that is closed under convolution and weak limits and contains the laws of all mixtures of distributions of random variables of the form $\varepsilon X + a$ where X has a negative exponential distribution, $\varepsilon = \pm 1$, and $a \in \mathbb{R}$. In terms of the Lévy–Khintchine representation, $\mathcal{B}(*)$ is characterized as the subfamily of $\mathcal{ID}(*)$ having characteristic triplet $(a, 0, \rho)$ where $a \in \mathbb{R}$ and ρ has a completely monotone density u, that is,

$$u(x) = \begin{cases} \int_0^\infty e^{-x\xi} \nu(\mathrm{d}x), & x > 0, \\ \int_{-\infty}^0 e^{x\xi} \nu(\mathrm{d}x), & x < 0, \end{cases}$$

where ν is a measure on the real line.

Lévy class $\mathcal{L}(*)$ This is the class of self-decomposable laws, introduced by Lévy and studied and applied extensively by many authors. Basic references for the theory of self-decomposability are Jurek and Mason (1993) and Sato (1999), while some of the applications are discussed in Barndorff-Nielsen (1998*a,b*), Barndorff-Nielsen and Shephard (2001*a,b*, 2005) and Barndorff-Nielsen and Schmiegel (2004).

By definition, $\mathcal{L}(*)$ is the class of distributions such that if X has law in $\mathcal{L}(*)$ then for all $c \in (0,1)$ there exists a random variable $X^{(c)}$, independent of X, with

$$X \overset{L}{=} cX + X^{(c)}. \tag{12.5}$$

In terms of the Lévy–Khintchine representation, $\mathcal{L}(*)$ is characterised as the subfamily of $\mathcal{ID}(*)$ for which the Lévy measure ρ has a density u that is increasing on $\mathbb{R}_{<0}$ and decreasing on $\mathbb{R}_{>0}$.

Example 3   The generalized inverse Gaussian and normal inverse Gaussian laws are all self-decomposable, as follows from the characterization just mentioned and the formulae for their Lévy densities, given in Examples 1 and 2.

We return briefly to these subclasses of the class of infinitely divisible laws in Section 12.7.

12.2.4   Lévy processes

A Lévy process is a stochastic process, starting at zero and stochastically continuous, with càdlàg sample paths, and having independent and stationary increments. The càdlàg assumption is innocuous, as any process having the other properties listed does possess a càdlàg modification. To any infinitely divisible distribution μ corresponds a Lévy process $Z = \{Z_t\}_{t\geq 0}$ such that the law of Z_1 is μ. We speak of the Lévy process generated by μ; this is uniquely determined in the distributional sense.

Let $Z = \{Z_t\}_{t\geq 0}$ be a Lévy process. For any $t > 0$ the random variable Z_t is infinitely divisible. Let (a, b, ρ) denote the characteristic triplet of Z_1. The key tool for understanding the nature of Lévy processes is the *Lévy–Ito representation* for Z,

$$Z_t = at + \sqrt{b}W_t + \int_0^t \int_{\mathbb{R}} x\{N(\mathrm{d}x, \mathrm{d}s) - \mathbf{1}_{[-1,1]}(x)\rho(\mathrm{d}x)\mathrm{d}s\},$$

where W is a Brownian motion and N is a Poisson basis—that is, a Poisson random measure—on $\mathbb{R} \times \mathbb{R}_{\geq 0}$, independent of W and with mean measure $\rho(\mathrm{d}x)\mathrm{d}t$.

When the Brownian part is absent from Z—that is, $b = 0$—we speak of a *pure Lévy process*. The following classification of the pure Lévy processes is important, not least from the applied point of view. Such processes may be divided according to the two criteria of finite or infinite activity and finite or infinite variation, as follows:

- finite activity and finite variation, if $\int_0^\infty \rho(\mathrm{d}x) < \infty$;
- infinite activity and finite variation, if $\int_0^\infty 1 \wedge |x| \rho(\mathrm{d}x) < \infty$ but $\int_0^\infty \rho(\mathrm{d}x) = \infty$;
- infinite activity and infinite variation, if $\int_0^\infty 1 \wedge |x| \rho(\mathrm{d}x) = \infty$.

Infinite activity means that the process exhibits infinitely many jumps in any time interval, however small. Finite variation means that the sample paths of Z are of finite variation with probability one, that is, for any $t > 0$ we have $\sup \sum |Z(t_j) - Z(t_{j-1})| < \infty$ where the supremum is over all subdivisions of $[0, t]$. In a variety of modelling contexts where volatility is manifest, such as in mathematical finance, data strongly point to the use of infinite activity processes—and in fact towards higher activity than at the bordering region between infinite activity and finite or infinite variation.

Example 4 When $a = b = 0$, the class of processes with finite activity and finite variation is identical to the class of compound Poisson processes. The generalized inverse Gaussian Lévy processes display infinite activity and finite variation. The normal inverse Gaussian Lévy process displays infinite activity and infinite variation, as follows from the fact that $K(x) \sim x^{-1}$ for $x \downarrow 0$; cf. eqn (12.3).

12.2.5 OU processes

The definition of self-decomposability, eqn (12.5), may be expressed equivalently as follows: X has law in $\mathcal{L}(*)$ if and only if for all $\lambda > 0$ there exists a Lévy process Z for which the stochastic differential equation

$$\mathrm{d}X_t = -\lambda X_t \mathrm{d}t + \mathrm{d}Z_{\lambda t}$$

has a stationary solution $\{X_t\}_{t \in \mathbb{R}}$ with $X_t \overset{L}{=} X$. The solution, termed an *Ornstein–Uhlenbeck-type (OU) process* is representable as

$$X_t = \int_{-\infty}^t \mathrm{e}^{-\lambda(t-s)} \mathrm{d}Z_{\lambda s},$$

where Z has been extended to a process with independent and stationary increments on the whole real line. The scaling by λ in the time for Z ensures that the one-dimensional marginal law of X does not change if λ is changed. The process Z is called the *background driving Lévy process*.

This construction works, in particular, for positive self-decomposable random variables, in which case the background driving Lévy process is a subordinator, that is, a Lévy process with non-negative increments.

When the process X is square integrable it has autocorrelation function $r(u) = \exp(-\lambda u)$, for $u > 0$.

Let $\mathcal{LB}(*)$ be the subclass of self-decomposable laws for which the background driving Lévy process belongs to the Bondesson class $\mathcal{B}(*)$. Barndorff-Nielsen *et al.* (2004d) show that $\mathcal{LB}(*)$ equals the Thorin class, that is, $\mathcal{LB}(*) = \mathcal{T}(*)$.

12.2.6 Lévy bases

A *Lévy basis* Z is an infinitely divisible independently scattered random measure on some Borel subset $\mathcal{S}$ of $\mathbb{R}^d$. To such a basis corresponds a Lévy–Khintchine type formula

$$C\{\zeta \ddagger Z(A)\} = ia(A)\zeta - \frac{1}{2}b(A)\zeta^2 + \int_{\mathbb{R}}\{e^{i\zeta x} - 1 - i\zeta c(x)\}\nu(dx; A), \quad (12.6)$$

where A denotes a Borel subset of $\mathcal{S}$, a is a signed measure, b is a measure, and $\nu(\cdot, A)$ is a Lévy measure. We refer to (a, b, ν) as the *characteristic triplet* of the Lévy basis Z and to ν as a *generalized Lévy measure*.

Lévy bases play an important role in many contexts. Some applications are indicated in Sections 12.3, 12.5, and 12.6.

It is informative to think of ν as factorised as

$$\nu(dx; d\sigma) = U(dx; \sigma)\mu(d\sigma),$$

and to consider $U(dx; \sigma)$ as the Lévy measure of an infinitely divisible random variable $Z'(\sigma)$, called the *spot mark* at site $\sigma \in \mathcal{S}$. Assuming for simplicity that $a = b = 0$, we then have the following key formula for the cumulant function of integrals $f \bullet Z = \int_{\mathcal{S}} f(\sigma)Z(d\sigma)$:

$$C\{\zeta \ddagger f \bullet Z\} = \int C\{\zeta f(\sigma) \ddagger Z'(\sigma)\}\mu(d\sigma).$$

This result follows essentially from the following formal calculation (which is in fact rigorous when f is a simple function) using the independent scattering property of Z and product integration:

$$\exp\left[C\{\zeta \ddagger f \bullet Z\}\right] = E\left[\{\exp\{i\zeta f \bullet Z\}\right]$$

$$= E\left[\prod_{\sigma \in \mathcal{S}} \exp\{i\zeta f(\sigma)Z(d\sigma)\}\right]$$

$$= \prod_{\sigma \in \mathcal{S}} E\left[\exp\{i\zeta f(\sigma)Z(d\sigma)\}\right]$$

$$= \prod_{\sigma \in \mathcal{S}} \exp\left[C\{\zeta f(\sigma) \ddagger Z'(\sigma)\}\mu(d\sigma)\right]$$

$$= \exp\left[\int_{\mathcal{S}} C\{\zeta f(\sigma) \ddagger Z'(\sigma)\}\mu(d\sigma)\right].$$

There is also an analogue for Lévy bases of the Lévy–Ito formula. In the case $a = b = 0$ it reads, in infinitesimal form,

$$Z(d\sigma) = \int_{\mathbb{R}} x\{N(dx; d\sigma) - \nu(dx; d\sigma)\}, \quad (12.7)$$

where N is a Poisson field on $\mathbb{R} \times \mathcal{S}$ with $E\{N(dx; d\sigma)\} = \nu(dx; d\sigma)$.

For more details and references to the literature, see Pedersen (2003) and Barndorff-Nielsen and Shephard (2005).

When $a = b = 0$ and $\nu(\mathrm{d}x; \mathrm{d}\sigma)$ is of the form $\nu(\mathrm{d}x; \mathrm{d}\sigma) = U(\mathrm{d}x)\mathrm{d}\sigma$, we say that the Lévy basis is *homogeneous*.

Example 5 (supOU processes) Suppose $\mathcal{S} = \mathbb{R}_{\geq 0} \times \mathbb{R}_{>0}$ with points $\sigma = (t, \xi)$ and

$$\nu(\mathrm{d}x; \mathrm{d}\sigma) = \rho(\mathrm{d}x)\mathrm{d}t\pi(\mathrm{d}\xi) \tag{12.8}$$

for some Lévy measure ρ and probability measure π, and define a *supOU process* $X = (X_t)_{t \geq 0}$ by

$$X_t = \int_{\mathbb{R}_{>0}} \mathrm{e}^{-\xi t} \int_{-\infty}^{\xi t} \mathrm{e}^s Z(\mathrm{d}s, \mathrm{d}\xi),$$

where Z is a Lévy basis with $a = b = 0$ and ν given by eqn (12.8). The process X is a superposition of independent OU processes with individual decay rates ξ. The case of finite superpositions, used for instance in Barndorff-Nielsen and Shephard (2001*b*), is included by letting π be a finite discrete probability measure.

Assuming square integrability, the autocorrelation function of X is

$$r(t) = \int_0^\infty \mathrm{e}^{-x\xi}\pi(\mathrm{d}\xi).$$

When π is the gamma law $\Gamma(2\bar{H}, 1)$ we have $r(|\tau|) = (1 + |\tau|)^{-2\bar{H}}$. In particular, the process X then exhibits a second-order long-range dependence if $H \in (\frac{1}{2}, 1)$ where $H = 1 - \bar{H}$ (Barndorff-Nielsen 1997).

12.3 Tempospatial modelling

12.3.1 A general class of tempospatial processes

Tempospatial models describe dynamic developments that take place in both time and space. For simplicity we here restrict attention to the case where space is one-dimensional.

A general and flexible class of tempospatial processes $X = \{X_t(\sigma)\}_{(t,\sigma) \in \mathbb{R}^2}$, building on the concept of Lévy bases, is of the form

$$X_t(\sigma) = \int_{-\infty}^t \int_{A_t(\sigma)} h_{t-s}(\rho; \sigma)Z(\mathrm{d}\rho \times \mathrm{d}s), \tag{12.9}$$

where h is a non-negative function, with $h_u(\rho; \sigma) = 0$ for $u < 0$, and Z is a Lévy basis, the *background driving Lévy basis*; finally, $A_t(\sigma)$ is an *ambit set*. The idea of ambit sets is illustrated in Figure 12.1. They determine which points in time and space contribute to the value of $X_t(\sigma)$, via the underlying Poisson field $N(\mathrm{d}x; \mathrm{d}s, \mathrm{d}\sigma)$; cf. formula (12.7). Such processes were introduced in Barndorff-Nielsen and Schmiegel (2004) for the study of aspects of turbulence.

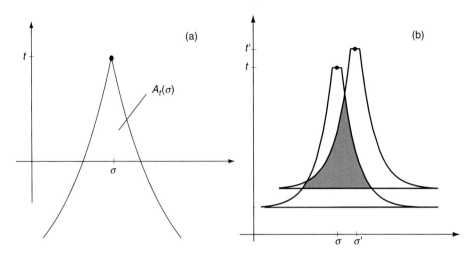

FIG. 12.1. Examples of ambit sets.

We shall refer to processes of the type of eqn (12.9) as *generalized shot noise* processes. It is also of interest to consider log generalized shot noise processes, which are positive processes $Y = \{Y_t(\sigma)\}$ for which $X_t(\sigma) = \log Y_t(\sigma)$ is of type (12.9). Dependence structures of such processes may be studied by means of *n-point correlators*

$$c(a_1, m_1; \ldots; a_n, m_n) = \frac{\mu(a_1, m_1; \ldots; a_n, m_n)}{\mu(a_1, m_1) \cdots \mu(a_n, m_n)}, \qquad (12.10)$$

where

$$\mu(a_1, m_1; \ldots; a_n, m_n) = \mathrm{E}\{Y_{t_1}(\sigma_1)^{m_1} \cdots Y_{t_n}(\sigma_n)^{m_n}\},$$

and $a_i = (t_i, \sigma_i)$ and $m_i \in \mathbb{R}$ for $i = 1, \ldots, n$. This definition of correlators is naturally adapted to the multiplicativity of the log generalized shot noise process since it allows for a cancellation of independent factors in the numerator and the denominator. An example of this is discussed in Section 12.6.2.

12.3.2 OU$_\wedge$ processes

The concept of OU processes on the real line may be extended to tempospatial settings by choosing the function h in eqn (12.9) to be simply $h_u(\rho; \sigma) = \mathrm{e}^{-\lambda u}$ for $u \geq 0$, in which case

$$X_t(\sigma) = \int_{-\infty}^{t} \int_{A_t(\sigma)} \mathrm{e}^{-\lambda(t-s)} Z(\mathrm{d}\rho \times \lambda \mathrm{d}s). \qquad (12.11)$$

It can then be shown (Barndorff-Nielsen and Schmiegel 2004) that, for all $\sigma \in \mathbb{R}$, $\{X_t(\sigma)\}_{t \in \mathbb{R}}$ is a stationary Markov process, decomposable as

$$X_t(\sigma) = \mathrm{e}^{-\lambda t} X_0(\sigma) + U_t(\sigma) + V_t(\sigma),$$

where $X_0(\sigma)$, $\{U_t(\sigma)\}_{t \geq 0}$, and $\{V_t(\sigma)\}_{t \geq 0}$ are independent.

12.3.3 OU$_\wedge$ Cox processes

In eqn (12.11) one may, in particular, choose the Lévy basis to be positive. The resulting generalized shot noise process may then be used, for instance, as intensity process for a Cox process. This provides an alternative to the log Gaussian processes discussed by Møller *et al.* (1998); see also Møller and Waagepetersen (2003). Certain advantages accrue due to the linear dependence of $X_t(\sigma)$ on the innovations $Z(\mathrm{d}\rho \times \lambda \mathrm{d}s)$, which in particular allows for explicit calculation of cumulant functions. The situation is similar to that in finance, where volatilities may be modelled either as log Gaussian or as OU or supOU processes (Barndorff-Nielsen and Shephard 2005).

12.4 Time change and chronometers

Time change for stochastic processes, through substituting ordinary time by an increasing stochastic process, is a topic of considerable current interest. If X is the original stochastic process and the increasing process determining the new time is T, then the time-changed process is $Y = X \circ T$ with $Y_t = X_{T_t}$. It is useful to have a name for the type of time-change processes we shall consider: a stochastic process $T = \{T_t\}_{t \geq 0}$ is said to be a *chronometer*—or a time-change—if it starts at 0, is increasing, stochastically continuous, and càdlàg.

A classical topic in the theory of Lévy processes is time change by subordination, that is, the chronometer T is itself a Lévy process—T is a subordinator—and T and X are independent. Then the time-changed process Y is again a Lévy process. This topic was initiated by Bochner (1949, 1955); some recent references are Bertoin (1996, 1997) and Barndorff-Nielsen *et al.* (2001*b*).

Example 6 Time changes of Brownian motion with drift are of particular interest. Then $Y = X \circ T$ where $X_t = B_t^{(\beta)} = \beta t + B_t$ with B a Brownian motion.

Let $Y_t^{(\mu)} = \mu t + Y_t$, with Y as just described. The *normal inverse Gaussian Lévy process* is the Lévy process $Y^{(\mu)} = \{Y_t^{(\mu)}\}_{t \geq 0}$ for which $Y_t^{(\mu)}$ has law NIG$(\alpha, \beta, t\mu, t\delta)$. In this case, Y is subordinated Brownian motion, with T being the *inverse Gaussian subordinator*, that is, the Lévy process for which T_t has law IG$(t\delta, t\gamma)$. More generally, when T is a subordinator with T_1 having law GIG$(\lambda, \delta, \gamma)$, one obtains the class of *generalized hyperbolic Lévy processes*. This has been extensively studied and applied in finance, see, for instance, Eberlein and Özkan (2003) and references therein. The class also contains the general variance gamma, or normal gamma, Lévy process introduced in finance by Carr *et al.* (1998); see also Geman *et al.* (2002).

Time change in finance other than subordination is considered in Section 12.5. For some discussions of time change in quantum physics and in turbulence see Chung and Zambrini (2003) and Barndorff-Nielsen *et al.* (2004*e*), respectively. Time change in a broad mathematical sense is treated in Barndorff-Nielsen and Shiryaev (2005). Barndorff-Nielsen *et al.* (2004*c*) have recently discussed conditions on X and T that imply that the time-changed process Y is infinitely divisible.

12.5   Lévy processes/bases and time change in finance

In recent years Lévy processes have entered the field of finance as an important modelling tool.

Using suitable Lévy processes instead of Brownian motion already provides a significant improvement in describing the behaviour of financial time series; cf. Example 6. However, this still fails to account for the significant timewise dependence structure of the series. To incorporate this one may consider time changes of Lévy processes using chronometers that, in contrast to subordinators, have non-independent increments.

Chronometers determined by integration of a stationary volatility process are of particular relevance; see for instance Barndorff-Nielsen and Shephard (2001*b*, 2004*a*, 2005), Barndorff-Nielsen *et al.* (2002), Carr *et al.* (2003), Cont and Tankov (2004), and references therein.

Example 7   Let $X = B^{(\beta)}$ be Brownian motion with drift β, and let T be an independent *integrated OU* or *integrated supOU* process, i.e.

$$T_t = \int_0^t V_s \mathrm{d}s, \qquad (12.12)$$

where V is a positive OU or supOU process. In the former case, the chronometer has the simple form

$$T_t = \lambda^{-1}\{Z_{\lambda t} - X_t + X_0\}$$
$$= \varepsilon(t; \lambda)X_0 + \int_0^t \varepsilon(t - s; \lambda)\mathrm{d}Z_s,$$

where Z is a subordinator and $\varepsilon(t; \lambda) = \lambda^{-1}(1 - \mathrm{e}^{-\lambda t})$. The linear dependence of T on the innovations of the background driving Lévy process Z allows for considerable analytic tractability, and a similar effect holds in the more general integrated supOU case. See Barndorff-Nielsen and Shephard (2001*a*, 2005) and Nicolato and Venardos (2003). Another case of interest takes V as the square root, or Cox–Ingersoll–Ross, process: the stationary solution of a stochastic differential equation of the form

$$\mathrm{d}V_t = (\mu - \lambda V_t)\mathrm{d}t + \sigma\sqrt{V_t}\mathrm{d}\tilde{B}_t,$$

where $\tilde{B}$ is a Brownian motion. In these cases, Y may be re-expressed in terms of another Brownian motion W, as

$$Y_t = \beta T_t + \int_0^t \sqrt{V_s}\mathrm{d}W_s,$$

showing that Y is of the type known in finance as a stochastic volatility model.

The general form of a one-dimensional continuous stochastic volatility model is

$$Y_t = \int_0^t a_s \mathrm{d}s + \int_0^t \sqrt{V_s} \mathrm{d}W_s,$$

having the form of a Brownian semimartingale, and the object of main interest is the *integrated volatility*—or *integrated variance*—process T, defined as in eqn (12.12). This is equal to the quadratic variation of Y, namely

$$T_t = [Y]_t \quad \left(= \text{p-lim} \sum (Y_{t_j} - Y_{t_{j-1}})^2 \right), \qquad (12.13)$$

where the limit is for finer and finer subdivision $0 < t_1 < \cdots < t_n \le t$; see eqn (11.1) Thus T is in principle calculable from observation of Y.

For Y observed at discrete time points the sum on the right-hand side of eqn (12.13) provides an estimate of T. The statistics of such estimates has been a subject of intense interest in econometrics since the recent advent of high-frequency data from the financial markets; see for instance Andersen *et al.* (2005*a*), Barndorff-Nielsen and Shephard (2004*a,c*), and references therein.

In the general time-change setting $Y = X \circ T$, it is of especial interest to ask to what extent information on the chronometer T can be obtained from observing the time-changed process Y only. This question is considered for Brownian subordination in Geman *et al.* (2002); their work has been extended by Winkel (2001) to time change of Brownian motion with general chronometers.

12.6 Lévy processes/bases and time change in turbulence

12.6.1 Introduction

The present section briefly summarizes two recent investigations in turbulence. In the first, the normal inverse Gaussian laws are used to study the distributions of velocity differences in three different large datasets. The second uses log generalized shot noise processes to model tempospatial structures in high Reynolds number turbulence. Frisch (1995) is an excellent and highly readable account of basis aspects of turbulence.

12.6.2 Normal inverse Gaussian laws and universality

The normal inverse Gaussian law provides excellent fits to velocity increments in high Reynolds number turbulence. In Barndorff-Nielsen *et al.* (2004*e*) this is demonstrated for three large datasets, from quite different experimental settings and with quite different Reynolds numbers. It is, moreover, shown that all three datasets can be parsimoniously described in terms of a few parameters via a time change from physical time t to a new time $\delta(t)$, which makes key features of the data collapse onto a single curve. Here $\delta(t)$ is the maximum likelihood estimate of the scale parameter of the normal inverse Gaussian distribution. The analysis has revealed several universal features of moderate to high Reynolds number turbulence, extending far beyond the Kolmogorov inertial subrange.

12.6.3   Log generalized shot noise processes and energy dissipation

Let $V_t(\sigma)$ denote the velocity of the turbulent field at position σ and time t. For simplicity we consider a one-dimensional position co-ordinate, corresponding to the fact that recorded measurements of $V_t(\sigma)$ often concern only the changes along the mean direction of the actual three-dimensional velocity.

Two key quantities in turbulence studies, and in particular in Kolmogorov's theory of isotropoic and homogeneous turbulence, are, up to a proportionality constant, the 'surrogate' energy dissipation rate

$$\varepsilon_t(\sigma) = \left\{ \frac{\partial}{\partial \sigma} V_t(\sigma) \right\}^2, \tag{12.14}$$

and its coarse-grained version

$$\bar{\epsilon}_t(\sigma) = \frac{1}{|B|} \int_B \epsilon_t(\sigma) \mathrm{d}\sigma,$$

where the integral is over some small region B around σ. For a detailed discussion of these, see Frisch (1995). The energy-dissipation rate expresses the intermittency or volatility of the turbulent field and plays a role very similar to that of stochastic volatility in finance—compare eqns (12.13) and (12.14). For the study of scaling properties and timewise and spatial dependence structures of the energy dissipation, modelling this as a log generalized shot noise process, rather than as a positive generalized shot noise process, offers considerable advantages. Accordingly, one considers processes $\{Y_t(\sigma)\}_{t\in\mathbb{R}}$ where $Y_t(\sigma) = \exp X_t(\sigma)$ with $\{X_t(\sigma)\}_{t\in\mathbb{R}}$ of generalized shot noise type.

A relatively simple case that is nonetheless of considerable interest has h in eqn (12.9) identically equal to 1 and ambit sets of the form

$$A_t(\sigma) = \{(\rho, s) : -\infty < s < t, \rho \in C_{s-t}(\sigma)\},$$

with

$$C_s(\sigma) = \begin{cases} \emptyset, & s < -T, \\ [\sigma - g(s), \sigma + g(s)], & -T \le s \le 0, \end{cases}$$

for some $T > 0$ and a non-negative decreasing function g on $[-T, 0]$; see part (b) of Figure 12.1.

Suppose, moreover, that the Lévy basis Z in eqn (12.9) is homogeneous, in the sense defined in Section 12.2.5, and that, for some $t_\eta \in (0, T)$,

$$g(s) = \frac{1}{2(T+s)}, \qquad t_\eta - T \le s \le 0,$$

while g is left arbitrary on $[-T, t_\eta - T]$, subject to the restrictions that it is decreasing on $[-T, 0]$ and that the volume $|A_t(\sigma)|$ of $A_t(\sigma)$ is finite. Let K denote the kumulant function of the spot marks, which all have the same distribution, due to the assumption of homogeneity of Z. It may then be shown that the time wise two-point correlator, given by eqn (12.10), has the form

$$c\{(\sigma,t),m;(\sigma,t'),m'\} = \left(\frac{T}{|t-t'|}\right)^{K(m+m')-K(m)-K(m')},$$

which is scaling in $|t-t'|$, and that the spacewise two-point correlator,

$$c\{(\sigma,t),m;(\sigma',t),m'\} \propto |\sigma-\sigma'|^{-\{K(m+m')-K(m)-K(m')\}},$$

is scaling in $|\sigma-\sigma'|$. This kind of correlation structure is observed for the energy dissipation in turbulent flows and is closely related to the multifractal and intermittent nature of the coarse-grained energy dissipation field. For further results and details, see Barndorff-Nielsen and Schmiegel (2004).

The modelling approach outlined here has turned out to be useful in the completely different setting of certain biological growth patterns, in particular cancer growth (Barndorff-Nielsen *et al.* 2004b).

12.7 Upsilon mappings

12.7.1 Introduction

Upsilon mappings are regularizing one-to-one mappings of the class of Lévy measures into itself, with associated one-to-one mappings of $\mathcal{ID}(*)$ into but not onto itself. The original Upsilon mapping was introduced by Barndorff-Nielsen and Thorbjørnsen (2004a) and this and generalizations have been studied in Barndorff-Nielsen and Thorbjørnsen (2004a,c), Barndorff-Nielsen *et al.* (2004d), and Barndorff-Nielsen and Peres-Abreu (2005). In Barndorff-Nielsen and Lindner (2004) the multivariate version of the Upsilon mapping, introduced in Barndorff-Nielsen *et al.* (2004d), is used to construct multivariate infinitely divisible laws with certain desired properties, via the concept of Lévy copulas.

12.7.2 The mappings Υ_0 and Υ

For any Lévy measure ρ we define $\tilde{\rho} = \Upsilon_0(\rho)$ by

$$\tilde{\rho}(\mathrm{d}x) = \int_0^\infty \rho(\xi^{-1}\mathrm{d}x)e^{-\xi}\mathrm{d}\xi.$$

This construction does yield a Lévy measure, and the mapping Υ_0 is one-to-one. However, the mapping is not onto. In fact, the mapping has a strong regularizing effect, as appears from the following result.

Theorem 12.1 *For any Lévy measure ρ the Lévy measure $\tilde{\rho} = \Upsilon(\rho)$ has a Lévy density $\tilde{r}$ satisfying*

$$\tilde{r}(x) = \begin{cases} \int_0^\infty \xi e^{-x\xi}\omega(\mathrm{d}\xi), & x > 0, \\ \int_{-\infty}^0 |\xi| e^{-x\xi}\omega(\mathrm{d}\xi), & x < 0, \end{cases}$$

where ω is the image of ρ by the mapping $x \mapsto x^{-1}$.

With Υ_0 in hand we may introduce the one-to-one mapping

$$\Upsilon: \quad \mathcal{ID}(*) \to \mathcal{ID}(*)$$
$$(a, b, \rho) \to (\tilde{a}, 2b, \tilde{\rho}),$$

where

$$\tilde{a} = a + \int_0^\infty \int_{\mathbb{R}} x\{1_{[-1,1]}(x) - 1_{[-\xi,\xi]}(x)\}\rho(\xi^{-1}\mathrm{d}x)\mathrm{e}^{-\xi}\mathrm{d}\xi.$$

This mapping has the key property

$$C_{\Upsilon(\mu)}(\zeta) = \int_0^\infty C_\mu(\zeta\xi)\mathrm{e}^{-\xi}\mathrm{d}\xi. \tag{12.15}$$

Using the characterizations of $\mathcal{L}(*)$, $\mathcal{T}(*)$ and $\mathcal{B}(*)$ given in Section 12.2.2, it can be shown that $\Upsilon(\mathcal{S}(*)) = \mathcal{S}(*)$, $\Upsilon(\mathcal{L}(*)) = \mathcal{T}(*)$ and $\Upsilon(\mathcal{ID}(*)) = \mathcal{B}(*)$. For the proof, see Barndorff-Nielsen and Thorbjørnsen (2004c) and Barndorff-Nielsen et al. (2004d).

There is, moreover, a stochastic interpretation of Υ:

Theorem 12.2 *Let $\mu \in \mathcal{ID}(*)$ and let X_t be the Lévy process for which X_1 has law μ. Then the integral*

$$Y = \int_0^1 |\log(1-t)|\mathrm{d}X_t \tag{12.16}$$

is well defined as the limit in probability for $n \to \infty$ of the stochastic integrals $\int_0^{1-1/n} |\log(1-t)|\mathrm{d}X_t$, and the law of Y is $\Upsilon(\mu)$.

12.7.3 The mappings Υ_0^α and Υ^α

It is possible to define a smooth family of mappings $\Upsilon^\alpha : \mathcal{ID}(*) \to \mathcal{ID}(*)$, for $\alpha \in [0,1]$, that interpolates between Υ ($\Upsilon^0(\mu) = \Upsilon(\mu)$) and the identity mapping on $\mathcal{ID}(*)$ ($\Upsilon^1(\mu) = \mu$), or—in the context of Section 12.8 below—between the classical Lévy–Khintchine formula and the free Lévy–Khintchine formula. For every α, Υ^α has properties similar to Υ. The Mittag–Leffler function takes a natural role in this. For details beyond those given below, see Barndorff-Nielsen and Thorbjørnsen (2004a,c).

Mittag–Leffler function The Mittag–Leffler function of negative real argument and index $\alpha > 0$ is

$$E_\alpha(-t) = \sum_{k=0}^\infty \frac{(-t)^k}{\Gamma(\alpha k + 1)}, \quad t < 0. \tag{12.17}$$

In particular we have $E_1(-t) = \mathrm{e}^{-t}$, and if we define E_0 by setting $\alpha = 0$ on the right-hand side of eqn (12.17) then $E_0(-t) = (1+t)^{-1}$.

The Mittag–Leffler function is infinitely differentiable and completely monotone if and only if $0 < \alpha \leq 1$. Hence for $0 < \alpha \leq 1$ it is representable as a Laplace transform and, in fact, we have (Feller 1971, p. 453) that

$$E_\alpha(-t) = \int_0^\infty e^{-tx}\zeta_\alpha(x)dx, \qquad 0 < \alpha < 1, \tag{12.18}$$

where

$$\zeta_\alpha(x) = \alpha^{-1}x^{-1-1/\alpha}\sigma_\alpha(x^{-1/\alpha}), \tag{12.19}$$

and σ_α denotes the density function of the positive stable law with index α and Laplace transform $\exp(-\theta^\alpha)$. Defining $\zeta_\alpha(x)$ as e^{-x} for $\alpha = 0$ and as the Dirac density at 1 when $\alpha = 1$ then eqn (12.18) is valid for all $\alpha \in [0, 1]$.

For $0 < \alpha < 1$, the function $\zeta_\alpha(x)$ is simply the probability density obtained from $\sigma_\alpha(y)$ by the transformation $y = x^{-1/\alpha}$. In other words, if we denote the distribution functions determined by ζ_α and σ_α by Z_α and S_α, respectively, then

$$Z_\alpha(x) = 1 - S_\alpha(x^{-1/\alpha}).$$

The mapping Υ_0^α We define Υ_0^α as the mapping $\rho \mapsto \tilde{\rho}_\alpha$, defined on the class of Lévy measures on $\mathbb{R}$, and determined by

$$\tilde{\rho}_\alpha(dt) = \int_0^\infty \rho(x^{-1}dt)\zeta_\alpha(x)dx. \tag{12.20}$$

Using eqn (12.19), this may be re-expressed as

$$\tilde{\rho}_\alpha(dt) = \int_0^\infty \rho(x^\alpha dt)\sigma_\alpha(x)dx.$$

Now, as for Υ_0, we have

Theorem 12.3 *The mapping Υ_0^α sends Lévy measures to Lévy measures. Furthermore, letting ω denote the image of the Lévy measure ρ under the mapping $x \mapsto x^{-1}$, we have that for any Lévy measure ρ the Lévy measure $\tilde{\rho}_\alpha = \Upsilon_0^\alpha(\rho)$ is absolutely continuous with respect to Lebesgue measure and the density $\tilde{r}_\alpha$ of $\tilde{\rho}_\alpha$ is the function on $\mathbb{R}\backslash\{0\}$ given by*

$$\tilde{r}_\alpha(t) = \begin{cases} \int_0^\infty s\zeta_\alpha(st)\omega(ds), & t > 0, \\ \int_{-\infty}^0 |s|\zeta_\alpha(st)\omega(ds), & t < 0. \end{cases}$$

The mapping Υ^α Based on Υ_0^α we can now define Υ^α as the mapping from $\mathcal{ID}(*)$ into $\mathcal{ID}(*)$ that sends a probability measure μ with characteristic triplet

(a, b, ρ) to the law $\Upsilon(\mu)$ in $\mathcal{ID}(*)$ having characteristic triplet $(a_\alpha, c_\alpha b, \tilde{\rho}_\alpha)$, where $\tilde{\rho}_\alpha = \Upsilon_0^\alpha(\rho)$ is defined by eqn (12.20),

$$c_\alpha = \frac{2}{\Gamma(2\alpha + 1)}, \qquad 0 \le \alpha \le 1,$$

and

$$a_\alpha = \frac{a}{\Gamma(\alpha + 1)} + \int_0^\infty \int_{\mathbb{R}} t\{\mathbf{1}_{[-1,1]}(t)\} - \mathbf{1}_{[-x^{-1}, x^{-1}]}(t)\}\rho(x^{-1}\mathrm{d}t)\zeta_\alpha(x)\mathrm{d}x.$$

Theorem 12.4 *The cumulant function of $\Upsilon^\alpha(\mu)$ may be represented as*

$$C_\alpha(\zeta) = i\frac{a}{\Gamma(\alpha + 1)}\zeta - c_\alpha b\zeta^2 + C_{0\alpha}(\zeta),$$

where

$$C_{0\alpha}(\zeta) = \int_{\mathbb{R}} \{E_\alpha(i\zeta t) - 1 - i\zeta\frac{t}{\Gamma(\alpha + 1)}\mathbf{1}_{[-1,1]}(t)\}\rho(\mathrm{d}t).$$

12.7.4 Stochastic representation

Let μ again denote an element of $\mathcal{ID}(*)$ and let X_t be the Lévy process generated by μ, that is, the Lévy process such that X_1 has law μ.

The mapping Υ^α may be given a stochastic interpretation similar to the representation (12.16) for Υ. Specifically, letting R_α denote the inverse of the distribution function Z_α, we have

Theorem 12.5 *The stochastic integral*

$$Y = \int_0^1 R_\alpha(s)X_s$$

is well defined for all $\alpha \in [0, 1]$, and the law of Y is $\Upsilon^\alpha(\mu)$.

12.8 Quantum stochastics

12.8.1 Introduction

The probabilistic aspects of quantum stochastics constitute a highly developed and mature field. In this, various new concepts of 'independence' and 'infinite divisibility', with associated 'Lévy processes', have recently arisen, the most widely studied being *free independence*—or freeness—and *free infinite divisibility*.

Concise introductions to the more statistical aspects of quantum stochastics are given in Holevo (2001*b*) and Barndorff-Nielsen *et al.* (2001*c*). The concept of *quantum instruments* has a key role in this, and *infinitely divisible quantum instruments* provide a bridge to quantum stochastic processes. These infinite divisibility themes are outlined below.

12.8.2 Freeness

The concept of freeness or free independence arose out of the study of von Neumann algebras. However, it was soon realized that this concept has an interesting interpretation in terms of the limit behaviour of large random matrices, in which Wigner's semicircular law takes a role similar to that of the Gaussian law under classical central limit theory. In a certain sense, free independence can be seen as the vestige of classical stochastic independence remaining after taking products of independent random matrices and passing to the limit by letting the dimension of the matrices go to infinity. For a more detailed introduction to free independence and free infinite divisibility than the brief account given here, see Barndorff-Nielsen and Thorbjørnsen (2002c).

Free infinite divisibility Out of von Neumann's formulation of the mathematical basis of quantum theory has grown the major and very active mathematical field of operator algebras, including the study of 'non-commutative probability'. Speaking briefly, an algebraic probability space is a pair $(\mathcal{A}, \phi)$ where $\mathcal{A}$ is an algebra, typically non-commutative with a unit element e, and ϕ is a positive linear functional on $\mathcal{A}$ such that $\phi(e) = 1$. Given an element a of $\mathcal{A}$ we may think of $\phi(a)$ as the mean value of a.

The idea of *free independence* is due to Voiculescu and hinges on the definition of a new type of 'convolution' of probability measures, denoted by $\boxplus$ instead of the classical $*$. In analogy with the classical case one says that a probability measure μ is *freely infinitely divisible* if for all $n \in \mathbb{N}$ there exists a probability measure μ_n such that $\mu = \mu_n^{\boxplus n}$.

The free convolution $\boxplus$ can be described in terms of an analogue of the classical cumulant transform C_μ called the *free cumulant transform* and denoted $\mathcal{C}_\mu$. This is defined, quite intricately, as

$$\mathcal{C}_\mu(z) = z\phi_\mu(z^{-1}),$$

where $\phi_\mu(z) = F_\mu^{\mathrm{inv}}(z) - z$, and $F_\mu^{\mathrm{inv}}(z)$ is the inverse function of $F_\mu(z) = 1/G_\mu(z)$, with $G_\mu(z)$ denoting the Cauchy or Stieltjes transform of μ,

$$G_\mu(z) = \int_{\mathbb{R}} \frac{1}{z - x}\mu(\mathrm{d}x).$$

It was shown by Voiculescu *et al.* (Voiculescu 2000) that $\phi_{\mu\boxplus\nu} = \phi_\mu + \phi_\nu$ and this is equivalent to $\mathcal{C}_{\mu\boxplus\nu} = \mathcal{C}_\mu + \mathcal{C}_\nu$. We have introduced the form $\mathcal{C}_\mu$ to make the analogy to the classical case as close as possible.

Voiculescu *et al.* have further proved that there is a free analogue of the Lévy–Khintchine formula, that we may express in terms of $\mathcal{C}$ as

$$\mathcal{C}_\mu(z) = az - bz^2 + \int_{\mathbb{R}} \left(\frac{1}{1 - zx} - 1 - zx\mathbf{1}_{[-1,1]}(x) \right) \rho(\mathrm{d}x),$$

where, as for the classical Lévy–Khintchine formula, $a \in \mathbb{R}$, $b \geq 0$ and ρ is a Lévy measure. We refer to (a, b, ρ) as the *characteristic triplet*, classical or free, as the case may be, of μ.

Connections between classical and free infinite divisibility   A direct connection between classical and free infinite divisibility was established by Bercovici and Pata (1999), who introduced a one-to-one mapping from the class $\mathcal{ID}(*)$ onto the class $\mathcal{ID}(\boxplus)$ of freely infinitely divisible probability measures. This *Bercovici–Pata mapping* is denoted by Λ. In the notation introduced above the mapping can be simply described as sending a probability measure μ with classical characteristic triplet (a, b, ρ) to the probability measure having free characteristic triplet (a, b, ρ).

We may now define a hierarchy within the class $\mathcal{ID}(\boxplus)$ of freely independent laws, analogous to the hierarchy (12.4):

$$\mathcal{G}(\boxplus) \subset \mathcal{S}(\boxplus) \subset \mathcal{T}(\boxplus) \subset \mathcal{L}(\boxplus), \ \mathcal{B}(\boxplus) \subset \mathcal{ID}(\boxplus),$$

where $\mathcal{G}(\boxplus) = \Lambda(\mathcal{G}(*))$, $\mathcal{S}(\boxplus) = \Lambda(\mathcal{S}(*))$, and so forth. The free counterpart of the Gaussian distributions, $\mathcal{G}(\boxplus)$, consists of the Wigner, or semi-circle, laws. The free stable laws and their domains of free attraction were studied in Bercovici and Pata (1999), and the free self-decomposable laws were introduced in Barndorff-Nielsen and Thorbjørnsen (2002c), not as here via Λ but through defining free stability and free self-decomposability from free analogues of their classical probabilistic meanings.

Barndorff-Nielsen and Thorbjørnsen (2002c) also established several algebraic and topological properties of Λ, in particular that Λ is a homeomorphism with respect to weak convergence, that is, convergence in law. These properties were instrumental in demonstrating that the two approaches to defining $\mathcal{L}(\boxplus)$ give identical results, and for developing some new properties of free Lévy processes, including results on free stochastic integration and a free analogue of the Lévy–Ito decomposition (Barndorff-Nielsen and Thorbjørnsen 2002a,b,c, 2004b).

Another type of connection between classical and free infinite divisibility is established via the mapping Υ discussed in Section 12.7. In fact, the key relation (12.15) extends to

$$C_{\Upsilon(\mu)}(\zeta) = \int_0^\infty C_\mu(\zeta\xi) e^{-\xi} \mathrm{d}\xi = \mathcal{C}_{\Lambda(\mu)}(i\zeta),$$

showing that the free cumulant function of $\Lambda(\mu)$ can, in spite of the complicated definition of the general free cumulant transform, be obtained from the classical cumulant function of μ via a simple integral and that the free cumulant function $\mathcal{C}_{\Lambda(\mu)}$ is, in fact, equal to a classical cumulant function, namely $C_{\Upsilon(\mu)}$.

Some related types of independence   In addition to free independence, two further independence concepts, called *monotone* and *antimonotone independence* have recently been introduced and studied in an algebraic framework that also encompasses classical (or tensor), free, and *boolean*, independence (Ben Ghorbal and Schürmann 2002, Franz 2003, Muraki 2003). In particular, associated to monotone independence there exists a non-commutative convolution concept, and for this the analogue of the Gaussian law in the classical case and of the Wigner law in the free case is the arcsine law.

12.8.3 Quantum measurements and quantum instruments

In comparison to the algebraic approach to probabilistic structures related to quantum theory, indicated in Section 12.8.1, the theory of quantum instruments is more analytic and, generally speaking, closer to present developments in quantum physics itself. At the same time the theory of quantum instruments, the foundation of which was laid by Davies and Lewis (1970), Davies (1976) and Kraus (1971, 1983), provides a bridge between von Neumann's original formulation of quantum physics, on the one hand, and probability and statistics based on Kolmogorov's axioms on the other.

According to von Neumann, the state of a quantum physical system $\mathcal{S}$ is represented by a linear operator ρ on a Hilbert space $\mathcal{H}$, where ρ is self-adjoint and non-negative with unit trace. In the modern formulation of quantum measurements, a measurement of $\mathcal{S}$ is represented by a *positive operator-valued measure* M defined on some measure space $(\Omega, \mathcal{F})$. More specifically, for each $B \in \mathcal{F}$, $M(B)$ is a positive linear self-adjoint operator on $\mathcal{H}$, and M is σ-additive, with $M(\Omega) = I$, the identity operator on $\mathcal{H}$. The result of the measurement is an outcome ω in Ω, whose distribution is the probability law

$$\pi(B; \rho) = \mathrm{pr}\{\omega \in B; \rho\} = \mathrm{tr}\{\rho M(B)\}.$$

In general, the outcome ω is not observed directly; only the value x of some random variate X on $(\Omega, \mathcal{F})$ is recorded.

The act of performing the measurement changes the state ρ of $\mathcal{S}$ to a new, in general unknown, state $\sigma = \sigma(\omega; \rho)$ that depends on ω and ρ. More generally, the posterior state of the quantum system conditioned by the event $\omega \in B$ is written as $\sigma(B; \rho)$ and is related to the probability measure π by

$$\sigma(B; \rho) = \frac{\int_B \sigma(\omega; \rho) \pi(d\omega; \rho)}{\pi(B; \rho)}.$$

When the positive operator-valued measure M is dominated by a σ-finite measure μ on $(\Omega, \mathcal{F})$, π has a density

$$p(\omega; \rho) = \frac{\pi(d\omega; \rho)}{\mu(d\omega)}.$$

Because of the basic tenets of quantum physics, π and σ cannot be prescribed arbitrarily but have to be linked to or—better expressed—must arise from a mathematical object known as a quantum instrument. There are two—dual— versions of this: *state instruments* and *operator instruments*. To define these, let $\mathcal{B}(\mathcal{H})$ be the linear space of bounded linear self-adjoint operators on $\mathcal{H}$ and let $\mathcal{T}(\mathcal{H})$ denote the set of trace class operators on $\mathcal{H}$, which includes all the states ρ of $\mathcal{S}$. Then we can make the following definitions.

Definition 12.6   *A state instrument is a mapping*

$$\mathcal{M} : \mathcal{F} \times \mathcal{T}(\mathcal{H}) \to \mathcal{T}(\mathcal{H})$$
$$(B, \kappa) \to \mathcal{M}(B)[\kappa]$$

satisfying the conditions that $\mathrm{tr}\{\mathcal{M}(\Omega)[\kappa]\} = \mathrm{tr}\{\kappa\}$; *that* $\mathcal{M}(\cdot)[\kappa]$ *is a* σ-additive measure on $(\Omega, \mathcal{F})$, *convergence being understood in the norm of* $\mathcal{T}(\mathcal{H})$; *and for which* $\mathcal{M}(B)[\cdot]$ *is a positive normal linear map.*

Definition 12.7   *An operator instrument is a mapping*

$$\mathcal{N} : \mathcal{F} \times \mathcal{B}(\mathcal{H}) \to \mathcal{B}(\mathcal{H})$$
$$(B, Y) \to \mathcal{M}(B)[Y]$$

satisfying the conditions that $\mathcal{N}(\Omega)[I] = I$; *that* $\mathcal{N}(\cdot)[Y]$ *is a* σ-additive measure on $(\Omega, \mathcal{F})$, *convergence being understood in the weak-$*$ topology; and for which* $\mathcal{N}(B)[\cdot]$ *is a positive normal bounded linear map.*

The duality of $\mathcal{M}$ and $\mathcal{N}$ is expressed in the relation

$$\mathrm{tr}\{Y\mathcal{M}(B)[\kappa]\} = \mathrm{tr}\{\kappa\mathcal{N}(B)[Y]\}.$$

In terms of $\mathcal{M}$ and $\mathcal{N}$ the positive operator-valued measure M, the probability measure π and the posterior state σ are given by

$$M(B) = \mathcal{N}(B)[I],$$
$$\pi(B; \rho) = \mathrm{tr}\{\rho\mathcal{N}(B)[I]\} = \mathrm{tr}\{\mathcal{M}(B)[\rho]\},$$
$$\sigma(\omega; \rho) = \frac{\mathrm{d}\mathcal{M}(\cdot)[\rho]}{\mathrm{d}\pi(\cdot; \rho)},$$

and we have

$$\int_B \mathrm{tr}\{\sigma(\omega; \rho)Y\}\pi(dw, \rho) = \mathrm{tr}\{\rho\mathcal{N}(B)[Y]\}.$$

Generic examples of quantum instruments are

$$\mathcal{N}(B)[X] = \sum_i \int_B V_i(x)^* X V_i(x)\nu_i(\mathrm{d}x),$$

and the dual

$$\mathcal{M}(B)[\tau] = \sum_i \int_B V_i(x)\tau V_i(x)^*\nu_i(\mathrm{d}x),$$

where the ν_i are σ-finite measures on $(\Omega, \mathcal{F})$ and the V_i are measurable $\mathcal{B}(\mathcal{H})$-valued functions on Ω.

Continuous-time quantum measurements The theory of continuously moni-
tored quantum systems is an area of great current interest, particularly in
quantum optics where very small quantum systems are now studied intensively.
Quantum state and operator instruments play an important role in this theory. Convolution semigroups of quantum instruments and associated instrument
processes are built from the concept of infinitely divisible quantum instruments.
This in turn leads to the construction of quantum stochastic process representations of the semigroups, a procedure known as *dilation*. A related approach con-
structs quantum measurement processes via classical stochastic calculus. For the
study and certain applications of these types of quantum stochastic processes,
specially designed Monte Carlo methods—'quantum Monte Carlo'—have been
constructed; see, for example, Mølmer and Castin (1996) and Barndorff-Nielsen
et al. (2001*c*, §9). The following remarks briefly indicate a few key aspects of
this. For more information, see Holevo (2001*b*, Ch. 4), Holevo (2001*a*), Barchielli
and Holevo (1995), and reference therein.

As is the case for classical infinite divisibility, Fourier transformation —the
use of characteristic functions—is a key tool. Suppose for simplicity that the
outcome space Ω is $\mathbb{R}$ equipped with the Borel σ-algebra. The *characteristic
function of an operator instrument* $\mathcal{N}$ is defined by

$$\Phi(\zeta)[X] = \int_{\mathbb{R}} e^{i\zeta x} \mathcal{N}(\mathrm{d}x)[X].$$

For completely positive instruments there is a characterisation of the Bochner–
Khintchine type, which in fact reduces to the Bochner–Khintchine theorem when
$\dim \mathcal{H} = 1$.

Infinite divisibility of quantum instruments is defined in analogy with the
classical probabilistic concept, and the infinitely divisible quantum instruments
are characterized by a generalization of the Lévy–Khintchine formula. Writing
the cumulant function of such an instrument as

$$\mathcal{K}(\zeta) = \log \Phi(\zeta),$$

we have that $\mathcal{K}$ is of the form

$$\mathcal{K}(\zeta) = \mathcal{K}_0 + \mathcal{K}_1(\zeta) + \mathcal{K}_2(\zeta),$$

where

$$\mathcal{K}_0[X] = i[H, X] + \sum (A_s X A_s - A_s^2 \circ X) + \sum \lambda_r (U_r^* X U_r - X),$$

$$\mathcal{K}_1(\zeta)[X] = \sigma^2 \left\{ L^* X L - L^* L \circ X + i\lambda(L^* X + XL) - \frac{1}{2}\lambda^2 X \right\},$$

$$\mathcal{K}_2(\zeta)[X] = \int_{\mathbb{R}\backslash\{0\}} \left\{ \mathfrak{J}(\mathrm{d}x)[X]e^{i\lambda x} - \mathfrak{J}(\mathrm{d}x)[I] \circ X - i[X, H(\mathrm{d}x)] - X\frac{i\lambda x}{1+x^2}\mu(\mathrm{d}x) \right\}.$$

Here A_s, U_r, and L are operators, $\mathfrak{J}(\mathrm{d}x)[\cdot]$ and $H(\mathrm{d}x)$ are operator-valued mappings, and $\lambda > 0$, $\sigma \geq 0$ are real numbers. See Holevo (2001b, §4.2.5) for details.

Given a continuous-time instrument process, it is in important cases possible to set up a stochastic differential equation, of a classical type, describing a dilation of the timewise evolution—concordant with the instrument process—of a quantum state $\psi(t)$ in a Hilbert space $\mathcal{H}$, in such a way that the dilated system behaves in mean like $\psi(t)$. A generic example of such a stochastic differential equation has the 'Lévy–Ito form'

$$\mathrm{d}\Psi(t) = -K\Psi(t_-)\mathrm{d}t + L\Psi(t_-)\mathrm{d}W(t) + \int_{\mathbb{R}\setminus\{0\}} L(y)\Psi(t_-)\Pi(\mathrm{d}y\mathrm{d}t),$$

where K, L and $L(y)$ are operators on $\mathcal{H}$, while W is Brownian motion and Π is a Poisson field. Associated to this there may be classical stochastic differential equations for the recorded measurement results. For a detailed account, see Barchielli and Holevo (1995).

Acknowledgements

Helpful comments from Alex Lindner, Jürgen Schmiegel, Neil Shephard, Steen Thorbjørnsen and two referees are gratefully acknowledged.

Bibliography

Abbott, A. and Hrycak, A. (1990), 'Measuring resemblence in sequence data: An optimal matching analysis of musicians' careers', *American Journal of Sociology* **96**, 144–185.

Abramowitz, M. and Stegun, I. A., (eds) (1964), *Handbook of Mathematical Functions*, Dover, New York.

Agresti, A. (2002), *Categorical Data Analysis*, 2nd, Wiley, New York.

Alcock, C., Alves, D. R., Becker, A., Bennett, D., Cook, K. H., Drake, A., Freeman, K., Geha, M., Griest, K., Kovacs, G., Lehner, M., Marshall, S., Minniti, D., Nelson, C., Peterson, B., Popowski, P., Pratt, M., Quinn, P., Rodgers, A., Stubbs, C., Sutherland, W., Vandehei, T. and Welch, D. L. (2003), 'The Macho project Large Magellanic Cloud variable star inventory. XI. Frequency analysis of the fundamental-mode RR Lyrae stars', *Astrophysical Journal* **598**, 597–609.

Aldous, D. J. (1981), 'Representations for partially exchangeable arrays of random variables', *Journal of Multivariate Analysis* **11**, 581–598.

Aldous, D. J. and Eagleson, G. K. (1978), 'On mixing and stability of limit theorems', *Annals of Probability* **6**, 325–331.

Andersen, T. G. and Bollerslev, T. (1997), 'Intraday periodicity and volatility persistence in financial markets', *Journal of Empirical Finance* **4**, 115–158.

Andersen, T. G. and Bollerslev, T. (1998), 'Deutsche mark-dollar volatility: Intraday activity patterns, macroeconomic announcements, and longer run dependencies', *Journal of Finance* **53**, 219–265.

Andersen, T. G., Bollerslev, T. and Diebold, F. X. (2003*c*), Some like it smooth, and some like it rough: Untangling continuous and jump components in measuring, modeling and forecasting asset return volatility. Unpublished paper: Economics Dept, Duke University.

Andersen, T. G., Bollerslev, T. and Diebold, F. X. (2005*a*), Parametric and nonparametric measurement of volatility, *in Handbook of Financial Econometrics*, North Holland, Amsterdam. To appear.

Andersen, T. G., Bollerslev, T., Diebold, F. X. and Labys, P. (2001), 'The distribution of exchange rate volatility', *Journal of the American Statistical Association* **96**, 42–55. Correction published in 2003, volume 98, page 501.

Andersen, T. G., Bollerslev, T., Diebold, F. X. and Labys, P. (2003*a*), 'Modeling and forecasting realized volatility', *Econometrica* **71**, 579–625.

Andersen, T. G., Bollerslev, T., Diebold, F. X. and Vega, C. (2003*b*), 'Micro effects of macro announcements: Real-time price discovery in foreign exchange', *American Economic Review* **93**, 38–62.

Andersen, T. G., Bollerslev, T. and Meddahi, N. (2005*b*), 'Correcting the errors: A note on volatility forecast evaluation based on high-frequency data and realized volatilities', *Econometrica* **73**, 279–296.

Anderson, R. M., Cox, D. R. and Hillier, H. C. (1989), 'Epidemiological and statistical aspects of the AIDS epidemic: Introduction', *Philosophical Transactions of the Royal Society of London, series B* **325**, 39–44.

Anderson, R. M. and May, R. M. (1991), *Infectious Diseases of Humans: Dynamics and Control*, Clarendon Press, Oxford.

Anderson, R. M., Medley, G. F., May, R. M. and Johnson, A. M. (1986), 'A preliminary study of the transmission dynamics of the human immunodeficiency virus (HIV), the causative agent of AIDS', *IMA Journal of Mathematics Applied in Medicine and Biology* **3**, 229–263.

Andersson, H. (1998), 'Limit theorems for a random graph epidemic model', *Annals of Applied Probability* **8**, 1331–1349.

Andersson, H. and Britton, T. (2000), *Stochastic Epidemic Models and their Statistical Analysis*, Springer, New York. Lecture Notes in Statistics, no. 151.

Andersson, H. and Djehiche, B. (1998), 'A threshold limit theorem for the stochastic logistic epidemic', *Journal of Applied Probability* **35**, 662–670.

Andrews, D. W. K. (1991), 'Asymptotic normality of series estimators for nonparametric and semiparametric models', *Econometrica* **59**, 307–345.

Anonymous (1984), 'Review of mortality results in randomized trials in early breast cancer', *Lancet* **ii**, 1205.

Armitage, P. (2003), 'Fisher, Bradford Hill, and randomization', *International Journal of Epidemiology* **32**, 925–928.

Arrigoni, F. and Pugliese, A. (2002), 'Limits of a multi-patch SIS epidemic model', *Journal of Mathematical Biology* **45**, 419–440.

Athreya, J. S. and Fidkowski, L. M. (2002), 'Number theory, balls in boxes, and the asymptotic uniqueness of maximal discrete order statistics', *Integers* **A3**, 5 pp. (electronic).

Austin, D. J., Bonten, M. J. M., Weinstein, R. A., Slaughter, S. and Anderson, R. M. (1999), 'Vancomycin-resistant enterococci in intensive-care hospital settings: Transmission dynamics, persistence and the impact of infection control programs', *Proceedings of the National Academy of Science* **96**, 6908–6913.

Back, K. (1991), 'Asset pricing for general processes', *Journal of Mathematical Economics* **20**, 371–395.

Bailey, N. T. J. (1975), *The Mathematical Theory of Infectious Diseases and its Applications*, Griffin, London.

Ball, F. (1999), 'Stochastic and deterministic models for SIS epidemics among a population partitioned into households', *Mathematical Biosciences* **156**, 41–67.

Ball, F. and Clancy, D. (1993), 'The final size and severity of a generalised stochastic multitype epidemic model', *Advances in Applied Probability* **25**, 721–736.

Ball, F. and Clancy, D. (1995), 'The final outcome of an epidemic model with several different types of infective in a large population', *Journal of Applied Probability* **32**, 579–590.

Ball, F. and Lyne, O. (2001), 'Stochastic multi-type SIR epidemics among a population partitioned into households', *Advances in Applied Probability* **33**, 99–123.

Ball, F. and Lyne, O. (2002), 'Optimal vaccination policies for stochastic epidemics among a population of households', *Mathematical Biosciences* **177**, 333–354.

Ball, F., Mollison, D. and Scalia-Tomba, G. (1997), 'Epidemics with two levels of mixing', *Advances in Applied Probability* **7**, 46–89.

Ball, F. and Neal, P. (2002), 'A general model for stochastic SIR epidemics with two levels of mixing', *Mathematical Biosciences* **180**, 73–102.

Ball, F. and Neal, P. (2003), 'The great circle epidemic model', *Stochastic Processes and their Applications* **107**, 233–268.

Bandi, F. M. and Russell, J. R. (2003), Microstructure noise, realized volatility, and optimal sampling. Unpublished paper presented at the Realized Volatility Conference, Montreal, 8th November, 2003.

Bankes, S. C. (2002), 'Agent-based modeling: A revolution?', *Proceedings of the National Academy of Sciences* **99**, 7199–7200.

Barabasi, A.-L. and Albert, R. (1999), 'Emergence of scaling in random networks', *Science* **286**, 509–512.

Barabasi, A.-L., Albert, R. and Jeong, H. (1999), 'Mean-field theory for scale-free random networks', *Physica A* **272**, 173–187.

Barbour, A. D. (1972), 'The principle of diffusion of arbitrary constants', *Journal of Applied Probability* **9**, 519–541.

Barbour, A. D. (1974), 'On a functional central limit theorem for Markov population processes', *Advances in Applied Probability* **6**, 21–39.

Barbour, A. D. (1975), 'The duration of the closed stochastic epidemic', *Biometrika* **62**, 477–82.

Barbour, A. D. and Kafetzaki, M. (1993), 'A host-parasite model yielding heterogeneous parasite loads', *Journal of Mathematical Biology* **31**, 157–176.

Barbour, A. D. and Mollison, D. (1990), Epidemics and random graphs, *in* J.-P. Gabriel, C. Lefèvre and P. Picard, (eds) *Stochastic Processes in Epidemic Theory. Lecture Notes in Biomathematics*, vol. 86, Springer, Berlin, pp. 86–89.

Barbour, A., Heesterbeek, J. and Luchsinger, C. (1996), 'Threshold and initial growth rates in a model of parasitic infection', *Advances in Applied Probability* **6**, 1045–1074.

Barchielli, A. and Holevo, A. S. (1995), 'Constructing quantum stochastic measurement processes via classical stochastic calculus', *Stochastic Processes and their Applications* **58**, 293–317.

Barndorff-Nielsen, O. E. (1986), 'Inference on full or partial parameters based on the standardized signed log likelihood ratio', *Biometrika* **73**, 307–322.

Barndorff-Nielsen, O. E. (1997), 'Superposition of Ornstein–Uhlenbeck type processes', *Theory of Probability and its Applications* **46**, 175–194.

Barndorff-Nielsen, O. E. (1998*a*), Probability and statistics: Self-decomposability, finance and turbulence, *in Probability towards 2000 (New York, 1995)*, Springer, New York, pp. 47–57.

Barndorff-Nielsen, O. E. (1998*b*), 'Processes of normal inverse Gaussian type', *Finance and Stochastics* **2**, 41–68.

Barndorff-Nielsen, O. E., Blæsild, P. and Schmiegel, J. (2004*e*), 'A parsimonious and universal description of turbulent velocity increments', *European Physical Journal, series B* **41**, 345–363.

Barndorff-Nielsen, O. E. and Cox, D. R. (1979), 'Edgeworth and saddle-point approximations with statistical applications (with discussion)', *Journal of the Royal Statistical Society, series B* **41**, 279–312.

Barndorff-Nielsen, O. E. and Cox, D. R. (1989), *Asymptotic Techniques for Use in Statistics*, Chapman & Hall, London.

Barndorff-Nielsen, O. E. and Cox, D. R. (1994), *Inference and Asymptotics*, Chapman & Hall, London.

Barndorff-Nielsen, O. E. and Cox, D. R. (1996), 'Prediction and asymptotics', *Bernoulli* **2**, 319–340.

Barndorff-Nielsen, O. E., Gill, R. D. and Jupp, P. E. (2001*c*), On quantum statistical inference. Research Report 2001–19. MaPhySto, University of Aarhus.

Barndorff-Nielsen, O. E., Graversen, S. E., Jacod, J., Podolskij, M. and Shephard, N. (2004*a*), A central limit theorem for realised power and bipower variations of continuous semimartingales. Festschrift in Honour of A. N. Shiryaev. Forthcoming.

Barndorff-Nielsen, O. E., Jensen, J. L. and Sørensen, M. (1990), 'Parametric modelling of turbulence', *Philosophical Transactions of the Royal Society of London, series A* **332**, 439–455.

Barndorff-Nielsen, O. E. and Lindner, A. (2004), Some aspects of Lévy copulas. Research Report (Submitted).

Barndorff-Nielsen, O. E., Maejima, M. and Sato, K. (2004*c*), Infinite divisibility and time change. *Journal of Theoretical Probablity*, forthcoming.

Barndorff-Nielsen, O. E., Maejima, M. and Sato, K. (2004*d*), Some classes of multivariate infinitely divisible distributions admitting stochastic integral representation. *Bernoulli*, forthcoming.

Barndorff-Nielsen, O. E., Mikosch, T. and Resnick, S. I., (eds) (2001*a*), *Lévy Processes: Theory and Applications*, Birkhäuser, Boston, MA.

Barndorff-Nielsen, O. E., Nicolato, E. and Shephard, N. (2002), 'Some recent developments in stochastic volatility modelling', *Quantitative Finance* **2**, 11–23.

Barndorff-Nielsen, O. E., Pedersen, J. and Sato, K. (2001*b*), 'Multivariate subordination, self-decomposability and stability', *Advances in Applied Probability* **33**, 160–187.

Barndorff-Nielsen, O. E. and Peres-Abreu, V. (2005), Matrix subordinators and Upsilon transformation. (In preparation).

Barndorff-Nielsen, O. E. and Schmiegel, J. (2004), 'Lévy-based tempo-spatial modelling; with applications to turbulence.', *Russian Mathematical Surveys (Uspekhi Acad. Nauk)* **59**, 65–90.

Barndorff-Nielsen, O. E. and Shephard, N. (2001*a*), Modelling by Lévy processes for financial econometrics, *in* O. E. Barndorff-Nielsen, T. Mikosch and S. I. Resnick, (eds), *Lévy Processes: Theory and Applications*, Birkhäuser, Boston, MA, pp. 283–318.

Barndorff-Nielsen, O. E. and Shephard, N. (2001*b*), 'Non-Gaussian Ornstein–Uhlenbeck-based models and some of their uses in financial economics (with discussion)', *Journal of the Royal Statistical Society, series B* **63**, 167–241.

Barndorff-Nielsen, O. E. and Shephard, N. (2002), 'Econometric analysis of realised volatility and its use in estimating stochastic volatility models', *Journal of the Royal Statistical Society, series B* **64**, 253–280.

Barndorff-Nielsen, O. E. and Shephard, N. (2003), Econometrics of testing for jumps in financial economics using bipower variation. *Journal of Financial Econometrics*, forthcoming.

Barndorff-Nielsen, O. E. and Shephard, N. (2004*a*), 'Econometric analysis of realised covariation: high frequency covariance, regression and correlation in financial economics', *Econometrica* **72**, 885–925.

Barndorff-Nielsen, O. E. and Shephard, N. (2004*b*), A feasible central limit theory for realised volatility under leverage. Unpublished discussion paper: Nuffield College, Oxford.

Barndorff-Nielsen, O. E. and Shephard, N. (2004*c*), 'Power and bipower variation with stochastic volatility and jumps (with discussion)', *Journal of Financial Econometrics* **2**, 1–48.

Barndorff-Nielsen, O. E. and Shephard, N. (2005), *Continuous Time Approach to Financial Volatility*, Cambridge University Press, Cambridge.

Barndorff-Nielsen, O. E. and Shiryaev, A. N. (2005), 'Change of time and change of measure'. (To appear.).

Barndorff-Nielsen, O. E. and Thorbjørnsen, S. (2002*a*), 'Lévy laws in free probability', *Proceedings of the National Academy of Sciences* **99**, 16568–16575.

Barndorff-Nielsen, O. E. and Thorbjørnsen, S. (2002*b*), 'Lévy processes in free probability', *Proceedings of the National Academy of Sciences* **99**(26), 16576–16580.

Barndorff-Nielsen, O. E. and Thorbjørnsen, S. (2002*c*), 'Self-decomposability and Lévy processes in free probability', *Bernoulli* **8**, 323–366.

Barndorff-Nielsen, O. E. and Thorbjørnsen, S. (2004*a*), 'A connection between free and classical infinite divisibility', *Infinite Dimensional Analysis, Quantum Probability, and Related Topics* **7**, 573–590.

Barndorff-Nielsen, O. E. and Thorbjørnsen, S. (2004*b*), 'The Lévy–Ito decomposition in free probability', *Probability Theory and Related Fields* . (To appear).

Barndorff-Nielsen, O. E. and Thorbjørnsen, S. (2004*c*), Regularising mappings of Lévy measures. (Submitted).

Barndorff-Nielsen, O. E., Vedel Jensen, E. B., Jónsdóttir, K. Ýr and Schmiegel, J. (2004), Spatio-temporal modelling—with a view to biological growth. Research Report 2004–12, T. N. Thiele Centre, Aarhus.

Bartlett, M. S. (1949), 'Some evolutionary stochastic processes', *Journal of the Royal Statistical Society, series B* **11**, 211–229.

Bartlett, M. S. (1956), Deterministic and stochastic models for recurrent epidemics, *in* 'Proceedings of the Third Berkeley Symposium on Mathematical Statistics and Probability', Vol. 4, University of California Press, Berkeley, pp. 81–109.

Bartlett, M. S. (1957), 'Measles periodicity and community size', *Journal of the Royal Statistical Society, series A* **120**, 48–70.

Bartlett, M. S. (1960), *Stochastic Population Models in Ecology and Epidemiology*, Methuen, London.

Bartlett, M. S. (1990), 'Chance or chaos (with discussion)', *Journal of the Royal Statistical Society, series A* **153**, 321–348.

Bates, D. M. and Watts, D. G. (1988), *Nonlinear Regression Analysis and Its Applications*, Wiley, New York.

Beck, N. (2000), 'Political methodology: a welcoming discipline', *Journal of the American Statistical Association* **95**, 651–654.

Becker, N. G. (1989), *Analysis of Infectious Disease Data*, Chapman & Hall, London.

Becker, N. G. and Britton, T. (1999), 'Statistical studies of infectious disease incidence', *Journal of the Royal Statistical Society, series B* **61**, 287–308.

Becker, N. G., Britton, T. and O'Neill, P. D. (2003), 'Estimating vaccine effects on transmission of infection from household outbreak data', *Biometrics* **59**, 467–475.

Becker, R. A., Chambers, J. M. and Wilks, A. R. (1988), *The New S Language: A Programming Environment for Data Analysis and Graphics*, Pacific Grove, CA: Wadsworth and Brooks/Cole.

Begun, J. M., Hall, W. J., Huang, W. M. and Wellner, J. (1983), 'Information and asymptotic efficiency in parametric-nonparametric models', *Annals of Statistics* **11**, 432–452.

Behnke, J. M., Bajer, A., Sinski, E. and Wakelin, D. (2001), 'Interactions involving intestinal nematodes of rodents: experimental and field studies', *Parasitology* **122**, 39–49.

Ben Ghorbal, A. and Schürmann, M. (2002), 'Non-commutative notions of stochastic independence', *Mathematical Proceedings of the Cambridge Philosophical Society* **133**, 531–561.

Beran, R. J. (1974), 'Asymptotically efficient adaptive rank estimators in location models', *Annals of Statistics* **2**, 63–74.

Bercovici, H. and Pata, V. (1999), 'Stable laws and domains of attraction in free probability', *Annals of Mathematics* **149**, 1023–1060.

Berger, J. O. and Wolpert, R. L. (1988), *The Likelihood Principle*, Vol. 6 of *Lecture Notes—Monograph Series*, 2nd edn, Institute of Mathematical Statistics, Hayward, California.

Berkson, J. (1950), 'Are there two regressions?', *Journal of the American Statistical Association* **45**, 164–180.

Bernardo, J. M. and Smith, A. F. M. (1994), *Bayesian Theory*, Wiley, New York.

Bernoulli, D. (1760), 'Essai d'une nouvelle analyse de la mortalité causée par la petite vérole et des advantages de l'inculation pour la prévenir', *Mémoires Mathématiques et Physiques de l'Académie Royale de Sciences de Paris*, pp. 1–45.

Bertoin, J. (1996), *Lévy Processes*, Cambridge University Press, Cambridge.

Bertoin, J. (1997), Subordinators: Examples and applications, *in* 'Lectures on Probability Theory and Statistics', Springer, Berlin, pp. 1–91. Ecole d'Eté de Probabilities de Saint-Flour XXVII.

Besag, J. E. (1974), 'Spatial interaction and the statistical analysis of lattice systems (with discussion)', *Journal of the Royal Statistical Society, series B* **34**, 192–236.

Besag, J. E. (2002), 'Discussion of 'What is a statistical model?' by P. McCullagh', *Annals of Statistics* **30**, 1267–1277.

Besag, J. E., York, J. and Mollié, A. (1991), 'Bayesian image restoration, with two applications in spatial statistics (with discussion)', *Annals of the Institute of Statistical Mathematics* **43**, 1–59.

Bickel, P. J. (1982), 'On adaptive estimation', *Annals of Statistics* **10**, 647–671.

Bickel, P. J. (1993), Estimation in semiparametric models, *in* C. R. Rao, (ed.), *Multivariate Analysis: Future Directions*, North-Holland, Amsterdam, pp. 55–73.

Bickel, P. J., Klaassen, C. A. J., Ritov, Y. and Wellner, J. A. (1993), *Efficient and Adaptive Estimation for Semiparametric Models*, Johns Hopkins University Press, Baltimore.

Bickel, P. J. and Kwon, J. (2001), 'Inference in semiparametric models: Some questions and one answer', *Statistica Sinica* **11**, 863–960.

Bickel, P. J. and Ritov, Y. (1988), 'Estimating squared density derivatives: Sharp best order of convergence estimates', *Sankhya* **50**, 136–146.

Bickel, P. J. and Ritov, Y. (2000), 'Non- and semiparametric statistics: Compared and contrasted', *Journal of Statistical Plannning and Inference* **91**, 209–228.

Bickel, P. J., Ritov, Y. and Stoker, T. M. (2004), Tailor-made tests for goodness-of-fit to semiparametric hypotheses, Technical report, Department of Statistics, The Hebrew University of Jerusalem.

Billard, L., Medley, G. F. and Anderson, R. M. (1990), The incubation period for the AIDS virus, *in* J.-P. Gabriel, C. Lefèvre and P. Picard, (eds), *Stochastic Processes in Epidemic Theory. Lecture Notes in Biomathematics*, vol. 86', Springer, Berlin, pp. 21–35.

Bird, S. M., Cox, D. R., Farewell, V. T., Goldstein, H., Holt, T. and Smith, P. C. (2005), 'Performance indicators: Good, bad and ugly', *Journal of the Royal Statistical Society, series A* **168**, 1–27.

Blanks, R. G., Moss, S. M., McGahan, C. E., Quinn, M. J. and Babb, P. J. (2000), 'Effect of NHS breast screening programme on mortality from breast cancer in England and Wales, 1990–8: Comparison of observed with predicted mortality', *British Medical Journal* **321**, 665–669.

Blossfeld, H.-P., Hamerle, A. and Mayer, K. U. (1989), *Event History Analysis. Statistical Theory and Applications in the Social Sciences*, Hillsday, NJ: Erlbaum.

Blundell, R. and Powell, J. L. (2003), Endogeneity in nonparametric and semiparametric regression models, *in* M. Dewatripont, L.-P. Hansen and S. J. Turnovsky, (eds), *Advances in Economics and Econometrics: Theory and Applications*, Vol. 2, Cambridge University Press, Cambridge, pp. 312–357.

Bochner, S. (1949), 'Diffusion equation and stochastic processes', *Proceedings of the National Academy of Sciences* **85**, 369–370.

Bochner, S. (1955), *Harmonic Analysis and the Theory of Probability*, University of California Press.

Bodmer, W. (2003), 'R.A. Fisher, statistician and geneticist extraordinary: A personal view', *International Journal of Epidemiology* **32**, 938–942.

Bolker, B. M. (2003), 'Combining endogenous and exogenous spatial variability in analytical population models', *Theoretical Population Biology* **64**, 255–270.

Bollobás, B. (1985), *Random Graphs*, Academic Press, London.

Bondesson, L. (1981), 'Classes of infinitely divisible distributions and densities', *Zeitschrift für Wahrscheinlichkeitstheorie und verwandte gebiete* **57**, 39–71.

Bondesson, L. (1992), *Generalized Gamma Distributions and Related Classes of Distributions and Densities*, Springer, Berlin.

Borg, I. and Groenen, P. (1997), *Modern Multidimensional Scaling. Theory and Applications*, Springer-Verlag, New York.

Bottomley, C., Isham, V. and Basáñez, M.-G. (2005) Population biology of multispecies helminth infection: Interspecific interactions and parasite distribution. *Parasitology*, in press.

Bouton, C., Henry, G., Colantuoni, C. and Pevsner, J. (2003), Dragon and Dragon View: Methods for the annotation, analysis and visualization of large-scale gene expression data, *in* G. Parmigiani, E. S. Garrett, R. A. Irizarry and S. L. Zeger, (eds), *The Analysis of Gene Expression Data*, Springer, New York.

Box, G. E. P. and Cox, D. R. (1964), 'An analysis of transformations (with discussion)', *Journal of the Royal Statistical Society, series B* **26**, 211–246.

Box, J. F. (1978), *R. A. Fisher. The Life of a Scientist*, Wiley, New York.

Bradford Hill, A. (1937), *Principles of Medical Statistics*, Arnold, London.

Bradford Hill, A. (1965), 'The environment and disease: Association or causation?', *Proceedings of the Royal Society of Medicine* **58**, 295–300.

Bradley, K. M., Bydder, G. M., Budge, M. M., Hajnal, J. V., White, S. J., Ripley, B. D. and Smith, A. D. (2002), 'Serial brain MRI at 3–6 month intervals as a surrogate marker for Alzheimer's disease.', *British Journal of Radiology* **75**, 506–513.

Brazzale, A. R. (2000), Practical Small-Sample Parametric Inference, PhD thesis, Department of Mathematics, Ecole Polytechnique Fédérale de Lausanne.

Breiman, L. (2001), 'Statistical modeling: The two cultures (with discussion)', *Statistical Science* **16**, 199–215.

Breiman, L. and Friedman, J. (1997), 'Predicting multivariate responses in multiple linear regression (with discussion)', *Journal of the Royal Statistical Society, series B* **59**, 3–37.

Breiman, L., Friedman, J. H., Olshen, R. A. and Stone, C. J. (1984), *Classification and Regression Trees*, Wadsworth and Brooks/Cole, Pacific Grove, California.

Breslow, N. E. and Clayton, D. G. (1993), 'Approximate inference in generalized linear mixed models', *Journal of the American Statistical Association* **88**, 9–25.

Brillinger, D. R. (1983), A generalized linear model with Gaussian regressor variables, in P. J. Bickel, K. A. Doksum and J. L. Hodges, (eds), *A Festschrift for Erich L. Lehmann*, Wadsworth and Brooks/Cole, Pacific Grove, California, pp. 97–114.

Brillinger, D. R. (1986), 'The natural variability of vital rates and associated statistics (with discussion)', *Biometrics* **42**, 693–734.

Brøns, H. (2002), 'Discussion of "What is a statistical model?" by P. McCullagh', *Annals of Statistics* **30**, 1279–1283.

Browne, M. W. (2000), 'Psychometrics', *Journal of the American Statistical Association* **95**, 661–665.

Bruce, M. C., Donnelly, C. A., Alpers, M. P., Galinski, M. R., Barnwell, J. W., Walliker, D. and Day, K. P. (2000), 'Cross-species interactions between malaria parasites in humans', *Science* **287**, 845–848.

Carr, P., Chang, E. and Madan, D. B. (1998), 'The variance gamma process and option pricing', *European Finance Review* **2**, 79–105.

Carr, P., Geman, H., Madan, D. B. and Yor, M. (2003), 'Stochastic volatility for Lévy processes', *Mathematical Finance* **13**, 345–382.

Carr, P. and Wu, L. (2004), 'Time-changed Lévy processes and option pricing', *Journal of Financial Economics* pp. 113–141.

Caspi, A., Sugden, K., Moffitt, T. E., Taylor, A., Craig, I. W., Harrington, H., McClay, J., Mill, J., Martin, J., Braithwaite, A. and Poulton, R. (2003), 'Influence of life stress on depression: Moderation by a polymorphism in the 5-HTT gene', *Science* **301**, 386–389.

Chalmers, I. (2003), 'Fisher and Bradford Hill: Theory and pragmatism?', *International Journal of Epidemiology* **32**, 922–924.

Chamberlain, G. (1986), 'Asymptotic efficiency in semiparametric models with censoring', *Journal of Econometrics* **32**, 189–218.

Chamberlain, G. (1987), 'Asymptotic efficiency in estimation with conditional moment restrictions', *Journal of Econometrics* **34**, 305–334.

Chambers, J. M. (1998), *Programming with Data. A Guide to the S Language*, Springer-Verlag, New York.

Chambers, J. M. and Hastie, T. J., (eds) (1992), *Statistical Models in S*, Chapman & Hall, New York.

Chambers, R. L. and Skinner, C. J. (2003), *Analysis of Survey Data*, Chichester: Wiley.

Chan, M. S. and Isham, V. S. (1998), 'A stochastic model of schistosomiasis immuno-epidemiology', *Mathematical Biosciences* **151**, 179–198.

Chen, H. (1988), 'Convergence rates for parametric components in a partly linear model', *Annals of Statistics* **16**, 136–146.

Chetelat, G. and Baron, J. C. (2003), 'Early diagnosis of Alzheimer's disease: contribution of structural neuroimaging', *Neuroimage* **2**, 525–541.

Chib, S. (1995), 'Marginal likelihood from the Gibbs output', *Journal of the American Statistical Association* **90**, 1313–21.

Chib, S. and Jeliazkov, I. (2001), 'Marginal likelihood from the Metropolis–Hastings output', *Journal of the American Statistical Association* **96**, 270–281.

Choi, E. and Hall, P. (1999), 'Nonparametric approach to analysis of space-time data on earthquake occurrences', *Journal of Computational and Graphical Statistics* **8**, 733–748.

Christensen, B. J. and Prabhala, N. R. (1998), 'The relation between implied and realized volatility', *Journal of Financial Economics* **37**, 125–150.

Chung, K. L. and Zambrini, J.-C. (2003), *Introduction to Random Time and Quantum Randomness*, World Scientific, Singapore.

Claeskens, G. (2004), 'Restricted likelihood ratio lack of fit tests using mixed spline models', *Journal of the Royal Statistical Society, series B* **66**, 909–926.

Clancy, D. (1994), 'Some comparison results for multitype epidemic models', *Journal of Applied Probability* **31**, 9–21.

Clancy, D. (1996), 'Carrier-borne epidemic models incorporating population mobility', *Mathematical Biosciences* **132**, 185–204.

Clancy, D. (1999*a*), 'Optimal interventions for epidemic models with general infection and removal rate functions', *Journal of Mathematical Biology* **39**, 309–331.

Clancy, D. (1999*b*), 'Outcomes of epidemic models with general infection and removal rate functions at certain stopping times', *Journal of Applied Probability* **36**, 799–813.

Clancy, D. and Pollett, P. K. (2003), 'A note on the quasi-stationary distributions of birth-death processes and the SIS logistic epidemic', *Journal of Applied Probability* **40**, 821–825.

Clapp, T. C. and Godsill, S. J. (1999), Fixed-lag smoothing using sequential importance sampling, *in* J. M. Bernardo, J. O. Berger, A. P. Dawid and A. F. M. Smith, (eds), *Bayesian Statistics 6*, Oxford University Press, Oxford, pp. 743–752.

Clayton, D. G. (1996), Generalized linear mixed models, *in* W. R. Gilks, S. Richardson and D. J. Spiegelhalter, (eds), *Markov Chain Monte Carlo in Practice*, Chapman & Hall, London, pp. 275–301.

Clayton, D. G. and Kaldor, J. (1987), 'Empirical Bayes estimates of age-standardised relative risks for use in disease mapping', *Biometrics* **43**, 671–681.

Clogg, C. C. (1992), 'The impact of sociological methodology on statistical methodology (with discussion)', *Statistical Science* **7**, 183–207.

Colantuoni, C., Henry, G., Bouton, C., Zeger, S. L. and Pevsner, J. (2003), Snomad: Biologist-friendly web tools for the standardization and normalization of microarray data, *in* G. Parmigiani, E. S. Garrett, R. A. Irizarry and S. L. Zeger, (eds), *The Analysis of Gene Expression Data*, Springer, New York.

Collins, F. S., Green, E. D., Guttmacher, A. E. and Guyer, M. S. (2003), 'A vision for the future of genomics research', *Nature* **422**, 835–847.

Cont, R. and Tankov, P. (2004), *Financial Modelling with Jump Processes*, Chapman & Hall/CRC, London.

Coombs, C. H. (1964), *A Theory of Data*, Wiley, New York.

Cornell, S. J. and Isham, V. S. (2004), 'Ultimate extinction in a bisexual, promiscuous Galton–Watson metapopulation', *Australian and New Zealand Journal of Statistics* **46**, 87–98.

Cornell, S. J., Isham, V. S. and Grenfell, B. T. (2004), 'Stochastic and spatial dynamics of nematode parasites in ruminants', *Proceedings of the Royal Society of London, Series B* **271**, 1243–1250.

Cornell, S. J., Isham, V. S., Smith, G. and Grenfell, B. T. (2003), 'Spatial parasite transmission, drug resistance and the spread of rare genes', *Proceedings of the National Academy of Science* **100**, 7401–7405.

Cornfield, J. (1951), 'A method for estimating comparative rates from clinical data: application to cancer of the lung, breast and cervix', *Journal of the National Cancer Institute* **11**, 1269–1275.

Corsi, F., Zumbach, G., Müller, U. and Dacorogna, M. (2001), 'Consistent high-precision volatility from high-frequency data', *Economic Notes* **30**, 183–204.

Cox, D. R. (1949), 'Theory of drafting of wool slivers', *Proceedings of the Royal Society of London, series A* **197**, 28–51.

Cox, D. R. (1955*a*), 'Some statistical methods connected with series of events (with discussion)', *Journal of the Royal Statistical Society, series B* **17**, 129–164.

Cox, D. R. (1955*b*), 'A use of complex probabilities in the theory of stochastic processes', *Proceedings of the Cambridge Philosophical Society* **51**, 313–319.

Cox, D. R. (1958*a*), *Planning of Experiments*, Wiley, New York.

Cox, D. R. (1958*b*), 'Some problems connected with statistical inference', *Annals of Mathematical Statistics* **29**, 359–372.

Cox, D. R. (1958*c*), 'The regression analysis of binary sequences (with discussion)', *Journal of the Royal Statistical Society, series B* **20**, 215–242.

Cox, D. R. (1961), Tests of separate families of hypotheses, *in* 'Proceedings of the 4th Berkeley Symposium on Mathematics, Probability and Statistics', Vol. 1, University of California Press, Berkeley, pp. 105–123.

Cox, D. R. (1962), *Renewal Theory*, Chapman & Hall, London.

Cox, D. R. (1966), 'Notes on the analysis of mixed frequency distributions', *British Journal of Mathematical Psychology* **19**, 39–47.

Cox, D. R. (1972), 'Regression models and life tables (with discussion)', *Journal of the Royal Statistical Society, series B* **34**, 187–220.

Cox, D. R. (1972a), The statistical analysis of point processes, *in* P. A. Lewis, (ed.), *Stochastic Point Processes*, Wiley, New York.

Cox, D. R. (1975), 'Partial likelihood', *Biometrika* **62**, 269–76.

Cox, D. R. (1981), 'Statistical analysis of time series: Some recent developments (with discussion)', *Scandivavian Journal of Statistics* **8**, 93–115.

Cox, D. R. (1984*a*), 'Interaction (with discussion)', *International Statistical Review* **52**, 1–31.

Cox, D. R. (1984*b*), Long range dependence: A review, *in* H. A. David and H. T. David, (eds), *Statistics: An Appraisal*, Iowa State University Press, pp. 55–74.

Cox, D. R. (1986), 'Comment on Holland, W., Statistics and causal inference', *Journal of the American Statistical Association* **81**, 945–960.

Cox, D. R. (1990), 'Role of models in statistical analysis', *Statistical Science* **5**, 169–174.

Cox, D. R. (1991), 'Long-range dependence, non-linearity and time irreversibility', *Journal of Time Series Analysis* **12**, 329–335.

Cox, D. R. (1992), 'Causality: Some statistical aspects', *Journal of the Royal Statistical Society, series A* **155**, 291–301.

Cox, D. R. (1995), 'The relation between theory and application in statistics (with discussion)', *Test* **4**, 207–261.

Cox, D. R. (1997), 'The current position of statistics: A personal view, (with discussion)', *International Statistical Review* **65**, 261–276.

Cox, D. R. (2000), 'Comment on "Causal inference without counterfactuals" by A. P. Dawid', *Journal of the American Statistical Association* **95**, 424–425.

Cox, D. R. (2004), 'The accidental statistician', *Significance* **1**, 27–29.

Cox, D. R., Fitzpatrick, R., Fletcher, A. E., Gore, S. M., Spiegelhalter, D. J. and Jones, D. R. (1992), 'Quality-of-life assessment: Can we keep it simple? (with discussion)', *Journal of the Royal Statistical Society, series A* **155**, 353–393.

Cox, D. R. and Hall, P. (2002), 'Estimation in a simple random effects model with nonnormal distributions', *Biometrika* **89**, 831–840.

Cox, D. R. and Hinkley, D. V. (1974), *Theoretical Statistics*, Chapman & Hall, London.

Cox, D. R. and Isham, V. (1986), 'The virtual waiting-time and related processes', *Advances in Applied Probability* **18**, 558–573.

Cox, D. R. and Isham, V. S. (1988), 'A simple spatial-temporal model of rainfall', *Proceedings of the Royal Society of London, series A* **415**, 317–328.

Cox, D. R. and Miller, H. D. (1965), *The Theory of Stochastic Processes*, Chapman & Hall, London.

Cox, D. R. and Reid, N. (1987), 'Parameter orthogonality and approximate conditional inference (with discussion)', *Journal of the Royal Statistical Society, series B* **49**, 1–39.

Cox, D. R. and Reid, N. (2004), 'A note on pseudolikelihood constructed from marginal densities', *Biometrika* **91**, 729–737.

Cox, D. R. and Snell, E. J. (1979), 'On sampling and the estimation of rare errors (Corr: **69**, 491)', *Biometrika* **66**, 125–132.

Cox, D. R. and Snell, E. J. (1981), *Applied Statistics: Principles and Examples*, Chapman & Hall, London.

Cox, D. R. and Solomon, P. J. (2003), *Components of Variance*, Chapman & Hall, London.

Cox, D. R. and Wermuth, N. (1996), *Multivariate Dependencies*, Chapman & Hall, London.

Cox, D. R. and Wermuth, N. (2001), 'Some statistical aspects of causality', *European Sociological Review* **17**, 65–74.

Cox, D. R. and Wermuth, N. (2004), 'Causality: a statistical view', *International Statistical Review* **72**, 285–305.

Cox, N. J., Frigge, M., Nicolae, D. L., Concanno, P., Harris, C. L., Bell, G. I. and Kong, A. (1999), 'Loci on chromosome 2 (NIDDM1) and 15 interact to increase susceptibility to diabetes in Mexican Americans', *Nature Genetics* **21**, 213–215.

Cox, T. F. and Cox, M. A. A. (2001), *Multidimensional Scaling*, 2nd edn, Chapman & Hall/CRC.

Dacorogna, M. M., Gencay, R., Müller, U. A., Olsen, R. B. and Pictet, O. V. (2001), *An Introduction to High-Frequency Finance*, Academic Press, San Diego.

Daley, D. J. and Gani, J. (1999), *Epidemic Modelling*, Cambridge University Press, Cambridge.

Daley, D. J. and Kendall, D. G. (1965), 'Stochastic rumours', *Journal of Institute of Mathematics and its Applications* **1**, 42–55.

Damien, P. and Walker, S. (2002), 'A Bayesian non-parametric comparison of two treatments', *Scandinavian Journal of Statistics* **29**, 51–56.

Daniels, H. E. (1991), A look at perturbation approximations for epidemics, *in* I. V. Basawa and R. L. Taylor, (eds), *Selected Proceedings of the Sheffield Symposium on Applied Probability*, IMS Monograph Series, Vol. 18, Hayward, CA, pp. 48–65.

Darby, S. C., McGale, P., Peto, R., Granath, F., Hall, P. and Ekbom, A. (2003), 'Mortality from cardiovascular disease more than 10 years after radiotherapy for breast cancer: nationwide cohort study of 90 000 Swedish women', *British Medical Journal* **326**, 256–257.

Darroch, J. and Seneta, E. (1967), 'On quasi-stationary distributions in absorbing continuous-time finite Markov chains', *Journal of Applied Probability* **4**, 192–196.

Datta, S., Longini, I. M. and Halloran, M. E. (1999), 'Efficiency of estimating vaccine efficacy for susceptibility and infectiousness: Randomization by individual versus household', *Biometrics* **55**, 792–798.

Davey Smith, G., Dorling, D., Mitchell, R. and Shaw, M. (2002), 'Health inequalities in Britain: continuing increases up to the end of the 20th century', *Journal of Epidemiology and Community Health* **56**, 434–435.

Davies, E. B. (1976), *Quantum Theory of Open Systems*, Academic Press, London.

Davies, E. B. and Lewis, J. T. (1970), 'An operational approach to quantum systems', *Communications in Mathematical Physics* **17**, 239–260.

Davison, A. C. (2003), *Statistical Models*, Cambridge University Press, Cambridge.

Davison, A. C. and Cox, D. R. (1989), 'Some simple properties of sums of random variables having long-range dependence', *Proceedings of the Royal Society of London, series A* **424**, 255–262.

Davison, A. C. and Wang, S. (2002), 'Saddlepoint approximations as smoothers', *Biometrika* **89**, 933–938.

Dawid, A. P. (2000), 'Causality without counterfactuals (with discussion)', *Journal of the American Statistical Association* **95**, 407–448.

De Finetti, B. (1974), *Theory of Probability: Volumes 1 and 2*, Wiley, New York.

De la Cal, J. (1989), 'On the three series theorem in number theory', *Annals of Probability* **17**, 357–361.

Deeming, T. J. (1975), 'Fourier analysis with unequally-spaced data', *Astrophysics and Space Science* **36**, 137–158.

Delattre, S. and Jacod, J. (1997), 'A central limit theorem for normalized functions of the increments of a diffusion process in the presence of round off errors', *Bernoulli* **3**, 1–28.

Denison, D. G. T., Holmes, C. C., Mallick, B. K. and Smith, A. F. M. (2002), *Bayesian Methods for Nonlinear Classification and Regression*, Wiley, Chichester.

DiBona, C., Ockman, S. and Stone, M., (eds) (1999), *Open Sources. Voices from the Open Source Revolution*, O'Reilly, Sebastopol, CA.

Dieckmann, U. and Law, R. (2000), Relaxation projections and the method of moments, *in* U. Dieckmann, R. Law and J. A. J. Metz, (eds), *The Geometry of Ecological Interactions: Simplifying Spatial Complexity*, Cambridge University Press, Cambridge, pp. 412–455.

Diekmann, O. and Heesterbeek, J. A. P. (2000), *Mathematical Epidemiology of Infectious Diseases*, Wiley, Chichester.

Dietz, K. (1976), The incidence of infectious diseases under the influence of seasonal fluctuations, *in Lecture Notes in Biomathematics*, vol. 11, Springer, Berlin, pp. 1–15.

Dietz, K. (1988), 'The first epidemic model: A historical note on P. D. En'ko', *Australian Journal of Statistics* **A3**, 56–65.

Dietz, K. (1989), 'A translation of the paper "On the course of epidemics of some infectious diseases" by P. D. En'ko', *International Journal of Epidemiology* **18**, 749–755.

Dietz, K. and Schenzle, D. (1985), Mathematical models for infectious disease statistics., *in* A. C. Atkinson and S. E. Fienberg, (eds), *A Celebration of Statistics: The ISI centenary volume*. Springer, New York, pp. 167–204.

Diggle, P. J., Heagerty, P., Liang, K.-Y. and Zeger, S. L. (2002), *Analysis of Longitudinal Data*, 2nd edn, Oxford University Press, Oxford.

Diggle, P. J., Moyeed, R. A. and Tawn, J. A. (1998), 'Model-based geostatistics (with discussion)', *Applied Statistics* **47**, 299–350.

Dobrushin, R. L. (1968), 'The description of a random field by means of conditional probabilities and conditions of its regularity', *Theory of Probability and its Applications* **13**, 197–224.

Dodds, S. and Watts, D. J. (2005), 'A generalized model of social and biological contagion', *Journal of Theoretical Biology*, **232**, 587–604.

Doll, R. (2003), 'Fisher and Bradford Hill: Their personal impact', *International Journal of Epidemiology* **32**, 929–931.

Dolton, P. (2002), 'Evaluation of economic and social policies', *Journal of the Royal Statistical Society, series A* **165**, 9–11.

Dominici, F., Daniels, M., Zeger, S. L. and Samet, J. (2002), 'Air pollution and mortality: estimating regional and national dose-response relationships', *Journal of the American Statistical Association* **97**, 100–111.

Dominici, F., McDermott, A., Zeger, S. L. and Samet, J. M. (2003), 'National maps of the effects of PM on mortality: Exploring geographical variation', *Environmental Health Perspectives* **111**, 39–43.

Donald, S. G. and Newey, W. (1994), 'Series estimation of semilinear models', *Journal of Multivariate Analyses* **50**, 30–40.

Donnelly, C. A., Woodroffe, R., Cox, D. R., Bourne, J., Gettinby, G., Le Fevre, A. M., McInerney, J. P. and Morrison, W. I. (2003), 'Impact of localised badger culling on tuberculosis incidence in British cattle', *Nature* **426**, 834–837.

Donoho, D. L. (1995), 'Nonlinear solution of linear inverse problems by wavelet-vaguelette decomposition', *Applied and Computational Harmonic Analysis* **2**, 101–126.

Doornik, J. A. (2001), *Ox: Object Oriented Matrix Programming, 3.0*, Timberlake Consultants Press, London.

Doucet, A., de Freitas, N. and Gordon, N. J., (eds) (2001), *Sequential Monte Carlo Methods in Practice*, Springer-Verlag, New York.

Drost, F. C., Klaassen, C. A. and Werker, B. J. M. (1994), 'Adaptiveness in time series models', *Asymptotic Statistics: Proceedings of the Fifth Prague Symposium* **5**, 203–211.

Duan, N. and Li, K. C. (1987), 'Distribution-free and link-free estimation for the sample selection model', *Journal of Econometrics* **35**, 25–35.

Duan, N. and Li, K. C. (1991), 'Slicing regression: A link-free regression method', *Annals of Statistics* **19**, 505–530.

Duerr, H. P., Dietz, K. and Eichner, M. (2003), 'On the interpretation of age-intensity profiles and dispersion patterns in parasitological surveys', *Parasitology* **126**, 87–101.

Duerr, H. P., Dietz, K., Schulz-Key, H., Büttner, D. W. and Eichner, M. (2002), 'Infection-associated immunosuppression as an important mechanism for parasite density regulation in onchcerciasis', *Transactions of the Royal Society of Tropical Hygiene and Medicine* **97**, 242–250.

Durrett, R. (1995), Spatial epidemic models, *in* D. Mollison, (ed.), *Epidemic Models: their Structure and Relation to Data*, Cambridge University Press, Cambridge, pp. 187–201.

Durrett, R. and Levin, S. (1994), 'Stochastic spatial models: a user's guide to ecological applications', *Philosophical Transactions of the Royal Society of London* **B343**, 329–350.

EBCTCG (1988), 'The effects of adjuvant tamoxifen and of cytotoxic therapy on mortality in early breast cancer: an overview of 61 randomized trials among 28,859 women', *New England Journal of Medicine* **319**, 1681–1692.

EBCTCG (1990), *Treatment of Early Breast Cancer*, Vol. 1: Worldwide Evidence, Oxford University Press, Oxford.

EBCTCG (1992), 'Systemic treatment of early breast cancer by hormonal, cytotoxic or immune therapy: 133 randomised trials involving 31,000 recurrences and 24,000 deaths among 75,000 women', *Lancet* **339**, 1–15 and 71–85.

EBCTCG (1995), 'Effects of radiotherapy and surgery in early breast cancer: An overview of the randomized trials', *New England Journal of Medicine* **333**, 1444–1455.

EBCTCG (1996), 'Ovarian ablation in early breast cancer: Overview of the randomised trials', *Lancet* **348**, 1189–1196.

EBCTCG (1998*a*), 'Polychemotherapy for early breast cancer: An overview of the randomised trials', *Lancet* **352**, 930–942.

EBCTCG (1998*b*), 'Tamoxifen for early breast cancer: An overview of the randomised trials', *Lancet* **351**, 1451–1467.

EBCTCG (2000), 'Favourable and unfavourable effects on long-term survival of radiotherapy for early breast cancer: An overview of the randomised trials', *Lancet* **355**, 1757–1770.

EBCTCG (2005), 'Effects of chemotherapy and hormonal therapy for early breast cancer on recurrence and 15-year survival: An overview of the randomised trials', *Lancet* **365**, 1687–1717.

Eberlein, E., Kallsen, J. and Kristen, J. (2001), Risk management based on stochastic volatility. FDM Preprint 72, University of Freiburg.

Eberlein, E. and Özkan, F. (2003), 'Time consistency of Lévy models', *Quantitative Finance* **3**, 40–50.

Ederington, L. and Lee, J. H. (1993), 'How markets process information: News releases and volatility', *Journal of Finance* **48**, 1161–1191.

Efromovich, S. (1994), Nonparametric curve estimation from indirect observations, *in* J. Sall and A. Lehman, (eds), *Computing Science and Statistics, Vol. 26: Computationally Intensive Statistical Methods*, Interface Foundation of America, pp. 196–200.

Efromovich, S. (1999), *Nonparametric Curve Estimation: Methods, Theory, and Applications*, Springer, New York.

Efron, B. (2001), 'Selection criteria for scatterplot smoothers', *Annals of Statistics* **29**, 470–504.

Elmore, R., Hall, P. and Neeman, A. (2003), 'A birational map out of the symmetrical power of A^{n+1}', Manuscript.

Engle, R. F., Granger, C. W. J., Rice, J. and Weiss, A. (1984), 'Semiparametric estimates of the relation between weather and electricity sales', *Journal of the American Statistical Association* **81**, 310–320.

Eraker, B., Johannes, M. and Polson, N. G. (2003), 'The impact of jumps in returns and volatility', *Journal of Finance* **53**, 1269–1300.

Erikson, R. and Goldthorpe, J. H. (1992), *The Constant Flux: A Study of Class Mobility in Industrial Societies*, Oxford: Clarendon Press.

Erwin, E., Obermayer, K. and Schulten, K. (1992), 'Self-organizing maps: ordering, convergence properties and energy functions', *Biological Cybernetics* **67**, 47–55.

Estevez-Perez, G., Lorenzo-Cimadevila, H. and Quintela-Del-Rio, A. (2002), 'Nonparametric analysis of the time structure of seismicity in a geographic region', *Annals of Geophysics* **45**, 497–511.

Evans, M. and Swartz, T. (2000), *Approximating Integrals via Monte Carlo and Deterministic Methods*, Oxford: Oxford University Press.

Feinberg, A. P. (2001), 'Cancer epigenetics take center stage', *Proceedings of the National Academy of Sciences* **98**, 392–394.

Feller, W. (1971), *An Introduction to Probability Theory*, 2nd edn, Wiley, New York. Volume II.

Ferguson, N. M., Donnelly, C. A. and Anderson, R. M. (2001*a*), 'The foot-and-mouth epidemic in Great Britain: Pattern of spread and impact of interventions', *Science* **292**, 1155–1160.

Ferguson, N. M., Donnelly, C. A. and Anderson, R. M. (2001*b*), 'Transmission intensity and impact of control policies on the foot-and-mouth epidemic in Great Britain', *Nature* **413**, 542–548.

Ferguson, N. M., Galvani, A. P. and Bush, R. M. (2003), 'Ecological and immunological determinants of influenza evolution', *Nature* **422**, 428–433.

Fienberg, S. E. (1989), Comments on 'Modeling considerations from a modeling perspective', *in* D. Kasprzyk, G. Duncan, G. Kalton and M. P. Singh, (eds), *Panel Surveys*, New York: Wiley, pp. 566–574.

Fienberg, S. E. (2000), 'Contingency tables and log-linear models: Basic results and new developments', *Journal of the American Statistical Association* **95**, 643–647.

Finkenstädt, B. F., Bjørnstad, O. and Grenfell, B. T. (2002), 'A stochastic model for extinction and recurrence of epidemics: estimation and inference for measles outbreaks', *Biostatistics* **3**, 493–510.

Firth, D. and Bennett, K. E. (1998), 'Robust models in probability sampling (with discussion)', *Journal of the Royal Statistical Society, series B* **60**, 3–21.

Fisher, B., Redmond, C., Legault-Poisson, S., Dimitrov, N.-V., Brown, A.-M., Wickerham, D. L., Wolmark, N., Margolese, R. G., Bowman, D. and Glass, A. G. (1990), 'Postoperative chemotherapy and tamoxifen compared with tamoxifen alone in the treatment of positive-node breast cancer patients aged 50 years and older with tumors responsive to tamoxifen: Results from the National Surgical Adjuvant Breast and Bowel Project B-16', *Journal of Clinical Oncology* **8**, 1005–1018.

Fisher, R. A. (1935), *The Design of Experiments*, Oliver and Boyd, Edinburgh.

Fox, J. (2002), *An R and S-Plus Companion to Applied Regression*, Thousand Oaks, CA: Sage.

Frangakis, C. E. and Rubin, D. B. (1999), 'Addressing complications of intention-to-treat analysis in the presence of all-or-none treatment-noncompliance and subsequent missing outcomes', *Biometrika* **86**, 365–379.

Frank, O. and Strauss, D. (1986), 'Markov graphs', *Journal of the American Statistical Association* **81**, 832–842.

Franklin, R. and Gosling, R. G. (1953), 'Molecular configuration in sodium thymonucleate', *Nature* **171**, 740–741.

Franz, U. (2003), 'Unification of Boolean, monotone, anti-monotone and tensor independence and Lévy processes', *Mathematische Zeitschrift* **243**, 779–816.

Fraser, D. A. S. (1990), 'Tail probabilities from observed likelihoods', *Biometrika* **77**, 65–76.

Fraser, D. A. S. (2003), 'Likelihood for component parameters', *Biometrika* **90**, 327–339.

Fraser, D. A. S. and Reid, N. (1995), 'Ancillaries and third order significance', *Utilitas Mathematica* **47**, 33–53.

Fraser, D. A. S., Reid, N. and Wu, J. (1999), 'A simple general formula for tail probabilities for frequentist and Bayesian inference', *Biometrika* **86**, 249–264.

Friedman, J. (1991), 'Multivariate adaptive regression splines (with discussion)', *Annals of Statistics* **19**, 1–141.

Friedman, J. H. and Stuetzle, W. (1981), 'Projection pursuit regression', *Journal of the American Statistical Association* **76**, 817–823.

Friedman, J. H., Stuetzle, W. and Schroeder, A. (1984), 'Projection pursuit density estimation', *Journal of the American Statistical Association* **79**, 599–608.

Frisch, U. (1995), *Turbulence: The Legacy of A. N. Kolmogorov*, Cambridge University Press, Cambridge.

Furnival, G. M. and Wilson Jnr., R. W. (1974), 'Regressions by leaps and bounds', *Technometrics* **16**, 499–511.

Gabriel, J.-P., Lefèvre, C. and Picard, P., (eds) (1990), *Stochastic Processes in Epidemic Theory*, number 86 *in Lecture Notes in Biomathematics*, Springer, Berlin.

Geman, H., Madan, D. B. and Yor, M. (2002), 'Stochastic volatility, jumps and hidden time changes', *Finance and Stochastics* **6**, 63–90.

Gentleman, R. and Carey, V. (2003), Visualization and annotation of genomic experiments, *in* G. Parmigiani, E. S. Garrett, R. A. Irizarry and S. L. Zeger, (eds), *The Analysis of Gene Expression Data*, Springer, New York.

Ghosh, J. K. and Ramamoorthi, R. V. (2003), *Bayesian Nonparametrics*, Springer, New York.

Ghosh, M., Natarajan, K., Stroud, T. W. F. and Carlin, B. P. (1998), 'Generalized linear models for small-area estimation', *Journal of the American Statistical Association* **93**, 273–282.

Ghysels, E., Harvey, A. C. and Renault, E. (1996), Stochastic volatility, *in* C. R. Rao and G. S. Maddala, (eds), *Statistical Methods in Finance*, North-Holland, Amsterdam, pp. 119–191.

Gibson, G. J. and Renshaw, E. (2001), 'Likelihood estimation for stochastic compartmental models using Markov chain methods', *Statistics and Computing* **11**, 347–358.

Gilks, W. R., Richardson, S. and Spiegelhalter, D. J., (eds) (1996), *Markov Chain Monte Carlo in Practice*, London: Chapman & Hall/CRC.

Gleason, A. M., Greenwood, E. and Kelly, L. M. (1980), *The William Lowell Putnam Mathematical Competition*, Mathematical Association of America, Washington, DC.

Glonek, G. F. V. and Solomon, P. J. (2004), 'Factorial and time course designs for cDNA microarray experiments', *Biostatistics* **5**, 89–111.

Goldstein, H. (1986), 'Multilevel mixed linear model analysis using iterative generalised least squares', *Biometrika* **73**, 43–56.

Gordon, N. J., Salmond, D. J. and Smith, A. F. M. (1993), 'A novel approach to nonlinear and non-Gaussian Bayesian state estimation', *IEE-Proceedings F* **140**, 107–113.

Granger, C. W. J. (1988), 'Some recent developments in a concept of causality', *Journal of Econometrics* **39**, 199–211.

Green, P. J. and Richardson, S. (2002), 'Hidden Markov models and disease mapping', *Journal of the American Statistical Association* **97**, 1055–1070.

Green, P. J. and Silverman, B. W. (1994), *Nonparametric Regression and Generalized Linear Models: A Roughness Penalty Approach*, Chapman & Hall, London.

Greenland, S., Robins, J. M. and Pearl, J. (1999), 'Confounding and collapsibility in causal inference', *Statistical Science* **14**, 29–46.

Greenwood, P. E. and Wefelmeyer, W. (1995), 'Efficiency of empirical estimators for Markov chains', *Annals of Statistics* **23**, 132–143.

Grenfell, B. T. (1992), 'Chance and chaos in measles dynamics', *Journal of the Royal Statistical Society, series B* **54**, 383–398.

Grenfell, B. T. and Dobson, A., (eds) (1995), *Ecology of Infectious Diseases in Natural Populations*, Cambridge University Press, Cambridge.

Guo, D., Wang, X. and Chen, R. (2003), 'Nonparametric adaptive detection in fading channels based on sequential Monte Carlo and Bayesian model averaging', *Annals of the Institute of Statistical Mathematics* **55**, 423–436.

Gupta, S., Swinton, J. and Anderson, R. M. (1991), 'Theoretical studies of the effects of heterogeneity in the parasite population on the transmission dynamics of malaria', *Proceedings of the Royal Society of London* **B256**, 231–238.

Gupta, S., Trenholme, K., Anderson, R. M. and Day, K. P. (1994), 'Antigenic diversity and the transmission dynamics of *plasmodium falciparum*', *Science* **263**, 961–963.

Hajek, J. (1970), 'A characterization of limiting distributions of regular estimates', *Zeitschrift für Wahrscheinlichkeitstheorie und verwandte gebiete* **14**, 323–330.

Halgreen, C. (1979), 'Self-decomposability of the generalized inverse Gaussian and hyperbolic distributions', *Zeitschrift für Wahrscheinlichkeitstheorie und verwandte gebiete* **47**, 13–17.

Hall, P. (2001), Statistical science—evolution, motivation and direction, *in* B. Engquist and W. Schmid, (eds), *Mathematics Unlimited—2001 and Beyond*, Springer, Berlin, pp. 565–575.

Hall, P. and Raimondo, M. (1998), 'On global performance of approximations to smooth curves using gridded data', *Annals of Statistics* **26**, 2206–2217.

Hall, P. and Zhou, X.-H. (2003), 'Nonparametric estimation of component distributions in a multivariate mixture', *Annals of Statistics* **31**, 201–224.

Halsey, A. H. and Webb, J. (2000), *Twentieth-Century British Social Trends*, Basingstoke: Macmillan.

Hamer, W. H. (1906), 'Epidemic disease in England', *The Lancet* **1**, 733–739.

Hampel, F. R. (1974), 'The influence function and its role in robust estimation', *Journal of the American Statistical Association* **62**, 1179–1186.

Handcock, M. S. (2003), Assessing degeneracy in statistical models of social networks, Technical report, Center for Statistics and the Social Sciences, University of Washington. Working Paper no. 39.

Hansen, P. R. and Lunde, A. (2003), An optimal and unbiased measure of realized variance based on intermittent high-frequency data. Unpublished paper, presented at the Realized Volatility Conference, Montreal, 7th November, 2003.

Hart, J. D. (1997), *Nonparametric Smoothing and Lack-of-Fit Tests*, Springer, New York.

Harvey, A. C. (1989), *Forecasting, Structural Time Series Models and the Kalman Filter*, Cambridge University Press, Cambridge.

Hasegawa, H. and Kozumi, H. (2003), 'Estimation of Lorenz curves: A Bayesian nonparametric approach', *Journal of Econometrics* **115**, 277–291.

Hastie, T. J. and Tibshirani, R. J. (1990), *Generalized Additive Models*, Chapman & Hall, London.

Hastie, T., Tibshirani, R., Eisen, M. B., Alizadeh, A., Levy, R., Staudt, L., Chan, W. C., Botstein, D. and Brown, P. (2000), 'Gene shaving as a method for identifying distinct sets of genes with similar expression patterns', *Genome Biology* **1**.

Hayes, L. J. and Berry, G. (2002), 'Sampling variability of the Kunst-Mackenbach relative index of inequality', *Journal of Epidemiology and Public Health* **56**, 762–765.

Head, J., Stansfeld, S. A. and Siegrist, J. (2004), 'The psychosocial work environment and alcohol dependence: a prospective study', *Occupational and Environmental Medicine* **61**, 219–214.

Heagerty, P. J. and Zeger, S. L. (2000), 'Marginalized multilevel models and likelihood inference (with discussion)', *Statistical Science* **15**, 1–26.

Heckman, J., Ichimura, H. and Todd, P. (1998), 'Matching as an econometric evaluation estimator', *Review of Economic Studies* **65**, 261–294.

Heckman, N. E. (1986), 'Spline smoothing in a partly linear model', *Journal of the Royal Statistical Society, series B* **48**, 244–248.

Heesterbeek, J. A. P. (1992), R_0. *PhD Thesis.*, CWI, Amsterdam.

Herbert, J. and Isham, V. (2000), 'Stochastic host-parasite population models', *Journal of Mathematical Biology* **40**, 343–371.

Hesterberg, T. C. (1996), 'Control variates and importance sampling for efficient bootstrap simulations', *Statistics and Computing* **6**, 147–157.

Hethcote, H. W. (2000), 'The mathematics of infectious diseases', *SIAM Review* **42**, 599–653.

Hirth, U. M. (1997), 'Probabilistic number theory, the GEM/Poisson-Dirichlet distribution and the arc-sine law', *Combinatorics, Probability and Computing* **6**, 57–77.

Hoem, J. M. (1989), The issue of weights in panel surveys of individual behavior, *in* D. Kasprzyk, G. Duncan, G. Kalton and M. P. Singh, (eds), *Panel Surveys*, New York: Wiley, pp. 539–565.

Hoffmann, M. and Lepski, O. (2002), 'Random rates in anisotropic regression (with discussion)', *Annals of Statistics* **30**, 325–396.

Holevo, A. S. (2001*a*), Lévy processes and continuous quantum measurements, *in* O. E. Barndorff-Nielsen, T. Mikosch and S. I. Resnick, (eds), *Lévy Processes: Theory and Applications*, Birkhäuser, Boston, MA, pp. 225–239.

Holevo, A. S. (2001*b*), *Statistical Structure of Quantum Theory*, Springer, Heidelberg.

Holland, P. (1986), 'Statistics and causal inference (with discussion)', *Journal of the American Statistical Association* **81**, 945–970.

Hortobagyi, G. N., Gutterman, J. U. and Blumenscein, G. R. (1978), Chemoimmunotherapy of advanced breast cancer with BCG, *in* W. D. Terry and D. Windhurst, (eds), *Immunotherapy of Cancer: Present Status of Trials in Man*, Raven Press, New York, pp. 155–168.

Huang, X. and Tauchen, G. (2003), The relative contribution of jumps to total price variation. Unpublished paper: Department of Economics, Duke University.

Huber, P. J. (1985), 'Projection pursuit (with discussion)', *Annals of Statistics* **13**, 435–525.

Ibragimov, I. A. and Has'minskii, R. Z. (1981), *Statistical Estimation: Asymptotical Theory*, Springer, New York.

Ihaka, R. and Gentleman, R. (1996), 'R: A language for data analysis and graphics', *Journal of Computational and Graphical Statistics* **5**, 299–314.

Indritz, J. (1963), *Methods in Analysis*, Macmillan, New York.

Irizarry, R. A., Gautier, L. and Cope, L. M. (2003), An R package for analysis of Affymetrix oligonucleotide arrays, *in* G. Parmigiani, E. S. Garrett, R. A. Irizarry and S. L. Zeger, (eds), *The Analysis of Gene Expression Data*, Springer.

Isham, V. (1988), 'Mathematical modelling of the transmission dynamics of HIV and AIDS: A review', *Journal of the Royal Statistical Society, series A* **151**, 5–30. With discussion, pp. 120–123.

Isham, V. (1991), 'Assessing the variability of stochastic epidemics', *Mathematical Biosciences* **107**, 209–224.

Isham, V. (1993*a*), Statistical aspects of chaos: A review, *in* O. E. Barndorff-Nielsen, D. R. Cox, J. L. Jensen and W. S. Kendall, (eds), *Networks and Chaos*, Chapman & Hall, London.

Isham, V. (1993*b*), 'Stochastic models for epidemics with special reference to AIDS', *Advances in Applied Probability* **3**, 1–27.

Isham, V. (1995), 'Stochastic models of host-macroparasite interaction', *Advances in Applied Probability* **5**, 720–740.

Isham, V. and Medley, G., (eds) (1996), *Models for Infectious Human Diseases: Their Structure and Relation to Data*, Cambridge University Press, Cambridge.

Jacod, J. (1994), 'Limit of random measures associated with the increments of a Brownian semimartingale'. Unpublished paper: Laboratoire de Probabilitiés, Université Pierre and Marie Curie, Paris.

Jacod, J. and Moral, P. D. (2001), Interacting particle filtering with discrete observations, *in* A. Doucet, N. de Freitas and N. J. Gordon, (eds), *Sequential Monte Carlo Methods in Practice*, Springer-Verlag, New York, pp. 43–77.

Jacod, J., Moral, P. D. and Protter, P. (2001), 'The Monte–Carlo method for filtering with discrete-time observations', *Probability Theory and Related Fields* **120**, 346–368.

Jacod, J. and Protter, P. (1998), 'Asymptotic error distributions for the Euler method for stochastic differential equations', *Annals of Probability* **26**, 267–307.

Jacquez, J. A. and Simon, C. P. (1990), 'AIDS: The epidemiological significance of two mean rates of partner change', *IMA Journal of Mathematics Applied in Medicine and Biology* **7**, 27–32.

Jacquez, J. A. and Simon, C. P. (1993), 'The stochastic SI model with recruitment and deaths. I. Comparisons with the closed SIS model', *Mathematical Biosciences* **117**, 77–125.

Jacquez, J. A., Simon, C. P. and Koopman, J. (1989), Structured mixing: Heterogeneous mixing by the definition of activity groups, *in* C. Castillo-Chavez, (ed.), *Mathematical and statistical approaches to AIDS epidemiology. Lecture notes in Biomathematics, 83*, Springer, Berlin, pp. 301–315.

Jeong, H., Mason, S. P., Barabási, A.-L. and Oltvai, Z. N. (2001), 'Lethality and centrality in protein networks', *Nature* **411**, 41.

Johannes, M., Polson, N. G. and Stroud, J. (2002), Nonlinear filtering of stochastic differential equations with jumps. Unpublished paper: Graduate School of Business, Columbia University.

Johannes, M., Polson, N. and Stroud, J. (2004), Practical filtering with parameter learning, *in* A. C. Harvey, S. J. Koopman and N. Shephard, (eds), *State Space and Unobserved Component Models: Theory and Applications. Proceedings of a Conference in Honour of James Durbin*, Cambridge University Press, Cambridge, pp. 236–247.

Johnson, A. M. (1996), Sources and use of empirical observations to characterise networks of sexual behaviour, *in* V. Isham and G. Medley, (eds), *Models for Infectious Human Diseases: Their Structure and Relation to Data*, Cambridge University Press, Cambridge, pp. 253–262.

Jones, J. H. and Handcock, M. S. (2003), 'An assessment of preferential attachment as a mechanism for human sexual network formation', *Proceedings of the Royal Society of London* **B270**, 1123–1128.

Jurek, Z. J. and Mason, J. D. (1993), *Operator-Limit Distributions in Probability Theory*, Wiley, New York.

Kalman, R. E. (1960), 'A new approach to linear filtering and prediction problems', *Journal of Basic Engineering* **82**, 35–45.

Kao, R. R. (2002), 'The role of mathematical modelling in the control of the 2001 FMD epidemic in the UK', *Trends in Microbiology* **10**, 279–286.

Karatzas, I. and Shreve, S. E. (1991), *Brownian Motion and Stochastic Calculus*, Vol. 113 of *Graduate Texts in Mathematics*, 2nd edn, Springer–Verlag, Berlin.

Karatzas, I. and Shreve, S. E. (1998), *Methods of Mathematical Finance*, Springer–Verlag, New York.

Kaufman, J. S., Kaufman, S. and Poole, C. (2003), 'Causal inference from randomized trials in social epidemiology', *Social Science and Medicine* **7**, 2397–2409.

Keeling, M. J. (1999*a*), 'Correlation equations for endemic diseases: externally imposed and internally generated heterogeneity', *Proceedings of the Royal Society of London* **B266**, 953–960.

Keeling, M. J. (1999*b*), 'The effects of local spatial structure on epidemiological invasions', *Proceedings of the Royal Society of London* **B266**, 859–867.

Keeling, M. J. (2000), 'Multiplicative moments and measures of persistence in ecology', *Journal of Theoretical Biology* **205**, 269–281.

Keeling, M. J. and Grenfell, B. T. (1997), 'Disease extinction and community size', *Science* **275**, 65–67.

Keeling, M. J. and Grenfell, B. T. (1998), 'Effect of variability in infection period on the persistence and spatial spread of infectious diseases', *Mathematical Biosciences* **147**, 207–226.

Keeling, M. J. and Grenfell, B. T. (2000), 'Individual-based perspectives on R_0', *Journal of Theoretical Biology* **203**, 51–61.

Keeling, M. J. and Grenfell, B. T. (2002), 'Understanding the persistence of measles: reconciling theory, simulation and observation', *Proceedings of the Royal Society of London* **B269**, 335–343.

Keeling, M. J., Woolhouse, M. E. J., May, R. M., Davies, G. and Grenfell, B. T. (2003), 'Modelling vaccination strategies against foot-and-mouth disease', *Nature* **421**, 136–142.

Keeling, M. J., Woolhouse, M. E. J., Shaw, D. J., Matthews, L., Chase-Topping, M., Haydon, D. T., Cornell, S. J., Kappey, J., Wilesmith, J. and Grenfell, B. T. (2001), 'Dynamics of the 2001 UK foot and mouth epidemic: Stochastic dispersal in a heterogeneous landscape', *Science* **294**, 813–817.

Kempthorne, O. (1952), *The Design and Analysis of Experiments*, Wiley, New York.

Kermack, W. O. and McKendrick, A. G. (1927), 'Contributions to the mathematical theory of epidemics, part I.', *Proceedings of the Royal Society of London* **A115**, 700–721.

Kerr, M. K. and Churchill, G. A. (2001), 'Experimental design for gene expression microarrays', *Biostatistics* **2**, 183–201.

Khintchine, A. Y. (1964), *Continued Fractions*, University of Chicago Press, Chicago.

Kilkenny, D., Reed, M. D., O'Donoghue, D., Kawaler, S., Mukadam, A., Kleinman, S., Nitta, A., Metcalfe, T. S., Provencal, J. L., Watson, T. K., Sullivan, D. J., Sullivan, T., Shobbrook, R., Jiang, X. ., Joshi, S., Ashoka, B. N., Seetha, S., Leibowitz, E., Ibbetson, P., Mendelson, H., Meistas, E., Kalytis, R., Alisauskas, D., P., M., van Wyk, F., Stobie, R. S., Marang, F., Zola, S., Krzesinski, J., Ogloza, W., Moskalik, P., Silvotti, R., Piccioni, A., Vauclair, G., Dolez, N., Chevreton, M., Dreizler, S., Schuh, S. L., Deetjen, J. L., Solheim, J. E., Perez, J. M. G., Ulla, A., Ostensen, R., Manteiga, M., Suarez, O., Burleigh, M., Kepler, S. O., Kanaan, A. and Giovannini, O. (2003), 'A whole earth telescope campaign on the pulsating subdwarf B binary system PG 1336-018 (NY Vir)', *Monthly Notices of the Royal Astronomical Society* **343**, 834–846.

Kim, S., Shephard, N. and Chib, S. (1998), 'Stochastic volatility: Likelihood inference and comparison with ARCH models', *Review of Economic Studies* **65**, 361–393.

Kingman, J. F. C. (1978), 'The representation of partition structures', *Journal of the London Mathematical Society* **18**, 374–380.

Kious, B. M. (2001), 'The Nuremberg Code: its history and implications', *Princeton Journal of Bioethics* **4**, 7–19.

Knott, M. and Bartholomew, D. J. (1999), *Latent Variable Models and Factor Analysis* (2nd edn.), London: Edward Arnold.

Kohonen, T. (1995), *Self-Organizing Maps*, Springer, Berlin.

Kong, A., McCullagh, P., Meng, X.-L., Nicolae, D. and Tan, Z. (2003), 'A theory of statistical models for Monte Carlo integration (with discussion)', *Journal of the Royal Statistical Society, series B* **65**, 585–618.

Koopman, J., Simon, C. P., Jacquez, J. A. and Park, T. S. (1989), Selective contact within structured mixing with an application to HIV transmission risk from oral and anal sex, *in* C. Castillo-Chavez, (ed.), *Mathematical and statistical approaches to AIDS epidemiology. Lecture notes in Biomathematics, 83*, Springer, Berlin, pp. 316–349.

Koshevnik, Y. A. and Levit, B. Y. (1976), 'On a nonparametric analogue of the information matrix', *Theory of Probability and its Applications* **21**, 738–753.

Kraus, K. (1971), 'General state changes in quantum theory', *Annals of Physics* **64**, 331–335.

Kraus, K. (1983), *States, Effects and Operations: Fundamental Notions of Quantum Theory*, Springer, Heidelberg.

Kretzschmar, M. (1989a), 'Persistent solutions in a model for parasitic infections', *Journal of Mathematical Biology* **27**, 549–573.

Kretzschmar, M. (1989b), 'A renewal equation with a birth-death process as a model for parasitic infections', *Journal of Mathematical Biology* **27**, 191–221.

Kruskal, J. B. (1964a), 'Multidimensional scaling by optimizing goodness-of-fit to a nonmetric hypothesis', *Psychometrika* **29**, 1–29.

Kruskal, J. B. (1964b), 'Non-metric multidimensional scaling: a numerical method', *Psychometrika* **29**, 115–129.

Kunst, A. E. and Mackenbach, J. P. (1995), *Measuring Socio-economic Inequalities in Health*, Copenhagen: World Health Organisation.

Kurtz, T. G. (1970), 'Solutions of ordinary differential equations as limits of pure jump Markov processes', *Journal of Applied Probability* **7**, 49–58.

Kurtz, T. G. (1971), 'Limit theorems for sequences of jump Markov processes approximating ordinary differential processes', *Journal of Applied Probability* **8**, 344–356.

Kurtz, T. G. (1981), *Approximation of Population Processes*, SIAM, Philadelphia.

Laio, F., Porporato, A., Ridolfi, L. and Rodriguez-Iturbe, I. (2001*a*), 'On the mean first passage times of processes driven by white shot noise', *Physical Review, series E* **63**, 36105.

Laio, F., Porporato, A., Ridolfi, L. and Rodriguez-Iturbe, I. (2001*b*), 'Plants in water-controlled ecosystems: Active role in hydrological processes and response to water stress. II. Probabilistic soil moisture dynamics', *Advances in Water Resources* **24**, 707–723.

Law, R. and Dieckmann, U. (2000), Moment approximations of individual-based models, *in* U. Dieckmann, R. Law J. A. J. Metz, (eds), *The Geometry of Ecological Interactions: Simplifying Spatial Complexity*, Cambridge University Press, Cambridge, pp. 252–270.

Lefèvre, C. (1990), Stochastic epidemic models for S-I-R infectious diseases: a brief survey of the recent general theory, *in* J.-P. Gabriel, C. Lefèvre and P. Picard, eds, *Stochastic Processes in Epidemic Theory. Lecture Notes in Biomathematics, vol. 86*, Springer, Berlin, pp. 1–12.

Levit, B. Y. (1975), 'On the efficiency of a class of non-parametric estimates', *Theory of Probability and its Applications* **20**, 723–740.

Levit, B. Y. (1978), 'Infinite dimensional informational inequalities', *Theory of Probability and its Applications* **23**, 371–377.

Li, H., Wells, M. and Yu, L. (2004), An MCMC analysis of time-changed Lévy processes of stock returns. Unpublished paper: Johnson Graduate School of Management, Cornell University, Ithaca, U.S.A.

Li, W.-H. (1997), *Molecular Evolution*, Sinauer, Sunderland, MA.

Liang, K.-Y., Chiu, Y.-F., Beaty, T. H. and Wjst, M. (2001), 'Multipoint analysis using affected sib pairs: Incorporating linkage evidence from unlinked regions', *Genetic Epidemiology* pp. 158–172.

Liang, K.-Y. and Zeger, S. L. (1995), 'Inference based on estimating functions in the presence of nuisance parameters (with discussion)', *Statistical Science* **10**, 158–172.

Lindley, D. V. (2002), 'Seeing and doing: The concept of causation (with discussion)', *International Statistical Review* **70**, 191–214.

Lindley, D. V. and Smith, A. F. M. (1972), 'Bayes estimates for the linear model (with discussion)', *Journal of the Royal Statistical Society, series B* **34**, 1–41.

Lipsitz, S. R., Ibrahim, J. and Zhao, L. P. (1999), 'A weighted estimating equation for missing covariate data with properties similar to maximum likelihood', *Journal of the American Statistical Association* **94**, 1147–1160.

Little, R. J. A. and Rubin, D. B. (2002), *Statistical Analysis with Missing Data* (2nd edn.), New York: Wiley.

Lloyd, A. L. (2001), 'Destabilisation of epidemic models with the inclusion of realistic distributions of infectious periods', *Proceedings of the Royal Society of London* **B268**, 985–993.

MacLachlan, G. J. and Peel, D. (2002), *Finite Mixture Models*, Wiley, New York.

Mancini, C. (2001), Does our favourite index jump or not? Dipartimento di Matematica per le Decisioni, Universita di Firenze.

Mancini, C. (2003), Estimation of the characteristics of jump of a general Poisson-diffusion process. Dipartimento di Matematica per le Decisioni, Universita di Firenze.

Maras, A. M. (2003), 'Adaptive nonparametric locally optimum Bayes detection in additive non-Gaussian noise', *IEEE Transactions on Information Theory* **49**, 204–220.

Marks, H. M. (2003), 'Rigorous uncertainty: Why R.A. Fisher is important', *International Journal of Epidemiology* **32**, 932–937.

Masoliver, J. (1987), 'First-passage times for non-Markovian processes: Shot noise', *Physical Review, series A* **35**, 3918–3928.

Matérn, B. (1986), *Spatial Variation*, Springer, New York.

McCullagh, P. (2000), 'Invariance and factorial models (with discussion)', *Journal of the Royal Statistical Society, series B* **62**, 209–256.

McCullagh, P. (2002), 'What is a statistical model? (with discussion)', *Annals of Statistics* **30**, 1225–1310.

McCullagh, P. and Nelder, J. A. (1989), *Generalized Linear Models*, 2nd edn, Chapman & Hall, London.

McKendrick, A. G. (1926), 'Applications of mathematics to medical problems', *Proceedings of the Edinburgh Mathematical Society* **44**, 98–130.

Medical Research Council (2001), *Medical Research Council Research Funding Strategy and Priorities, 2001–2004*, Medical Research Council, London.

Merton, R. C. (1980), 'On estimating the expected return on the market: An exploratory investigation', *Journal of Financial Economics* **8**, 323–361.

Milton, J. (2003), 'Spies, magicians, and Enid Blyton: How they can help improve clinical trials', *International Journal of Epidemiology* **32**, 943–944.

Møller, J., Syversveen, A. R. and Waagepetersen, R. P. (1998), 'Log Gaussian Cox processes', *Scandinavian Journal of Statistics* **25**, 451–482.

Møller, J. and Waagepetersen, R. (2003), *Statistical Inference and Simulation for Spatial Point Processes*, Chapman & Hall/CRC, London.

Mollison, D. (1991), 'Dependence of epidemic and population velocities on basic parameters', *Mathematical Biosciences* **107**, 255–287.

Mollison, D., ed. (1995), *Epidemic Models: their Structure and Relation to Data*, Cambridge University Press, Cambridge.

Mollison, D., Isham, V. and Grenfell, B. (1994), 'Epidemics: Models and data (with discussion)', *Journal of the Royal Statistical Society, series A* **157**, 115–149.

Mølmer, K. and Castin, Y. (1996), 'Monte Carlo wavefunctions', *Coherence and Quantum Optics* **7**, 193–202.

Moore, G. E. (1965), 'Cramming more components onto integrated circuits', *Electronics* **38**(8), 114–117.

Morris, M. (1991), 'A loglinear modelling framework for selective mixing', *Mathematical Biosciences* **107**, 349–377.

Morris, M. (1996), Behaviour change and non-homogeneous mixing, *in* V. Isham and G. Medley, (eds), *Models for Infectious Human Diseases: their Structure and Relation to Data*, Cambridge University Press, Cambridge, pp. 239–252.

Morris, M. and Kretzschmar, M. (1989), 'Concurrent partnerships and the spread of HIV', *AIDS* **11**, 641–648.

MRC Streptomycin in Tuberculosis Trials Committee (1948), 'Streptomycin treatment for pulmonary tuberculosis', *British Medical Journal* pp. 769–782.

Munch, R. (2003), 'Robert Koch', *Microbes and Infection* **5**, 69–74.

Muraki, N. (2003), 'The five independencies as natural products', *Infinite Dimensional Analysis, Quantum Probability, and Related Topics* **6**, 337–371.

Murphy, S. A. and van der Vaart, A. W. (2000), 'Semiparametric likelihood ratio inference', *Annals of Statistics* **25**, 1471–1509.

Nåsell, I. (1985), *Hybrid Models of Tropical Infections*, Lecture Notes in Biomathematics, no. 59, Springer, Berlin.

Nåsell, I. (1995), The threshold concept in stochastic epidemic and endemic models, *in* D. Mollison, (ed.), *Epidemic Models: Their Structure and Relation to Data,* Cambridge University Press, pp. 71–83.

Nåsell, I. (1996), 'The quasi-stationary distribution of the closed endemic SIS model', *Advances in Applied Probability* **28**, 895–932.

Nåsell, I. (1999), 'On the time to extinction in recurrent epidemics', *Journal of the Royal Statistical Society, series B* **61**, 309–330.

Nåsell, I. (2002*a*), Measles outbreaks are not chaotic, *in* C. Castillo-Chavez, (ed.), *Mathematical Approaches for Emerging and Reemerging Infectious Diseases II: Models, Methods and Theory. IMA Math. Appl. vol 126,* Springer, New York, pp. 85–114.

Nåsell, I. (2002*b*), 'Stochastic models of some endemic infections', *Mathematical Biosciences* **179**, 1–19.

Nåsell, I. (2003), 'Moment closure and the stochastic logistic model', *Theoretical Population Biology* **63**, 159–168.

National Institutes of Health (2000), 'Adjuvant Therapy for Breast Cancer', *NIH Consensus Statement* **17**, 1–35.

Nelder, J. A. (1977), 'A reformulation of linear models (with discussion)', *Journal of the Royal Statistical Society, series A* **140**, 48–77.

Nelder, J. A. and Wedderburn, R. W. M. (1972), 'Generalized linear models', *Journal of the Royal Statistical Society, series A* **135**, 370–384.

Newey, W. (1990), 'Semiparametric efficiency bounds', *Journal of Applied Econometrics* **5**, 99–135.

Newman, M. E. J. (2003), 'Properties of highly clustered networks', *Physical Review* **E68**, 026121.

Newman, M. E. J., Jensen, I. and Ziff, R. M. (2002), 'Percolation and epidemics in a two-dimensional small world', *Physical Review E* **65**, 021904.

Newman, M. E. J., Watts, D. J. and Strogatz, S. H. (2002), 'Random graph models of social networks', *Proceedings of the National Academy of Science* **99**, 2566–2572.

Neyman, J. and Scott, E. L. (1948), 'Consistent estimates based on partially consistent observations', *Econometrica* **16**, 1–32.

Nicolato, E. and Venardos, E. (2003), 'Option pricing in stochastic volatility models of the Ornstein–Uhlenbeck type', *Mathematical Finance* **13**, 445–466.

Noble, M., Smith, G., Penhale, B., Wright, G., Dibben, C., Owen, T. and Lloyd, M. (2000), *Measuring Multiple Deprivation at the Small Area Level: The Indices of Deprivation 2000*, London: Department of the Environment, Transport and the Regions, Regeneration Series.

Ogata, Y. (2001), 'Exploratory analysis of earthquake clusters by likelihood-based trigger models', *Journal of Applied Probability* **38A**, 202–212.

Oh, H. S., Nychka, D. and Brown, T. (2004), 'Period analysis of variable stars by robust smoothing', *Applied Statistics* **53**, 15–30.

O'Neill, P. D. (2002), 'A tutorial introduction to Bayesian inference for stochastic epidemic models using Markov chain Monte Carlo methods', *Mathematical Biosciences* **180**, 103–114.

O'Neill, P. D. and Roberts, G. O. (1999), 'Bayesian inference for partially observed stochastic epidemics', *Journal of the Royal Statistical Society, series A* **162**, 121–129.

Owen, A. B. (2001), *Empirical Likelihood*, Chapman & Hall/CRC, Boca Raton.

Pace, L. and Salvan, A. (1997), *Principles of Statistical Inference from a Neo-Fisherian Perspective*, World Scientific, Singapore.

Patil, G. P. (1963), 'Minimum variance unbiased estimation and certain problems of additive number theory', *Annals of Mathematical Statistics* **34**, 1050–1056.

Pattison, P. and Robins, G. (2002), 'Neighbourhood-based models of social networks', *Sociological Methodology* **32**, 301–337.

Pearl, J. (2000), *Causality: Models, Reasoning and Inference*, Cambridge University Press, Cambridge.

Pearl, J. (2002), 'Comments on "Seeing and doing: The concept of causation" by D. V. Lindley', *International Statistical Review* **70**, 207–214.

Pearl, J. (2003), 'Statistics and causal inference: A review', *Test* **12**, 281–345.

Pedersen, J. (2003), The Lévy–Ito decomposition of independently scattered random measures. Research Report 2003-2, MaPhySto, University of Aarhus.

Peto, R. (1996), 'Five years of tamoxifen—or more?', *Journal of the National Cancer Institute* **88**, 1791–1793.

Peto, R., Boreham, J., Clarke, M., Davies, C. and Beral, V. (2000), 'UK and USA breast cancer deaths down 25% at ages 20–69 years', *Lancet* **355**, 1822.

Pfanzagl, J. and Wefelmeyer, W. (1982), *Contributions to a General Asymptotic Statistical Theory*, Vol. 13 of *Lecture Notes in Statistics*, Springer, New York.

Pfeffermann, D. (1993), 'The role of sampling weights when modeling survey data', *International Statistical Review* **61**, 317–337.

Pierce, D. A. and Bellio, R. (2004), The effect of choice of reference set on frequency inferences. Preprint.

Pierce, D. A. and Peters, D. (1992), 'Practical use of higher order asymptotics for multiparameter exponential families (with discussion)', *Journal of the Royal Statistical Society, series B* **54**, 701–738.

Pinheiro, J. C. and Bates, D. M. (2000), *Mixed-Effects Models in S and S-PLUS*, Springer-Verlag, New York.

Pison, G., Rousseeuw, P. J., Filzmoser, P. and Croux, C. (2003), 'Robust factor analysis', *Journal of Multivariate Analysis* **84**, 145–172.

Pitman, E. J. G. (1937), 'Significance tests which may be applied to samples from any population', *Journal of the Royal Statistical Society, Supplement* **4**, 119–130.

Pitt, M. K. and Shephard, N. (1999), 'Filtering via simulation: auxiliary particle filter', *Journal of the American Statistical Association* **94**, 590–599.

Pitt, M. K. and Shephard, N. (2001), Auxiliary variable based particle filters, *in* N. de Freitas, A. Doucet and N. J. Gordon, (eds), *Sequential Monte Carlo Methods in Practice*, Springer-Verlag, New York, pp. 273–293.

Poole, D. and Raftery, A. E. (2000), 'Inference from deterministic simulation models: The Bayesian melding approach', *Journal of the American Statistical Association* **95**, 1244–1255.

Porporato, A., Daly, E. and Rodriguez-Iturbe, I. (2004), 'Soil water balance and ecosystem response to climate change', *American Naturalist* **164**, 625–633.

Porporato, A., Laio, F., Ridolfi, L., Caylor, K. K. and Rodriguez-Iturbe, I. (2003), 'Soil moisture and plant stress dynamics along the Kalahari precipitation gradient', *Journal of Geophysical Research* **108**, 4127–4134.

Porporato, A., Laio, F., Ridolfi, L. and Rodriguez-Iturbe, I. (2001), 'Plants in water-controlled ecosystems: Active role in hydrological processes and response to water stress. III. Vegetation water stress', *Advances in Water Resources* **24**, 725–744.

Porporato, A. and Rodriguez-Iturbe, I. (2002), 'Ecohydrology—a challenging multidisciplinary research perspective', *Hydrological Sciences Journal* **47**, 811–822.

Poterba, J. and Summers, L. (1986), 'The persistence of volatility and stock market fluctuations', *American Economic Review* **76**, 1124–1141.

Preston, C. J. (1973), 'The generalised Gibbs states and Markov random fields', *Advances in Applied Probability* **5**, 242–261.

Preston, D. L., Shimizu, Y., Pierce, D. A., Suyama, A. and Mabuchi, K. (2003), 'Studies of mortality of atomic bomb survivors. Report 13: Solid cancer and noncancer disease mortality 1950–1997', *Radiation Research* **160**, 381–407.

Pritchard, K. I., Paterson, A. H., Fine, S., Paul, N. A., Zee, B., Shepherd, L. E., Abu-Zahra, H., Ragaz, J., Knowling, M., Levine, M. N., Verma, S., Perrault, D., Walde, P. L., Bramwell, V. H., Poljicak, M., Boyd, N., Warr, D., Norris, B. D., Bowman, D., Armitage, G. R., Weizel, H. and Buckman, R. A. (1997), 'Randomized trial of cyclophosphamide, methotrexate, and fluorouracil chemotherapy added to tamoxifen as adjuvant therapy in postmenopausal women with node-positive estrogen and/or progesterone receptor-positive breast cancer: A report of the National Cancer Institute of Canada Clinical Trials Group. Breast Cancer Site Group', *Journal of Clinical Oncology* **15**, 2302–2311.

Protter, P. (1992), *Stochastic Integration and Differential Equations*, 2nd edn, Springer-Verlag, New York.

Quinlan, J. R. (1993), *C4.5: Programs for Machine Learning*, Morgan Kaufmann, San Mateo, CA.

Quinn, B. G. (1999), 'A fast efficient technique for the estimation of frequency: interpretation and generalisation', *Biometrika* **86**, 213–220.

Quinn, B. G. and Hannan, E. J. (2001), *The Estimation and Tracking of Frequency*, Cambridge University Press, Cambridge.

Quinn, B. G. and Thompson, P. J. (1991), 'Estimating the frequency of a periodic function', *Biometrika* **78**, 65–74.

Quinn, M., Babb, P., Brock, A., Kirby, E. and Jones, J. (2001), Cancer Trends in England and Wales, *in Studies on Medical and Population Subjects No. 66*, The Stationery Office, London, pp. 1950–1999.

Raftery, A. E. (2000), 'Statistics in sociology, 1950–2000', *Journal of the American Statistical Association* **95**, 654–661.

Raftery, A. E., Tanner, M. A. and Wells, M. T. (2001), *Statistics in the 21st Century*, Chapman & Hall, London.

Ramsay, J. O. and Silverman, B. W. (1997), *Functional Data Analysis*, Springer, New York.

Ramsay, J. O. and Silverman, B. W. (2002), *Applied Functional Data Analysis: Methods and Case Studies*, Springer, New York.

Rao, J. N. K. (2003), *Small Area Estimation*, New York: Wiley.

Read, J. M. and Keeling, M. J. (2003), 'Disease evolution on networks: the role of contact structure', *Proceedings of the Royal Society of London, Series B* **B270**, 699–708.

Reid, N. (1994), 'A conversation with Sir David Cox', *Statistical Science* **9**, 439–455.

Reid, N. (2003), 'Asymptotics and the theory of inference', *Annals of Statistics* **31**, 1695–1731.

Reid, N., Mukerjee, R. and Fraser, D. A. S. (2003), Some aspects of matching priors, *in* M. Moore, S. Froda and C. Léger, (eds), *Mathematical Statistics and Applications: Festschrift for C. van Eeden*, Vol. 42 of *Lecture Notes – Monograph Series*, Institute of Mathematical Statistics, Hayward, CA, pp. 31–44.

Renshaw, E. (1991), *Modelling Biological Populations in Space and Time*, Cambridge University Press, Cambridge.

Rényi, A. (1963), 'On stable sequences of events', *Sankyā, Series A* **25**, 293–302.

Rice, J. A. (1986), 'Convergence rates for partially splined estimates', *Statistics and Probability Letters* **4**, 203–208.

Rice, J. A. and Silverman, B. W. (1991), 'Estimating the mean and covariance structure nonparametrically when the data are curves', *Journal of the Royal Statistical Society, series B* **53**, 233–243.

Richardson, M. P. and Stock, J. H. (1989), 'Drawing inferences from statistics based on multi-year asset returns', *Journal of Financial Economics* **25**, 323–48.

Riley, S., Fraser, C., Donnelly, C. A., Ghani, A. C., Abu-Raddad, L. J., Hedley, A. J., Leung, G. M., Ho, L.-M., Lam, T.-H., Thach, T. Q., Chau, P., Chan, K.-P., Lo, S.-V., Leung, P.-Y., Tsang, T., Ho, W., Lee, K.-H., Lau, E. M. C., Ferguson, N. M. and Anderson, R. M. (2003), 'Transmission dynamics of the etiological agent of SARS in Hong Kong: Impact of public health interventions', *Science* **300**, 1961–66.

Ripley, B. D. (1993), Statistical aspects of neural networks, *in* O. E. Barndorff-Nielsen, J. L. Jensen and W. S. Kendall, (eds), *Networks and Chaos—Statistical and Probabilistic Aspects*, Chapman & Hall, London, pp. 40–123.

Ripley, B. D. (1994), 'Neural networks and related methods for classification (with discussion)', *Journal of the Royal Statistical Society, series B* **56**, 409–456.

Ripley, B. D. (1996), *Pattern Recognition and Neural Networks*, Cambridge University Press, Cambridge.

Ritov, Y. and Bickel, P. J. (1990), 'Achieving information bounds in non- and semi-parametric models', *Annals of Statistics* **18**, 925–938.

Ritter, H., Martinetz, T. and Schulten, K. (1992), *Neural Computation and Self-Organizing Maps. An Introduction*, Addison-Wesley, Reading, MA.

Robins, J. M. (2004), Optimal structural nested models for optimal decisions, *in* D. Y. Lin and P. Heagerty, (eds), *Proceedings of the 2nd Seattle Symposium on Survival Analysis*, Springer, New York.

Robins, J. M., Mark, S. D. and Newey, W. K. (1992), 'Estimating exposure effects by modelling the expectation of exposure conditional on confounders', *Biometrics* **48**, 479–495.

Robins, J. M. and Ritov, Y. (1997), 'Toward a course of dimensionality appropriate (CODA) asymptotic theory for semi-parametric models', *Statistics in Medicine* **26**, 285–319.

Robins, J. M. and Rotnitzky, A. (1992), Recovery of information and adjustment for dependent censoring using surrogate markers, *in* N. Jewell, K. Dietz and V. Farewell, (eds), *AIDS Epidemiology-Methodological Issues*, V. Farewell, Boston, USA, pp. 297–331.

Robins, J. M. and Rotnitzky, A. (2001), 'Discussion of: "Inference in semiparametric models: some questions and one answer"', *Statistica Sinica* **11**, 863–960.

Robins, J. M., Rotnitzky, A. and van der Laan, M. (2000), 'Discussion of: "On profile likelihood"', *Journal of the American Statistical Association* **95**, 431–435.

Robins, J. M. and van der Vaart, A. W. (2004), A unified approach to estimation in non-semiparametric models using higher order influence functions, Technical report, Department of Epidemiology, Harvard School of Public Health.

Robins, J. M. and Wasserman, L. (2000), 'Conditioning, likelihood, and coherence: A review of some foundational concepts', *Journal of the American Statistical Association* **95**, 1340–1346.

Robinson, G. K. (1991), 'That BLUP is a good thing: The estimation of random effects (with discussion)', *Statistical Science* **6**, 15–51.

Robinson, P. (1988), 'Root-N-consistent semiparametric regression', *Econometrica* **56**, 931–954.

Rodriguez-Iturbe, I. (2000), 'Ecohydrology: A hydrologic perspective of climate-soil-vegetation dynamics', *Water Resources Research* **36**, 3–9.

Rodriguez-Iturbe, I., Cox, D. R. and Isham, V. S. (1987), 'Some models for rainfall based on stochastic point processes', *Proceedings of the Royal Society of London, series A* **410**, 269–288.

Rodriguez-Iturbe, I., Cox, D. R. and Isham, V. S. (1988), 'A point process model for rainfall: Further developments', *Proceedings of the Royal Society of London, series A* **417**, 283–298.

Rodriguez-Iturbe, I. and Porporato, A. (2005), *Ecohydrology of Water-Controlled Ecosystems*, Cambridge University Press, Cambridge.

Rodriguez-Iturbe, I., Porporato, A., Ridolfi, L., Isham, V. and Cox, D. R. (1999), 'Probabilistic modelling of water balance at a point: The role of climate, soil and vegetation', *Proceedings of the Royal Society of London, series A* **455**, 3789–3805.

Rosenbaum, P. R. (2002), *Observational Studies* (2nd edn.), New York: Springer.

Rosenbaum, P. R. and Rubin, D. B. (1983), 'The central role of the propensity score in observational studies for causal effects', *Biometrika* **70**, 41–55.

Rosenblatt, M. (1952), 'Remarks on a multivariate transformation', *Annals of Mathematical Statistics* **23**, 470–2.

Ross, R. (1911), *The Prevention of Malaria,* 2nd *edn*, Murray, London.

Ross, R. (1916), 'An application of the theory of probabilities to the study of *a priori* pathometry, I', *Proceedings of the Royal Society of London* **A92**, 204–30.

Ross, R. and Hudson, H. P. (1917*a*), 'An application of the theory of probabilities to the study of *a priori* pathometry, II', *Proceedings of the Royal Society of London* **A93**, 212–25.

Ross, R. and Hudson, H. P. (1917*b*), 'An application of the theory of probabilities to the study of *a priori* pathometry, III', *Proceedings of the Royal Society of London* **A93**, 225–40.

Roth, K. F. (1955), 'Rational approximations to algebraic numbers', *Mathematika* **2**, 1–20, (corrigendum **2**, 168).

Rothman, K. J. and Greenland, S. (1998), *Modern Epidemiology*, 2nd edn, Lippincott Williams and Wilkins, Philadelphia.

Royal Statistical Society (1988), 'Royal Statistical Society Meeting on AIDS', *Journal of the Royal Statistical Society, series A* **151**, 1–136.

Royall, R. (1997), *Statistical Evidence: A Likelihood Paradigm*, Chapman & Hall, London.

Rubin, D. B. (1974), 'Estimating causal effects of treatments in randomized and non-randomized studies', *Journal of Educational Psychology* **66**, 688–701.

Rubin, D. B. (1976), 'Inference and missing data', *Biometrika* **63**, 581–590.

Rubin, D. B. (1987), *Multiple Imputation for Nonresponse in Surveys*, New York: Wiley.

Rubin, D. B. (1996), 'Multiple imputation after 18+ years', *Journal of the American Statistical Association* **91**, 473–489.

Rubin, D. B. (2001), 'Estimating the causal effects of smoking', *Statistics in Medicine* **20**, 1395–1414.

Rubin, D. B. (2002), 'The ethics of consulting for the tobacco industry', *Statistical Methods in Medical Research* **11**, 373–380.

Ruczinski, I., Kooperberg, C. and LeBlanc, M. (to appear), 'Logic regression', *Journal of Computational and Graphical Statistics* **11**.

Ruud, P. A. (1983), 'Sufficient conditions for the consistency of maximum likelihood estimation despite misspecification of distribution in multinomial discrete choice models', *Econometrica* **51**, 225–228.

Ruud, P. A. (1986), 'Consistent estimation of limited dependent variable models despite misspecification of distribution', *Econometrica* **32**, 157–187.

Rvachev, L. A. and Longini, I. M. (1985), 'A mathematical model for the global spread of influenza', *Mathematical Biosciences* **75**, 3–23.

Sasieni, P. (2003), 'Evaluation of the UK breast screening programmes', *Annals of Oncology* **14**, 1206–1208.

Sato, K. (1999), *Lévy Processes and Infinitely Divisible Distributions*, Cambridge University Press, Cambridge.

Scaccia, L. and Green, P. J. (2003), 'Bayesian growth curves using normal mixtures with nonparametric weights', *Journal of Computational and Graphical Statistics* **12**, 308–331.

Schafer, J. L. (1997), *Analysis of Incomplete Multivariate Data*, London: Chapman & Hall/CRC.

Schaffer, W. M. (1985), 'Can nonlinear dynamics elucidate mechanisms in ecology and epidemiology?', *IMA Journal of Mathematics Applied in Medicine and Biology* **2**, 221–252.

Scharfstein, D. O. and Robins, J. M. (2002), 'Estimation of the failure time distribution in the presence of informative censoring', *Biometrika* **89**, 617–634.

Scharfstein, D. O., Rotnitzky, A. and Robins, J. M. (1999), 'Adjusting for non-ignorable drop-out using semi-parametric non-response models', *Journal of the American Statistical Association* **94**, 1096–1146.

Schenzle, D. (1984), 'An age-structured model of the pre- and post-vaccination measles transmission', *IMA Journal of Mathematics Applied in Medicine and Biology* **1**, 169–191.

Schervish, M. J. (1995), *Theory of Statistics*, Springer, New York.

Schwert, G. W. (1989), 'Why does stock market volatility change over time?', *Journal of Finance* **44**, 1115–1153.

Schwert, G. W. (1990), 'Stock market volatility', *Financial Analysts Journal* **46**, 23–34.

Sellke, T. (1983), 'On the asymptotic distribution of the size of a stochastic epidemic', *Journal of Applied Probability* **20**, 390–394.

Sen, A. and Srivastava, M. (1990), *Regression Analysis: Theory, Methods, and Applications*, Springer, New York.

Severini, T. A. (2000*a*), *Likelihood Methods in Statistics*, Clarendon Press, Oxford.

Severini, T. A. (2000*b*), 'The likelihood ratio approximation to the conditional distribution of the maximum likelihood estimator in the discrete case', *Biometrika* **87**, 939–945.

Shephard, N. (1994), 'Partial non-Gaussian state space', *Biometrika* **81**, 115–131.

Shephard, N., (ed.) (2004), *Stochastic Volatility: Selected Readings*, Oxford University Press, Oxford.

Siegel, C. L. (1942), 'Iterations of analytic functions', *Annals of Mathematics* **43**, 607–612.

Silverman, B. W. (1981), 'Using kernel density estimates to investigate multimodality', *Journal of the Royal Statistical Society, series B* **43**, 97–99.

Simon, J. (1969), *Basic Research Methods in Social Science: The Art of Experimentation*, Random House, New York.

Simon, J. (1993), *Resampling: The New Statistics*, preliminary edn, Duxbury, Belmont, CA.

Simpson, J. A. and Weiner, E. S. C., (eds) (1989), *Oxford English Dictionary*, Oxford University Press, London.

Singpurwalla, N. D. (2002), 'On causality and causal mechanisms', *International Statistical Review* **70**, 198–206.

Skovgaard, I. M. (1996), 'An explicit large-deviation approximation to one-parameter tests', *Bernoulli* **2**, 145–166.

Skovgaard, I. M. (2001), 'Likelihood asymptotics', *Scandinavian Journal of Statistics* **28**, 3–32.

Small, C. G. and McLeish, B. G. (1994), *Hilbert Space Methods in Probability and Statistical Inference*, Wiley, New York.

Smith, J. Q. (1985), 'Diagnostic checks of non-standard time series models', *Journal of Forecasting* **4**, 283–91.

Snijders, T. A. B. (2001), 'The statistical evaluation of social network dynamics', *Sociological Methodology* **31**, 361–395.

Snijders, T. A. B. (2002), 'Markov chain Monte Carlo estimation of exponential random graph models', *Journal of Social Structure.* **3**, part 2.

Snijders, T. A. B. and Bosker, R. J. (1999), *Multilevel Analysis: An Introduction to Basic and Advanced Multilevel Modelling*, London: Sage.

Sobel, M. (2000), 'Causal inference in the social sciences', *Journal of the American Statistical Association* **95**, 647–651.

Soper, H. E. (1929), 'Interpretation of periodicity in disease prevalence', *Journal of the Royal Statistical Society* **92**, 34–73.

Spence, R. (2001), *Information Visualization*, Addison-Wesley, Harlow.

Spiegelhalter, D. J. (2002), 'Funnel plots for institutional comparison', *Quality and Safety in Health Care* **11**, 390–391.

Spielman, R. S., McGinnis, R. E. and Ewens, W. J. (1993), 'Transmission test for linkage disequilibrium: the insulin gene region and insulin-dependent diabetes mellitus (IDDM)', *American Journal of Human Genetics* **52**, 506–516.

Stark, J., Iannelli, P. and Baigent, S. (2001), 'A nonlinear dynamics perspective of moment closure for stochastic processes', *Nonlinear Analysis* **47**, 753–764.

Stein, C. (1956), Efficient nonparametric testing and estimation, *in* 'Proceedings of the Third Berkeley Symposium on Mathematical Statistics and Probability', Vol. 1, University of California Press, Berkeley, pp. 187–195.

Stein, M. L. (1999), *Interpolation of Spatial Data*, Springer, New York.

Stewart, L. A. and Parmar, M. K. B. (1993), 'Meta-analysis of the literature or of individual patient data: Is there a difference?', *Lancet* **341**, 418–220.

Stone, C. (1975), 'Adaptive maximum likelihood estimation of a location parameter', *Annals of Statistics* **3**, 267–284.

Storvik, G. (2002), 'Particle filters in state space models with the presence of unknown static parameters', *IEEE Transactions on Signal Processing,* **50**, 281–289.

Tardella, L. (2002), 'A new Bayesian method for nonparametric capture-recapture models in presence of heterogeneity', *Biometrika* **89**, 807–817.

Taylor, S. J. and Xu, X. (1997), 'The incremental volatility information in one million foreign exchange quotations', *Journal of Empirical Finance* **4**, 317–340.

Thorin, O. (1977), 'On infinite divisibility of the lognormal distribution', *Scandinavian Actuarial Journal* pp. 141–149.

Thorin, O. (1978), 'An extension of the notion of a generalized gamma convolution', *Scandinavian Actuarial Journal* pp. 121–148.

Tibshirani, R. J. (1996), 'Regression shrinkage and selection via the lasso', *Journal of the Royal Statistical Society, series B* **58**, 267–288.

Tikhonov, A. N. (1963), On the solution of incorrectly put problems and the regularisation method, *in* 'Outlines of the Joint Symposium on Partial Differential Equations (Novosibirsk, 1963)', Academy of Science, USSR, Siberian Branch, Moscow, pp. 261–265.

Titterington, D. M. and Cox, D. R., (eds) (2001), *Biometrika: 100 Years*, Oxford University Press, Oxford.

Titterington, D. M., Smith, A. F. M. and Makov, U. E. (1985), *Statistical Analysis of Finite Mixture Distributions*, Wiley, New York.

Troynikov, V. S. (1999), 'Probability density functions useful for parametrization of heterogeneity in growth and allometry data', *Bulletin of Mathematical Biology* **60**, 1099–1122.

Tukey, J. W. (1974), 'Named and faceless values: An initial exploration in memory of Prasanta C. Mahalanobis', *Sankhyā, series A* **36**, 125–176.

Tukey, J. W. (1977), *Exploratory Data Analysis*, Addison-Wesley, Reading, MA.

van de Vijver, M. J., He, Y. D., van 't Veer, L. J., Dai, H., Hart A. A. M., Voskuil, D. W., Schreiber, G. J., Peterse, J. L., Roberts, C., Marton, M. J., Parrish, M., Atsma, D., Witteveen, A., Glas, A., Delahaye, L., van der Velde, T., Bartelink, H., Rodenhuis, S., Rutgers, E. T., Friend, S. H. and Bernards, R. (2002), 'A gene-expression signature as a predictor of survival in breast cancer', *New England Journal of Medicine* **347**, 1999–2009.

van der Geer, S. (2000), *Empirical Processes in M-Estimation*, Cambridge University Press, Cambridge.

van der Laan, M. J. and Robins, J. M. (2003), *Unified Methods for Censored Longitudinal Data and Causality*, Springer, New York.

van der Vaart, A. W. (1988), *Statistical Estimation in Large Parameter Spaces*, CWI Tracts 44, Centrum voor Wiskunde en Informatica, Amsterdam.

van der Vaart, A. W. (1998), *Asymptotic Statistics*, Cambridge University Press.

van der Vaart, A. W. (2000), Semiparametric statistics, *in* P. Bernard, (ed.), *Lectures on Probability Theory and Statistics*, Springer, Berlin.

Vardi, Y. (1985), 'Empirical distributions in selection bias models', *Annals of Statistics* **13**, 178–203.

Venables, W. N. and Ripley, B. D. (1994), *Modern Applied Statistics with S-Plus*, Springer-Verlag, New York.

Venables, W. N. and Ripley, B. D. (2002), *Modern Applied Statistics with S*, 4th edn, Springer-Verlag, New York.

Versluis, M., Schmitz, B., von der Heydt, A. and Lohse, D. (2000), 'How snapping shrimp snap: Through cavitating bubbles', *Science* **289**, 2114–2117.

Voiculescu, D. (2000), Lectures on free probability, *in Lectures on Probability Theory and Statistics*, Springer, Heidelberg, pp. 283–349. Ecole d'Eté de Saint Flour XXVIII.

von Mering, C., Zdobnov, E. M. and Tsoka, S. (2003), 'Genome evolution reveals biochemical networks and functional modules', *Proceedings of the National Academy of Sciences of the USA* **100**, 15428–15433.

Wahba, G. (1985), 'A comparison of GCV and GML for choosing the smoothing parameter in the generalized spline smoothing problem', *Annals of Statistics* **13**, 1378–1402.

Wahba, G. (1990), *Spline Models for Observational Data*, SIAM, Philadelphia, PA.

Wakefield, J. (2004), 'Ecological inference for 2×2 tables (with discussion)', *Journal of the Royal Statistical Society, series A* **167**, 385–425.

Waterman, R. P. and Lindsay, R. G. (1996), 'Projected score methods for approximating conditional scores', *Biometrika* **83**, 1–14.

Watson, J. D. and Crick, F. H. C. (1953), 'Molecular structure of nucleic acids', *Nature* **171**, 737–738.

Watts, D. J. and Strogatz, S. H. (1998), 'Collective dynamics of "small-world" networks', *Nature* **393**, 440–442.

Welch, B. L. and Peers, H. W. (1963), 'On formulae for confidence points based on integrals of weighted likelihoods', *Journal of the Royal Statistical Society, series B* **25**, 318–329.

West, M. and Harrison, J. (1997), *Bayesian Forecasting and Dynamic Models*, Springer, New York.

Whittle, P. (1955), 'The outcome of a stochastic epidemic—a note on Bailey's paper', *Biometrika* **42**, 116–122.

Whittle, P. (1957), 'On the use of the normal approximation in the treatment of stochastic processes', *Journal of the Royal Statistical Society, series B* **19**, 268–281.

Wilkinson, L. (1999), *The Grammar of Graphics*, Springer-Verlag, New York.

Wilson, E. B. and Burke, M. H. (1942), 'The epidemic curve', *Proceedings of the National Academy of Science* **28**, 361–367.

Wilson, E. B. and Burke, M. H. (1943), 'The epidemic curve, II', *Proceedings of the National Academy of Science* **29**, 43–48.

Winkel, M. (2001), The recovery problem for time-changed Lévy processes. Research Report 2001-37, MaPhySto, University of Aarhus.

Witten, I. H. and Franke, E. (2000), *Data Mining. Practical Machine Learning Tools and Techniques with Java Implementations*, Morgan Kaufmann, San Francisco.

Wjst, M., Fischer, G., Immervoll, T., Jung, M., Saar, K., Rueschendorf, F., Reis, A., Ulbrecht, M. Gomolka, M., Weiss, E. H., Jaeger, L., Nickel, R., Richter, K., Kjellman, N. I. M., Griese, M., von Berg, A., Gappa, M., Riedel, F., Boehle, M., van Koningsbruggen, S., Schoberth, P., Szczepanski, R., Dorsch, W., Silbermann, M., Loesgen, S., Scholz, M., Bickeboller, H. and Wichmann, H. E. (1999), 'A genome-wide search for linkage to asthma. German Asthma Genetics Group', *Genomics* **58**, 1–8.

Wood, S. N., Kohn, R., Shively, T. and Jiang, W. (2002), 'Model selection in spline non-parametric regression', *Journal of the Royal Statistical Society, series B* **64**, 119–139.

Woodroofe, M. (1970), 'On choosing a delta-sequence', *Annals of Mathematical Statistics* **41**, 1665–1671.

Wulfsohn, M. S. and Tsiatis, A. A. (1997), 'A joint model for survival and longitudinal data measured with error', *Biometrics* **53**, 330–339.

Xie, Y. (1992), 'The log-multiplicative layer effect model for comparing mobility tables', *American Sociological Review* **57**, 380–95.

Yang, Y. H. and Speed, T. (2002), 'Design issues for cDNA microarray experiments', *Nature Reviews: Genetics* **3**, 579–588.

Yau, P., Kohn, R. and Wood, S. N. (2003), 'Bayesian variable selection and model averaging in high-dimensional multinomial nonparametric regression', *Journal of Computational and Graphical Statistics* **12**, 23–54.

Yu, J. (2004), 'On leverage in a stochastic volatility model', *Journal of Econometrics*. Forthcoming.

Zeger, S. L., Liang, K.-Y. and Albert, P. S. (1988), 'Models for longitudinal data: a generalized estmating equation approach', *Biometrics* **44**, 1049–1060.

Zerhouni, E. (2003), 'The U.S. National Institutes of Health Roadmap', *Science* **302** **(5642)**, 63–72.

Zhang, L., Mykland, P. and Aït-Sahalia, Y. (2003), A tale of two time scales: determining integrated volatility with noisy high-frequency data. Unpublished paper presented at the Realized Volatility Conference, Montreal, 8th November 2003.

Zhuang, J., Ogata, Y. and Vere-Jones, D. (2002), 'Stochastic declustering of space-time earthquake occurrences', *Journal of the American Statistical Association* **97**, 369–380.

Index of Names

Holt, T., 25, 165
Hornik, K., 206
Hortobagyi, G. N., 191
Howard S. V., 17
Hrycak, A., 164
Huang, W. M., 121, 122
Huang, X., 221
Huber, P. J., 9, 137
Huber-Carol, C., 24
Hudson, H. P., 29

Iannelli, P., 42
Ibragimov, I. A., 123
Ibrahim, J., 134
Ichimura, H., 163
Ihaka, R., 165, 206
Indritz, J., 149
Ingham, J., 12
Irizarry, R. A., 170
Isham, V. S., 17, 18, 19, 20, 22, 23, 24, 27, 28, 32, 33, 35, 40, 41, 42, 43, 55, 58, 60, 89
Ito, K., 234

Jacod, J., 216, 217, 220, 224, 230
Jacquez, J. A., 36, 43, 44
Jeliazkov, I., 228
Jensen, J. L., 233
Jeong, H., 46, 177
Jessip, W. N., 15
Jiang, W., 139
Johannes, M., 223, 224, 226, 231
Johnson, A. M., 44, 46
Johnson, N. L., 16, 18, 20
Jones, D. A., 24
Jones, D. R., 21, 158
Jones, J. H., 47
Jónsdóttir, K. Ýr, 247
Joyce, H., 140
Jupp, P. E., 250, 255
Jurek, Z. J., 238

Kafetzaki, M., 41
Kakou, A., 24
Kallsen, J., 223
Kaldor, J., 175
Kallianpur, G., 18
Kalman, R. E., 174

Kao, R. R., 51
Karatzas, I., 215
Kardoun, O. J. W. F., 25
Kaufman, J. S., 182
Kaufman, S., 182
Keeling, M. J., 28, 34, 37, 42, 44, 48, 49, 50, 51
Kelly, L. M., 150
Kempthorne, O., 8, 172
Kendall, D. G., 33
Kermack, W. O., 29
Kerr, D., 23
Kerr, M. K., 177
Khintchine, A. Y., 145, 223
Kilkenny, D., 144
Kim, S., 224, 229
Kingman, J. F. C., 91, 92
Kious, B. M., 168
Klaassen, C. A. J., 117, 119, 120, 126, 127, 129, 130, 131
Knott, M., 158
Koch, R., 181
Kohn, R., 139
Kohonen, T., 204
Kolmogorov, A. N., 233
Kooperberg, C., 170
Koopman, J., 44
Koshevnik, Y. A., 123
Kotz, S., 18, 20, 22, 23
Kozumi, H., 139
Kraus, K., 253
Kretzschmar, M., 40, 48
Krishnaiah, P. R., 18
Kristen, J., 223
Kruskal, J. B., 202, 203
Kruskal, W. H., 17
Kunst, A. E., 159
Kurtz, T. G., 33
Kwon, J., 117

Labys, P., 219
Laio, F., 56, 59, 61, 62, 65, 66, 67, 69, 70, 71, 72
Lauh, E., 16
Law, G. R., 24, 25
Law, R., 24, 51
LeBlanc, M., 170
LeCam, L. M., 14, 16

Index of Terms